Lecture Notes in Electrical Engineering

Volume 676

The book series *Lecture Notes in Electrical Engineering* (LNEE) publishes the latest developments in Electrical Engineering—quickly, informally and in high quality. While original research reported in proceedings and monographs has traditionally formed the core of LNEE, we also encourage authors to submit books devoted to supporting student education and professional training in the various fields and applications areas of electrical engineering. The series cover classical and emerging topics concerning:

- Communication Engineering, Information Theory and Networks
- Electronics Engineering and Microelectronics
- Signal, Image and Speech Processing
- Wireless and Mobile Communication
- Circuits and Systems
- Energy Systems, Power Electronics and Electrical Machines
- Electro-optical Engineering
- Instrumentation Engineering
- Avionics Engineering
- Control Systems
- Internet-of-Things and Cybersecurity
- Biomedical Devices, MEMS and NEMS

For general information about this book series, comments or suggestions, please contact leontina.dicecco@springer.com.

To submit a proposal or request further information, please contact the Publishing Editor in your country:

China

Jasmine Dou, Associate Editor (jasmine.dou@springer.com)

India, Japan, Rest of Asia

Swati Meherishi, Executive Editor (Swati.Meherishi@springer.com)

Southeast Asia, Australia, New Zealand

Ramesh Nath Premnath, Editor (ramesh.premnath@springernature.com)

USA, Canada:

Michael Luby, Senior Editor (michael.luby@springer.com)

All other Countries:

Leontina Di Cecco, Senior Editor (leontina.dicecco@springer.com)

**** Indexing: The books of this series are submitted to ISI Proceedings, EI-Compendex, SCOPUS, MetaPress, Web of Science and Springerlink ****

More information about this series at http://www.springer.com/series/7818

Zuber Patel · Shilpi Gupta · Nithin Kumar Y. B.
Editors

Advances in VLSI and Embedded Systems

Select Proceedings of AVES 2019

Springer

Editors
Zuber Patel
Department of Electronics Engineering
S. V. National Institute of Technology
Surat, Gujarat, India

Shilpi Gupta
Department of Electronics Engineering
S. V. National Institute of Technology
Surat, Gujarat, India

Nithin Kumar Y. B.
National Institute of Technology Goa
Ponda, Goa, India

ISSN 1876-1100 ISSN 1876-1119 (electronic)
Lecture Notes in Electrical Engineering
ISBN 978-981-15-6228-0 ISBN 978-981-15-6229-7 (eBook)
https://doi.org/10.1007/978-981-15-6229-7

This Springer imprint is published by the registered company Springer Nature Singapore Pte Ltd.
The registered company address is: 152 Beach Road, #21-01/04 Gateway East, Singapore 189721, Singapore

Committees

Organizing Committee

Organizing Chairs

Anand D. Darji, SVNIT, Surat
Rasika Dhavse, SVNIT, Surat
Zuber Patel, SVNIT, Surat
Pinalkumar Engineer, SVNIT, Surat

Program Chairs

Virendra Singh, IIT Bombay, India
R. M. Patrikar, VNIT Nagpur, India
Vineet Sahula, MANIT Jaipur, India

Web Chairs

Seena V., IIST, India, India
Mahendra Sakre, IIT Ropar, India

Hospitality Chairs

A. H. Lalluwadia, SVNIT Surat, India
Abhilash Mandloi, SVNIT Surat, India
Mehul Patel, SVNIT Surat, India

Registration Chairs

Shweta Shah, SVNIT Surat, India
K. P. Upla, SVNIT Surat, India
P. K. Shah, SVNIT Surat, India

Publicity and Social Media Chairs

Upena Dalal, SVNIT Surat, India
R. K. Chauhan, MMMUT Gorakhpur, India
Dipankar Saha, IIT Bombay, India

Sponsorship Chairs

Jignesh Sarvaiya, SVNIT Surat, India
P. N. Patel, SVNIT Surat, India
Manish kumar, MMMUT Gorakhpur, India

Publication Chairs

Zuber Patel, SVNIT, Surat
Shilpi Gupta, SVNIT, Surat
Nithin Kumar Y. B., NIT Goa, India

Exhibition Chairs

N. B. Kanirkar, SVNIT Surat, India
G. Santra, SVNIT Surat, India
Abhishek Agrahari, Mentor Graphics, India

Advisory Committee

S. N. Merchant, IIT Bombay, India
Roy P. Paily, IIT Guwahati, India
Samudravijaya K., IIT Guwahati, India
Avinash Keskar, VNIT Nagpur
Nirupam S. D., NTU, Singapore
Partik Sheth, APTIV, Dublin
Dileep Dhakal, X-FAB Semiconductor, Germany
Samitsubhro Banerji, Qualcomm. USA
Punj Pokharel, NVIDIA, USA
Binod Bhatta, Intel, Singapore
YatinVarachhia, Euphotic Labs, India
Leena Panchmatia, Apple, USA
Mihir Shah, IBM, USA
Taral Chhatbar, Intel, USA
Maulik Prabhudesai, Intel, USA
Arpit Gandhi, Intel, India
H. S. Jatana, SCL, Mohali, India
Sunita Verma, Meity, New Delhi, India

Technical Programme Committee

R. B. Deshmukh, VNIT Nagpur, India
Joycee M. Mekie, IIT Gandhinagar, India
Biswajit Mishra, DAIICT Gandhinagar, India
Hitesh Shrimali, IIT Mandi, India
Nihar Ranjan Mohapatra, IIT Gandhinagar, India
Brijesh Kumar, MMMUT Gorakhpur, India
Shree prakash Tiwari, IIT Jodhpur, India
S. Sivanantham, VIT Vellore, India
Arvind Rajawat, MANIT Bhopal, India
Ashwin Kothari, VNIT Nagpur, India
Rutu Parekh, DAIICT Gandhinagar, India
Santosh Kumar Vishvakarma, IIT Indore, India
Amit Joshi, MNIT Jaipur, India
Marshnil Dave, Lion Semiconductor, USA
Nitin Kale, Nanosniff, India
Vijay Nath, BIT Mesra, India

Message from General Chairs

Anand D. Darji

Zuber Patel

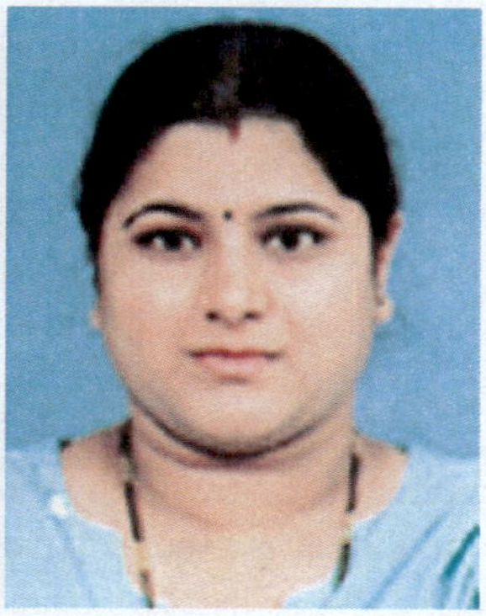

Rasika Dhavse

Pinalkumar Engineer

On behalf of organizing committee, we are very pleased and honored to welcome you to International Conference on Advances in VLSI and Embedded Systems (AVES-2019) and Electronics Engineering Department, S V National Institute of Technology (SVNIT), Surat. This conference is technically sponsored by Springer.

We believe that change is the only constant in semiconductor industry. AVES-2019 is the premier forum for the presentation of new advances and research results in related fields. It aims to bring together leading researchers, engineers, and scientists in the domain of VLSI Design and Embedded Systems from around the world. It consists of one-day pre-conference tutorials, a plenary talk and four keynote speeches on radical topics in VLSI and Embedded Systems under a common umbrella.

We are very pleased to invite Prof. Gaurav Sharma (University of Rochester, NY, USA) for presenting a Plenary talk on "Smart Light-Weight Body Worn Sensors for Health Analytics". We are grateful for Keynote speeches delivered by Prof. Virendra Singh (IIT Bombay) on "Computer Architecture Challenges in 21st Century", Prof. R. K. Chauhan (MMMUT, Gorakhpur) on "WBG Semiconductors for Next Generation Electronic Applications", Prof. Rutu Parekh for talk on "Design, modelling and Simulation of 3-D integrated Single Electron Transistor based circuits operating at room temperature", and H. S. Jatana (SCL, ISRO) on "Challenges in development of SOI-CMOS technology for ultra-low power applications". Two state-of-the-art pre-conference tutorials have been also prepared to meet the intellectual abilities and curiosity of the attendees.

AVES-2019 has truly been a team effort. We extend our special thanks to authors for noteworthy contribution and heartfelt thanks to members of organizing committee for their dedicated efforts and making AVES-2019 special and memorable. Very special thanks to TEQIP-III, NPIU Delhi, MMMUT Gorakhpur, Optimized Solutions Limited, and Cadre for their generous support toward AVES-2019.

Anand D. Darji
Zuber Patel
Rasika Dhavse
Pinalkumar Engineer
General Chairs, AVES 2019
Department of Electronics Engineering, SVNIT, Surat

Preface

It gives us immense pleasure to write the preface of the proceeding of Springer sponsored “International conference on Advances in VLSI and Embedded Systems (AVES-2019)”. The conference aims to bring together leading researchers, engineers, and scientists in the domain of VLSI Design and Embedded Systems from around the world. The conference has been organized in the Department of Electronics Engineering, Sardar Vallabhbhai National Institute of Technology (SVNIT), Surat (Gujarat State, India), during December 20–21, 2019.

With recent advancements in the CMOS and allied novel technologies, major research and development efforts are ongoing in the field of VLSI and Embedded Systems. Challenges faced by conventional CMOS technology are being overcome by new architectures, algorithms, implementation methodologies, re-configurability, devices, and processes. Research in this area spans many levels of abstraction: from basic design concepts to circuit and system design, fabrication, testing, and standardization.

The conference provides a forum where experts, researchers, academicians, industry professionals, experiences, and proficiencies gather to discuss the state-of-the-art topics. The theme of our conference was nothing but the next-generation requirements in Embedded Systems and VLSI design, i.e., Embedded Systems—Embedded systems for industrial automation and power electronics, Embedded systems for defence, healthcare, automotive, and signal processing, Technologies for secure embedded circuits and systems, IoT and wearable devices, High-performance embedded systems; VLSI Design—System-level design, Advances in digital design, Low-power design, Analog and mixed-signal design, FPGA design and reconfigurable systems, RF/Wireless systems, System-on-chip designs, Networks-on-chip Design; and Devices and Emerging Technology—Semiconductor device simulation, Device modeling, Fabrication and characterization,

Device circuit interactions, Sensors and transducers, MEMS/NEMS, Organic Electronics, Spintronics, 2-D material for electronic devices, Device physics design and circuits using non-silicon materials, Fabrication and packaging technologies.

Editorial Team

Surat, India — Zuber Patel
Surat, India — Shilpi Gupta
Ponda, India — Nithin Kumar Y. B.

Acknowledgements

The journey of the conference started many months ago in the Department of Electronics Engineering, SVNIT. The General chairs Mr. A. D. Darji, Mr. Z. M. Patel, Mrs. R. N. Dhavse, and Mr. P. J. Engineer initiated this task which was then carried out by various committees, who they started working diligently. During this journey, the whole-hearted support was received from the staff of Electronics Engineering Department of SVNIT and various people of TEQIP-III. We extend our special thanks to the authors for noteworthy contribution.

Our esteemed sponsors are—Springer, TEQIP-III, Cadre, and optimized solutions limited, without whose financial support this conference would not take its shape. We are very grateful to MMMUT Gorakhpur for joining hands with us for this task. There was extreme support from TPC team and reviewers, all the keynote speakers, invited paper speakers, session chairs, and registered participants. It is worth to express deep gratitude toward SVNIT authority people—Director, Registrar In-Charge, and administrative staff for providing the infrastructural support and approvals for various arrangements in the Institute premise.

Contents

Reusability and Scalability of an SoC Testbench in Mixed-Signal Verification—The Inevitable Necessity 1
Babun Chandra Pal

Automated Simulator for the Validation of Bio-Impedance Devices 17
S. Sruthi, Rasika Dhavse, and Jignesh N. Sarvaiya

Analysis of Memory-Based Real Fast Fourier Transform Architectures for Low-Area Applications 31
Rajasekhar Turaka and M. Satya Sai Ram

Optimization of MEMS-Based Capacitive Sensor with High-k Dielectric for Detection of Heavy Metal Ions 41
Dinesh Rotake and Anand D. Darji

Statistical Analysis of Vehicle Detection in the ITS Application for Monitoring the Traffic and Road Accident Using Internet of Things 55
Diya Vadhwani and Devendra Thakor

Qualitative and Quantitative Analysis of Parallel-Prefix Adders 71
Sudhanshu Janwadkar and Rasika Dhavse

A Novel Method of Multiplication with Ekanyunena Purvena 89
M. A. Sayyad and D. N. Kyatanavar

FPGA Design of SAR Type ADC Based Analog Input Module for Industrial Applications 97
G. Dhanabalan and T. Murugan

Need for Predictive Data Analytics in Cold Chain Management 115
Swati D. Kale and Shailaja C. Patil

FPGA-Based Implementation of Artifact Suppression and Feature Extraction 131
Shrikant Kumbhar, Anand D. Darji, and Harikrishna M. Singapuri

Test Time Reduction Using Power-Aware Dynamic Clock Allocation to Scan Vectors 145
Harikrishna Parmar and Usha Mehta

Impact of Spacers in Raised Source/Drain 14 nm Technology Node $In_{0.53}Ga_{0.47}As$-nFinFET on Short Channel Effects 159
Jay Pathak and Anand D. Darji

Impact of Multi-Metal Gate Stacks on the Performance of β-Ga_2O_3 MOS Structure 169
Narendra Yadava and R. K. Chauhan

Improved VLSI Architecture of Dual-CLCG for Pseudo-Random Bit Generator 175
Mangal Deep Gupta and R. K. Chauhan

Low Power Radix-8 Modulo $2^n + 1$ Multiplier Using Modified Weighted Method 183
Naveen Kumar Kabra and Zuber M. Patel

Design of Prominent Single-Precision 32-Bit Floating-Point Adder Using Single-Electron Transistor Operating at Room Temperature 201
Ankur Sharma, Rasika Dhavse, Yash Agrawal, and Rutu Parekh

Image Communication Using Quasi-Cyclic Low-Density Parity-Check (QC-LDPC) Code 211
Dharmesh Jayantibhai Patel, Pinalkumar Engineer, and Ninad Sunilkumar Bhatt

Smart Soldier Health Monitoring System Incorporating Embedded Electronics 223
Krishna Teja, Umang Patel, Parthkumar Patel, Yash Agrawal, and Rutu Parekh

A 3–7 GHz CMOS Power Amplifier Design for Ultra-Wide-Band Applications 235
Vishakha Purnanand Bhale and Upena Dalal

An Approach to Detect and Prevent Distributed Denial of Service Attacks Using Blockchain Technology in Cloud Environment 247
Vishwani Patel, Fenil Khatiwala, and Yashi Choksi

Variability Analysis of On-Chip Interconnect System Using Prospective Neural Network 259
Ajita Misra, Yash Agrawal, Vinay Palaparthy, and Rutu Parekh

Prospective Incorporation of Booster in Carbon Interconnects for High-Speed Integrated Circuits 273
Takshashila Pathade, Yash Agrawal, Rutu Parekh, and Mekala Girish Kumar

Island Engineering of Single-Electron Transistor for Room Temperature Operation 289
Raj Shah, Rutu Parekh, and Rasika Dhavse

About the Editors

Dr. Zuber Patel is Associate Professor in S. V. National Institute of Technology (SVNIT), Surat, Gujarat, India. He has done his PhD in wireless communication form SVNIT, Surat, and M.Tech. in VLSI design and microelectronics from IIT Bombay. His areas of research include HDL/FPGA based design, VLSI design of RISC microprocessors, FPGA/ASIC design of wireless transceiver hardware and embedded systems. He is one of the investigators of the prestigious SMDP-C2SD project funded by MeitY, Government of India. He has been a reviewer of numerous journals and an editor of notable conference proceedings.

Dr. Shilpi Gupta is Associate Professor in S. V. National Institute of Technology (SVNIT), Surat, Gujarat, India. She holds a PhD in wireless communication. Her areas of research include wireless communication, signal processing, orthogonal frequency division multiplexing (OFDM), multiple input multiple output (MIMO) systems, massive MIMO detection techniques, free space optics, and fiber optic sensors. She has published several articles in peer-reviewed journals and conference proceedings. She has also organized fifteen national level events including workshops, seminars, conferences, and training programmes. She is a member of many professional bodies like IETE, IEEE, IEICE and IJERIA.

Dr. Nithin Kumar Y. B. received the B.E. degree from Manipal Institute of Technology Manipal, India, and the M.Tech. degree from National Institute of Technology-Surathkal, and the Ph.D. degree from Indian Institute of Technology-Kharagpur, India. Currently he is working as an Assistant Professor at National Institute of Technology Goa, India. He is a recipient of young Indian researcher award from Government of Italy. He has published many research papers in reputed international journals/conferences and delivered several technical presentations at various international conferences/universities in India and abroad. He also filed 4 patents in the relevant area. His current research interests include analog and mixed signal design.

Reusability and Scalability of an SoC Testbench in Mixed-Signal Verification—The Inevitable Necessity

Babun Chandra Pal

Abstract The digital verification with Universal Verification Methodology (UVM) provides vast opportunities for automation. The momentous themes in the context of digital verification are constraint random stimulus generation, functional coverage, and self-checking test benches that make digital verification very robust. However, analog verification is still a manual process. Therefore, the demand is to transform analog verification techniques into an automated process in the context of digital-centric Mixed-Signal verification for more coverage. A new verification approach that simulates Mixed-Signal Design Under Test (DUT), which contains Register Transfer Logic (RTL), SPICE netlist, and Verilog-AMS models or Real Number Model (RNM) blocks has been demonstrated in the paper that describes the Mixed-Signal simulation environment and simulation techniques to achieve more simplistic yet very powerful strategies to address Mixed-Signal verification challenges.

Keyword Simulation speed up techniques · Mixed-signal simulation in digital kernel · Analog-centric mixed-signal with UVM · Digital-centric mixed-signal with UVM

1 Introduction

The ultimate goal of ASIC designs is to construct a chip based on highly precise specifications. These predefined specifications have to be inspected during and after design flow. In order to acquire the highest possible level of certainty in the functionality of the design it is of vital importance to use accurate and liable verification techniques.

In design of today's SoCs, functional verification complexity rises exponentially with hardware complexity doubling exponentially with time. Many novel, advanced,

---

No academic titles or descriptions of academic positions should be included in the addresses. The affiliations should consist of the author's institution town/city and country.

B. C. Pal (✉)
Einfochips, Kolkata, India
e-mail: babun.pal@einfochips.com

Z. Patel et al. (eds.), *Advances in VLSI and Embedded Systems*, Lecture Notes in Electrical Engineering 676, https://doi.org/10.1007/978-981-15-6229-7_1

and liable methods have been introduced for digital verification. However, up to 70% of design development time and resources are still dedicated to functional verification.

As SoCs became multifunctional, more analog blocks are integrated on a single chip. Therefore, Mixed-Signal verification difficulties became more visible to verification teams. Moreover, at chip-level verification, when block-level verification is accomplished successfully, it is required to develop a technique through which analog and digital sub-blocks of a design are considered holistically.

Universal Verification Methodology (UVM) by offering base class libraries, brings much automation to the digital verification world. In addition, this methodology is simulator vendor independent. It is possible to create UVM components and reuse them in different projects. Taking advantage of constraint random stimuli generation offered by UVM, the engineering effort has been turned into building automatic checkers instead of writing directed test. The typical approach in UVM-based verification flow is to define a test case and simulate the Design Under Test (DUT) with constraint random stimuli to put the design into corner cases. Automatic checkers ensure the correctness of the design functionality according to the generated stimuli. Furthermore, coverage mechanisms are used in order to measure the inspected specification of the DUT and to point out to the verification closure.

The main goal of this work is to utilize the provided facilities by UVM in Mixed-Signal verification environments. The proposed structure by this work highly improves the Mixed-Signal verification quality in terms of verification performance with minimum human effort devoted. It is possible to generate real value input stream and drive analog interfaces of the Design Under Test (DUT) using automated digital verification techniques like constraint random stimuli generation. Moreover, the utilized analog transaction-level modeling approach enhances full-chip verification procedure significantly.

In this work, a UVM-based Mixed-Signal verification environment is proposed.

2 SoC Verification Methodologies

During last decades, semiconductor industry has experienced immense advancements in terms of integrated circuit's capacity, complexity, and fabrication technology. New technological capabilities besides increasing demand for high quality, multifunctional ICs bring about integrating more and more diverse functions on a single chip. Moore's prediction has been realized and as complexity and capacity of electronic designs grow, production costs rise concurrently.

Costly semiconductor realization draws attention to obtain assurance of debugging all the bugs prior to the design fabrication. Expensive design development signifies the necessity of verification in all design steps. In the absence of verification specifically from the early stages of the design, presence of functional defects is not far-fetched.

2.1 Introduction to Verification

Design flow of a System-on-Chip involves diverse steps. Data "Transformation" when crossing the steps is an inevitable outcome.

The design flow tasks are performable taking advantage of numerous complicated Electronic Design Automation (EDA) tools. Large amount of data is required to be able to apply these advanced tools properly and this is not an unlikely occurrence to face several errors during the flow.

A general verification definition can be

> "Act of checking, testing, inspecting whether a product, service or system (or set thereof) meets its initial requirements and specifications."

The seriousness of verification has been frequently emphasized in many papers. A financial disaster might occur for a company if a bug is discovered in an already sold IC. Intel's Pentium floating-point division bug in 1994, imposed 475 million U.S dollar detriment [1].

Although verification has been a main investigation subject between designers in recent years, still there is wide gap between design capabilities and verification capabilities. Today, verification is the major bottleneck of designs and between 50–70% of project effort is dedicated to design verification.

2.2 Analog Circuit Simulation

Analog integrated circuits are simulated using analog simulators, like SPICE, over the last decades. Simulation Program with Integrated Circuit Emphasis (SPICE) is the most well-known tool for analog simulation. SPICE simulation before manufacturing the integrated circuits is a common task for designers to ensure the correctness of the design. The tool was first developed at the University of California at Berkeley as a class project about 40 years ago and since then it became a worldwide standard circuit simulator. SPICE simulation covers wide range of components from simple passive to complicated active devices like MOSFETs. By writing a text-description of a component, the SPICE simulation model of it is achievable and its behavior is predictable. SPICE like simulators are suitable for detailed design analysis, when the design is constructed in transistor level. Even for a small portion of a design SPICE simulation might take a week or more to be accomplished. SPICE simulation performance was improved to speed up the simulation at about 2–4 orders of magnitude over the last years. Although this improved simulation speed was achieved at the expense of losing simulation accuracy.

In order to provide higher levels of productivity, it is of significant importance to utilize various simulation tools during all steps of design development. For an analog circuit designer to gain better productivity it is crucial to model the circuit in a higher level of abstraction and verify its functionality. This is how designers can simply detect some functional defects prior to devoting time to design to the transistor level.

Moreover, in chip-level verification when the design is going to be checked in context of the whole system and verification aim is more to check system-level behavior or interconnectivity between blocks, detailed results of SPICE models are not requested.

3 Analog Transaction-Level Modeling

Today's design methodologies are highly affected by two different directions; in one hand, it is required to inspect electrical behavior of the design to avoid physical defects and later product re-spins. Detailed design representation is necessary in order to verify it at this level. On the other hand, time-to-market pressure forces designers to use abstract representation to be able to overcome the challenging task of complex Mixed-Signal circuit verification.

Today, between 70 and 80% of designs are Mixed-Signal designs and, therefore, Mixed-Signal verification takes enormous amount of time to be accomplished. To be able to use digital-like simulators and, therefore, speed up the verification process, it is crucial to consider analog and digital portions of the design holistically. This highlights the demand for novel strategies in order to verify an analog IP in the SoC context.

The aim of this paper is to reuse and scale a UVM SoC-level test bench for Mixed-Signal verification from project to project.

3.1 Mixed-Signal Verification Challenges

As it has been recorded, more than 50% of design re-spins at 65 nm and below are the overcome of Mixed-Signal designing. The consequent wasted time and high costs of product re-spin bring about the demand for new strategies for SoC verification. It is not possible any more simply assuming the practicality of old strategies in SoC-level Mixed-Signal verification. To have an overview of old Mixed-Signal SoC-level verification strategies, brief description of two of them is stated in this section.

Black Box Approach. One of the common methods to verify a mixed-signal design with dominant digital content is called black box verification. In this approach, a pre-verified analog block is integrated into a bigger design using its highly abstracted model. The abstract model contains the interface description only in many cases. This is how simulation speed degradation can be avoided but this problem arises that verification quality degrades instead. Today's Mixed-Signal SoCs are not only more complex in terms of added analog blocks, but also in terms of complicated interconnections and feedback loops between analog and digital portions of the design. It is not reliable any more to replace analog portions of the design with a black box model and perform SoC-level verification.

3.2 *Mixed-Signal Verification Facilitation*

SoC-level verification of Mixed-Signal ICs has been getting more attention in last years. Several approaches have been proposed by researchers in order to enhance the verification process. Since digital verification teams were able to build highly automated test benches, it was time to introduce more digital-like and consequently more automated techniques for Analog and Mixed-Signal (AMS) circuit verification. Although analog verification with SPICE is still a golden standard and cannot be ignored, to achieve better simulation speed, different levels of design abstraction are used to model an analog subsystem. Taking into account SPICE and fast-SPICE simulation next level of abstraction, analog behavioral modeling (with Verilog-AMS), improved simulation speed significantly. Furthermore, Real Number Modeling (RNM) and pure digital models are the other approaches to describe the analog sub-circuit in higher levels of abstraction and, therefore, to achieve higher simulation performance. Moreover, another crucial factor to choose an appropriate abstract model is the required effort to build a simulation environment. To clarify more a comparative chart is shown in Fig. 1.

The chart shows the distinction between mentioned approaches in terms of required effort for simulation setup in Fig. 1 and performance trade-off in Fig. 2. As depicted in this figure, although RNMs or pure digital models are less accurate models but less effort is required to build a simulation environment using these models than AMS models. Therefore, this is the most apparent advantage of these models in full-chip verification.

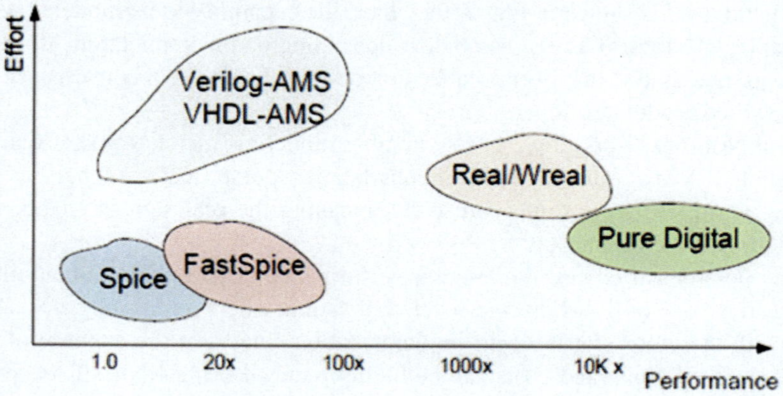

Fig. 1 A comparison between different modeling approaches in mixed-signal verification in terms of required effort and performance

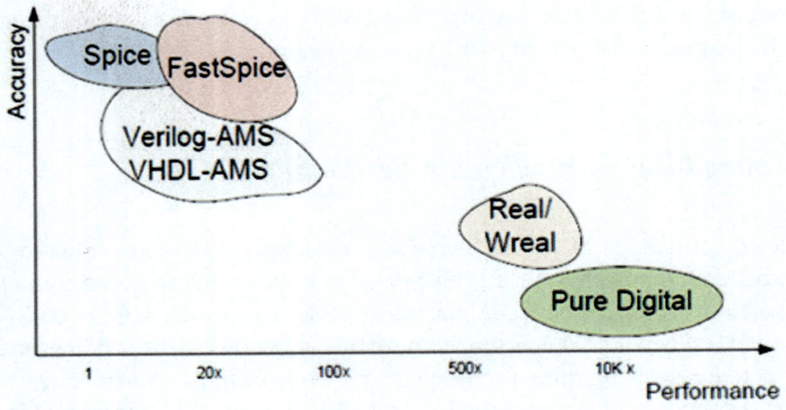

Fig. 2 A comparison between different modeling approaches in mixed-signal verification in terms of accuracy and performance

Analog Behavioral (Electrical) Modeling. In general, analog behavioral modeling approach is used to create a module, which encapsulates high-level behavioral description of an analog or Mixed-Signal subsystem. In order to enable designers to write these behavioral models, some specific languages were created like Verilog-AMS, Verilog-A, VHDL-AMS, etc. These Analog and Mixed-Signal (AMS) languages offer both continuous-time and event-driven semantics for verification task. Apparently, analog designers are the best choice to write the analog behavioral models. Although they are familiar with their own analog design, they usually lack Verilog or VHDL knowledge to write these models. Similarly, digital designers with enough knowledge of these languages are not able to write an accurate model since they know less about the analog circuits. As it is clearly depicted in Fig. 1, AMS model simulation needs most setup effort. In one hand, it is possible to write these models to gain more simulation speed, at the sacrifice of accuracy. On the other hand, as it is shown in Fig. 2, they have the capability to be modeled with the accuracy close to SPICE like simulators depending on the application. However, it is important to notice that over-idealized models can bring about convergence issues for unskilled model developers.

Real Number Modeling. "In real number modeling, also known as Real Value Modeling (RVM), values are continuous-floating-point (real) numbers as in the analog world. However, time is discrete, meaning the real signals change values based on discrete events".

It is possible to achieve simulation performance near digital simulation speed taking advantage of RNM. In chip-level verification, where it is usually at an abstract level and, therefore, not in depth implementation details, at the expense of losing some accuracy, time, and costs can be highly reduced using RNM. This approach is restricted to signal flow and, therefore, benefits from the absence of troublesome convergence issues. From Fig. 2, it is possible to notice that in comparison with

Table 1 Simulation performance comparison with master bias block

Parameters	Spice	Verilog-AMS	SV-RNM model
Simulator	Spectre	AMS engine	Digital simulator (e.g., NC-SIM)
Stimuli	Fixed directed value with toggling of digital control signal	Fixed directed value with toggling of digital control signal	Fixed directed value with toggling of digital control signal
Nature of simulation	Pure spice simulation with spectre engine	Mixed-signal simulation	Digital simulation with digital solver
Simulation time	5 ms	5 ms	5 ms
Elapsed time (wall clock)	265 s	40 s	11 s

AMS models, this approach requires less effort and provides higher performance when verification goal is apart from implementation details (accuracy).

Three HDL languages support RNM such as

- VHDL (with real type)
- SystemVerilog (with real type)
- Verilog-AMS (with wreal type).

Simulation performance comparable to what is already existing in digital verification techniques, makes it possible to apply highly automated digital techniques like verification planning, random test generation, coverage, and assertions in analog verification area by using real number models. This is the most well-liked verification capability, which can be achieved in Analog and Mixed-Signal verification.

3.3 Performance Comparison with Spice, Verilog-AMS Model, and System Verilog Model

Three experimental simulations have been performed with Master Bias Block using the Schematic, Verilog-AMS Model and the SV-RNM Model in Cadence Platform. The simulation details are captured in Table 1 as shown below.

3.4 Analog Transaction

A sample DUT with SPICE netlist or Verilog-AMS or RNM blocks and RTL blocks is presented in Fig. 3. In this case, the DUT has both digital and real-valued pins. The existing approach to drive real-valued pins of DUT is to hard code stimuli in the test (direct way). However, the proposed technique which leads to a way in which analog pins of the DUT are driven using constrained random stimulus generation.

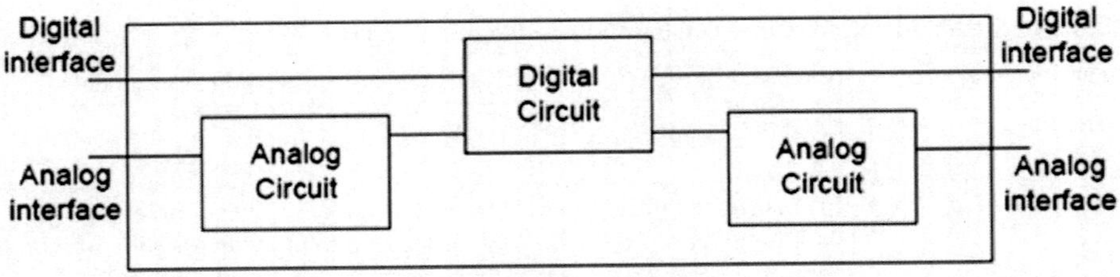

Fig. 3 A mixed-signal design with analog sub-circuits and digital RTL interfacing with each other

The proposed method to drive analog pins of depicted sample DUT in Fig. 3 resulted in a new concept, which is Analog Transaction. To clarify more on transaction concept, it is noticeable that transaction is a data structure containing parameters. In a UVM-based test bench, data fields within transaction are randomized and are used in driver, which implements the protocol separately and wiggles the DUT pins.

To be able to extend the concept of transaction in analog domain, a term replacement from "protocol" in digital world to "shape" in analog world is suggested. Analog signals can have different shapes, for example, harmonic, linear or cubic, spline, etc. To be able to describe or generate these analog shapes besides the name of the shape we have to define parameters. This brings about a request to use analog transactions to generate analog waves with defined parameters and consequently in TLM way [2]. In other words, in a UVM-based test bench, data structure including parameters is passed to the driver and according to the specific numerical algorithm driver reflects those parameters to generate the desired analog wave.

3.5 *Metric Driven Verification (MDV) Adoption to Analog Environment*

UVM, as a popular verification methodology, enhanced the verification process of complex SoCs. The most efficient aspect of this methodology is the application of metric-driven verification. In order to enhance analog verification one should think about MDV techniques and the possibility of adopting them to an analog environment [2]. In this section, some of the most important limitations are discussed briefly.

Analog vPlan and Coverage. Planning for verification (vPlan) is a beneficial technique that is, however, more complicated when talking about analog features. As vPlan must capture the defined features of the design, it has to be richer to adopt analog features as well. Moreover, measuring those analog features could be a sophisticated task. This is because analog properties including amplitude, gain, frequency, or some other similar values that are different from logic values naturally.

Simulation performance: In order to adopt MDV and, therefore, take advantage of its ability to a run large number of simulations automatically, analog designs have to be modeled using AMS languages or Real Number Model (RNM). This is because, the SPICE-level simulation is slow in comparison to digital simulation.

Constrained Random Stimulus. To explore different behaviors of DUT, it is required to generate both digital and analog random stimulus. Efficient randomization of analog inputs to the design is a major request when adopting MDV to analog environment.

Self-checking. A self-checking test bench can determine that the design behaves as planned or not. In digital designs, this is carried out by monitoring the output pins of DUT and comparing it with expected output in Scoreboard. This process is more complicated when checking an analog design.

4 Testbench Component Implementation

The goal of this work is the improvement of existing verification methodology and test bench structure of Mixed-Signal designs. In order to do so, a test bench structure is proposed in which analog interfaces of the depicted DUT in Fig. 3 have been driven taking advantage of transaction-level modeling. The proposed test bench structure shown in Fig. 4 is responsible for generating random stimulus in order to shape various analog waves and sending them to the DUT's analog interface automatically.

The aim is to generate more different analog waves in transaction-based manner in which the driver component receives transactions from sequencer and drives DUT pins. The driver reflects the received transaction and eventually generates the intended shape at the inputs of the DUT.

Moreover, generated components in the proposed verification environment can support and deliver transactions with various data structures. This enables the user to

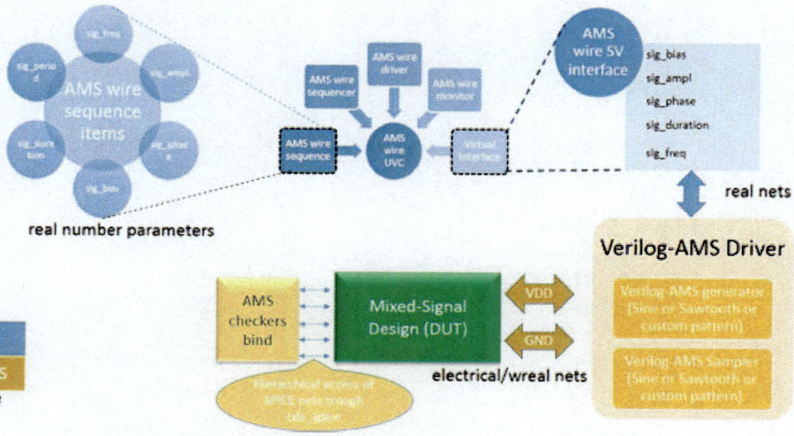

Fig. 4 Stimuli driver in an AMS testbench

implement the application-specific algorithm and reuse the generated SystemVerilog library in test bench environment.

To be able to use advanced verification techniques like randomization and, therefore, to avoid writing analog real-valued input stimulus in a directed way, transaction-level modeling is extended into analog domain.

The proposed technique generates analog stimulus through passing sequence items (data structures) to the driver as transactions are data structures containing particular parameters via sequencer and driver. Driver wiggles DUT's pins according to the specified protocol and by decoding parameters within a received transaction. Hence, "transactions offer abstraction of the protocol."

UVM sequence item class holds parameters, which are required to define an analog wave (like slope and value of a certain point in time for a linear signal). During run task of the driver, when it sends a request to the sequencer for a new sequence item, those parameters within sequence item will be randomized. As an example, generation of a ramp analog signal using Verilog-AMS is shown below.

Program Code.

```
module ramp_gen (enable, ramp_duration,ramp_val, o_ramp);
 input wire enable;
  input wire [31:0] ramp_duration;
  input wreal ramp_val;
  `ifdef UVC_ELECTRICAL_DRIVERS
    output  o_ramp;
    electrical o_ramp;
  `else
    output wreal o_ramp;
  `endif
  parameter real vdelta=1u from (0:inf); // voltage delta
  parameter real ttol=1n from (0:1m]; // time tolerance
  parameter real vtol=10n from (0:inf); // voltage
  // tolerance
  real ramp_real;
  real realX_ramp;
  real trise;
  real tfall;
  real noise_val;
  `ifndef UVC_ELECTRICAL_DRIVERS
    assign  o_ramp  =  (enable  ==  1)  ?  realX_ramp  :
`wrealZState;
  `endif
  analog initial begin
    ramp_real = 0;
    noise_val = 0;
  end
```

```
    initial  $display("inside ramp_gen module");

    always begin
      trise = ramp_duration;
      tfall= trise;
      @(ramp_duration);
    end
    always @(absdelta(ramp_real, vdelta, ttol, vtol ))
      realX_ramp = ramp_real;

    analog begin
      if(enable == 1) begin
      //noise_val =   (($random() % 1000) * 0.000001);
      //ramp_real       =       transition(ramp_val,       0.0,
  trise*1n,tfall*1n) + noise_val;
        ramp_real       =       transition(ramp_val,       0.0,
  trise*1n,tfall*1n);
        `ifdef UVC_ELECTRICAL_DRIVERS
          V(o_ramp) <+ ramp_real;
        `endif
      end
    end

  endmodule
```

The randomized spectrum of signal is delivered by the UVM sequence item class and the ramp signal is obtained through Verilog-AMS driver (either in "electrical" domain or in "wreal" domain).

As the verification is simulation based, thus a simulator (either a Mixed-Signal simulator or a digital solver) is a mandatory one.

For this purpose, let's consider Cadence Mixed-Signal Simulator (e.g., AMS Designer).

In a typical Mixed-Signal test bench, the Mixed-Signal DUT can be configured with SPICE netlist and Verilog-AMS models, mixed and matched. This configuration can be controlled from the AMS test bench configuration based upon the test scenarios, which is explained in the next section [3, 4].

4.1 Mixed-Signal Testbench Components

In this paper, the goal is to describe a UVM-based Mixed-Signal test bench through which analog input pins of a sample DUT are driven using the same strategy as digital input pins. Since different verification methodologies are used to verify digital and analog blocks, it is crucial to have a holistic view of both digital and analog sub-blocks of the design at chip-level verification.

The verification approach followed here is Command Line Based or Batch Mode [3, 4].

The various components of a Mixed-Signal test bench are shown in Fig. 5.

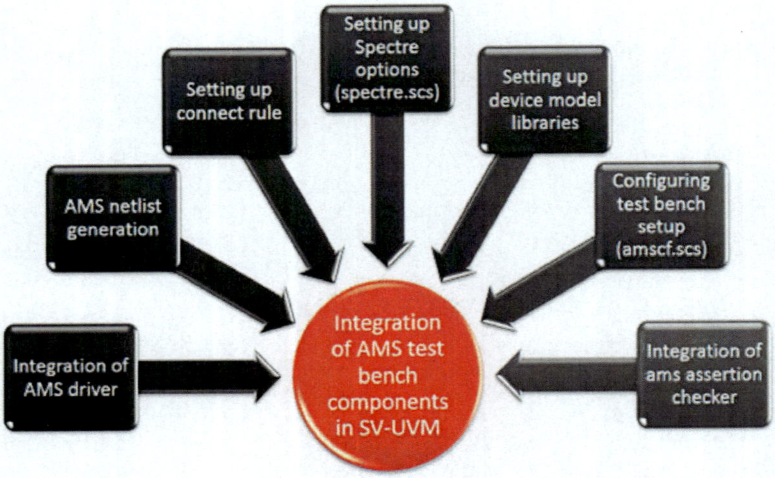

Fig. 5 Different AMS testbench components

AMS Netlist Generation. AMS netlist generation in the integration process where the netlist can be generated with the help of "runams" script which needs a config view of the design. The following command shows how to generate a Verilog-AMS netlist using a config view:

Program Code.

```
runams -lib <lib_name> -cell <cell_name> -view config -
netlist -rundir <run_dir_name>
```

The final netlist file will be generated beneath run_dir_name directory as netlist.vams.

The Verilog-AMS netlist can also be generated using ADE-L/XL in Virtuoso [3, 4].

Setting up Connect Rules. A connect module is required to be inserted into the design whenever module with ports of different disciplines are connected in order to map the signal values from one domain to the other. Connect modules are written in Verilog-AMS language. It is a good practice to use "-iereport" "-ieinfo" in the irun option for efficient debugging in a typical Cadence Flow. This will generate ams_ieinfo.log file which contains the information about the connect modules inserted.

Connect Rule can be set with the "mode = split" option to improve the accuracy of the simulation. Default option for mode is "merged" if not specified.

Insertion of connect modules with different attributes are demystified in Fig. 6.

Let's take some examples of setting up connect rules through "i.e. card".

Program Code.

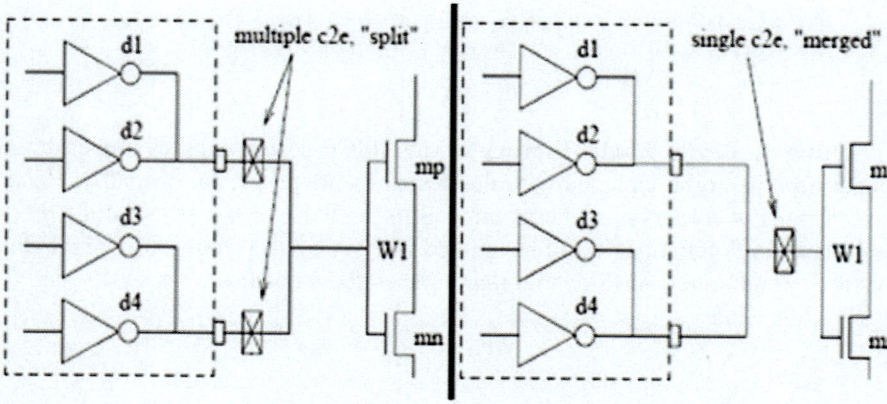

Fig. 6 Different attributes of connect module

```
amsd{
          ie vsup=1.8 tr=1n instport=<port name including
hierarchical path> mode=split
          ie vsup=1.2 tr=1n instport=<port name including
hierarchical path>
     }
```

In the above example:

- Parameter "vsup". "vsup" is the parameter to determine the supply voltage level (or, maximum voltage level) that a signal can reach to.
- Parameter "tr". "tr" is the parameter to determine rise time of the signal when getting converted into electrical domain.

Setting up Simulator (e.g., Spectre) Options. In order to obtain appropriate accuracy, accuracy control options (e.g., errpreset) must be used to obtain the faithful outputs. Hence, setting up proper simulation options is very crucial in AMS verification [3, 4].

A typical spectre.scs file, which looks like below

Program Code.

```
simulator lang=spectre
global 0
tran1 tran stop=10m errpreset=moderate relref=alllocal
```

For user friendly debugging, the simulation must be run with debug switch enabled in the simulation control file (e.g., spectre.scs). Typical example of such options using Spectre simulator is shown below.

Program Code.

```
myinfo info what=opppoint where=rawfile
spectrel options diagnose=detailed debugstepdrop=yes \
simstat=detailed debug=yes
```

Setting up Device Model Libraries. Any SPICE design requires proper device model libraries to be included with the design for its proper functionality. Hence, correct process corners must be mentioned in the field known as "section". For a typical corner simulation, it will be marked with "tt" in the section while including the library path in the AMS configuration file as shown below.

Program Code.

```
include "<hierarchy path>/<file>.scs" section=mos_tt
include "<hierarchy path>/<file>.scs" section=res_tt
include "<hierarchy path>/<file>.scs" section=bjt_tt
include "<hierarchy path>/<file>.scs" section=mim_tt
include "<hierarchy path>/<file>.scs" section=rtmom_tt
include "<hierarchy path>/<file>.scs" section=vpmom_tt
include "<hierarchy path>/<file>.scs" section=dio_dnw_tt
```

Note. The device model libraries may be included in the AMS configuration file (e.g., amscf.scs).

Setting up Test Bench Configuration File (e.g., amscf.scs). This file is the mixed-signal configuration file which includes the following:

- spectre.scs
- device model libraries
- amsd block for configuring connect rules using "i.e. card"
- amsd block for configuring SPICE subcircuit using portmap statement.

Typical example of such "portmap" statement configuring an analog subckt is shown below

Program Code.

```
amsd{
            Portmap subckt=<subckt> porttype=name
            Config cell=<cell name> use=spice
    }
```

Note. The mixed-signal test bench configuration file (e.g., amscf.scs) must be parsed with the "irun" or "xrun" file list.

AMS Assertion Checkers. These checkers are written in Verilog-AMS and they check voltage levels of electrical signal. In order to determine voltage level in electrical domain, a "cross" statement (Verilog-AMS) can be used and the checker can be bind with the electrical signal of DUT.

5 Conclusion

Based on the information described before, the aim of this work is to build a test bench which is fully reusable and scalable in different projects when other shapes of analog wave are required to drive a DUT.

To clarify more, UVM_seq_item class is implemented to hold parameters which are required to generate a specific analog wave. In described example to generate a ramp signal, signal parameters are needed. In one hand, parameters within UVM_seq_item class were defined to deliver signal's spectrum. In other words, sequence items are sent to driver from sequencer.

In this work, the main focus is to implement a UVM-based test bench in which transaction type, driver class are also implemented in a generic way. In this manner, it is possible to generate various analog signal shapes like harmonic, cubic spline, or any other shapes using the same structure. In many verification projects, it is required to generate different shapes of analog waves. To have a fully flexible test bench with the ability of generating different shapes of analog waves it is necessary to define an unconstrained transaction-level communication between UVM components. Desired analog output wave is generated eventually.

In a nutshell, a UVM SoC test bench can be leveraged for Mixed-Signal verification in the following manner as shown in Table 2.

Acknowledgments At First, I would like to express my sincere gratitude to Mr. Gopal Bhavsar for continuous support of my work, for his patience, motivation, and immense knowledge. His guidance helped me in all the time of work and writing of this paper.

Table 2 Digital integration with different modeling standard for mixed-signal simulation

Approach	Verilog-A	Verilog-AMS	VAMS wreal	System Verilog (2012)
Description	Verilog lookalike for SPICE	Superset of Verilog-A and Verilog-D	Verilog-AMS RVM subset.	Most recent approach for RVM
Evaluation	Continuous	Flexible	Discrete time	Discrete time
Digital integration	Via co-sims	True AMS, limited UVM/SV	True AMS, limited UVM/SV	Excellent integration
Modeling features	True analog modeling	True analog + mixed-signal interaction	Abstract signal chain only	Abstract signal chain but most permissive
Speed	Slow	Better	Close to digital	Fast, close to digital

Note. It is not recommended to use both Verilog-AMS and SystemVerilog modeling mixed and matched. A Verilog-AMS model for an analog block is always preferred over a SystemVerilog model when interacting with a SPICE netlist

Besides my advisor, I would like to thank the rest of my Analog Mixed-Signal team members for not only their insightful comments and encouragement, but also for the hard questions, which motivated me to hone my technical know-how in detail from various perspectives.

My sincere thanks also go to Mr. Saurabh Desai, who provided me an opportunity to join his team, and who gave access to open up a new center at Kolkata with Analog and Mixed-Signal work. Without his precious support it would not be possible to write this paper.

I thank my fellow colleagues for the stimulating discussions and for all the fun we have had in the last 3 years. Also I thank my friends at eInfochips. In particular, I am grateful to Mr. Snehal A. Patel for enlightening me the first glance of writing paper.

Last but not the least, I would like to thank my family for supporting me spiritually throughout writing this paper and my life in general.

References

1. FDIV Bug of Intel's Pentium Processor. https://olenick.com/index.php/olenick-blog/177-infamous-software-bugs-fdiv-bug.html
2. Universal Verification Methodology (UVM) 1.2 User's Guide, October, 2015. https://www.accellera.org/images//downloads/standards/uvm/uvm_users_guide_1.2.pdf
3. Virtuoso® AMS Designer Simulator User Guide Product Version 15.1 July 2015, Cadence
4. Virtuoso® Spectre® Circuit Simulator and Accelerated Parallel Simulator User Guide Product Version 14.1 October 2014, Cadence
5. Pant N, Harinarayan GS (2015) Comparing Verilog-AMS versus SPICE view usage for robust AMS verification of power management controller and mode transition, freescale semiconductor. https://www.edn.com/Pdf/ViewPdf?contentItemId=4438475
6. Harinarayan GS, Rana M, Pant N (2015) Verilog-AMS versus SPICE view: an SoC verification comparison. https://www.edn.com/Pdf/ViewPdf?contentItemId=4439239
7. Jain G, Harinarayan GS, Pant N, Rana M (2015) Mixed-mode verification of DDR, LCD, and memory sub-systems: Verilog-AMS versus SPICE. http://www.edn.com/Pdf/ViewPdf?contentItemId=4439299
8. Jain G, Rathi S, Rana M (2014) IC mixed-mode verification: the sandwiched-SPICE approach. http://www.edn.com/Pdf/ViewPdf?contentItemId=4436273
9. Kaundal N, Kamal K, Prakash SJ (2014) How an LCD controller drives an LCD glass. http://www.edn.com/Pdf/ViewPdf?contentItemId=4424795
10. Kundert K, Simulation of analog and mixed-signal circuits. https://kenkundert.com/docs/bctm98-MSsim.pdf
11. Kundert K, Simulation of analog and mixed-signal circuits. Cadence
12. Sarkinen G (2011) Plan & metric driven mixed-signal verification for medical devices, Medtronics Inc. https://dvcon.org/2011/proceedings/papers/03_3.pdf
13. Sutherland S, Fitzpatrick T (2012) Keeping up with chip—the Proposed SystemVerilog 2012 standard makes verifying ever-increasing design complexity more efficient. http://events.dvcon.org/2012/proceedings/papers/04_3.pdf
14. Eisawy A, Ams design configuration schemes, design topologies, Mentor Graphics. www.verificationacademy.com
15. Dr. Nanda M, Rao PR (2018) An approach for generating the self-checking test-bench. https://www.ijraset.com/fileserve.php?FID=17914
16. Nita I, Rapan A, Improving verification methodologies in digital circuits modeling. https://www.scientificbulletin.upb.ro/rev_docs_arhiva/fulla0f_605416.pdf

Automated Simulator for the Validation of Bio-Impedance Devices

S. Sruthi, Rasika Dhavse, and Jignesh N. Sarvaiya

Abstract The electrical behaviour of biological tissue can be studied by passing high frequency Alternating Current (AC) through the tissue using electrodes. The response of the tissue is measured by an acquisition system and the bio-impedance of tissue to the current applied is extracted. Bio-impedance-based diagnosis of diseases has gained popularity over the last decade due to its non invasive and user friendly nature. Bio-impedance simulators are used to validate and ensure the safety and accuracy of developed bio-impedance devices before they are used for real-time measurement in human subjects. In this paper, a fully automated bio-impedance simulator is developed. This work follows an embedded solution approach and uses off the shelf components. The propounded bio-impedance simulator is using a digital potentiometer as resistor bank and 16 channel multiplexer controlled capacitor bank to constitute the Frickse-Morse model of 2R-1C network. Arduino Uno controls digital potentiometer via Serial Peripheral Interface (SPI), LCD via Inter Integrated Circuit (I2C) and multiplexer via digital pins D7-D0. The implemented bio-impedance simulator is validated by using as a varying load of a precise constant current source where the current showed a variability of only 1.69 μA and the power consumed is 125 mW.

Keywords Bio-impedance simulator · 2R-1C network · Digital potentiometer · De-multiplexer

1 Introduction

Bio-impedance is the ability of biological tissue to impede current. Characteristics of any biological tissue can be understood by applying a high frequency low amplitude (mV voltage excitation or μA current excitation) AC signal to the tissue under study. Bio-impedance is obtained by acquiring, conditioning and processing this response of

S. Sruthi (✉) · R. Dhavse · J. N. Sarvaiya
Electronics Department, S.V. National Institute of Technology, Surat, Gujarat 395007, India
e-mail: sruthisindu@gmail.com

Z. Patel et al. (eds.), *Advances in VLSI and Embedded Systems*, Lecture Notes in Electrical Engineering 676, https://doi.org/10.1007/978-981-15-6229-7_2

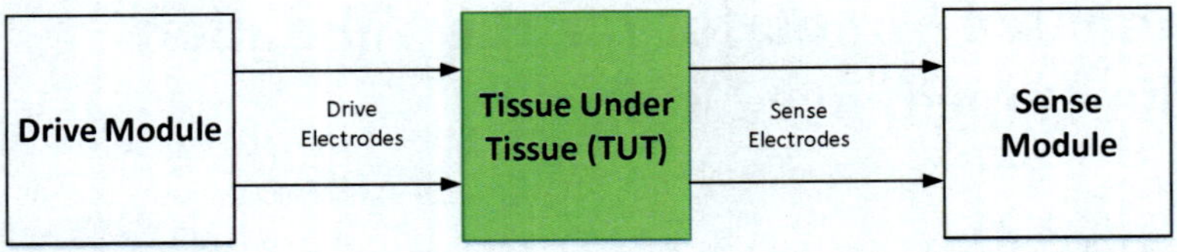

Fig. 1 Block diagram of bio-impedance measurement

the tissue after application of excitation [1]. Body composition analysis [2], study of cardiovascular abnormalities through impedance cardiography (ICG) [3], impedance glottography (IG) [4], impedance pneumography (IPG) [5], uro-lithosis [6], study of cell suspension in cancer [7], observation of post knee replacements, prevention of amputation of limps in diabetic patients [1], etc. are some of the applications where bio-impedance measurement is employed.

Unlike X-ray and Computed Tomography (CT) bio-impedance measurement does not require radiation exposure and presence of trained personnels. It is also a noninvasive procedure which involves no risks as compared to blood tests or pap smear tests and also does not involve complications like general anaesthesia. Due to these immense advantages over other diagnostic methods, it is necessary to develop cost effective, user friendly, indigenous, accurate and application-specific measuring device for bio-impedance measurement. Bio-impedance measurement comprises a drive module which is voltage or current excitation, a Tissue Under Test (TUT), electrodes and sense module which measures the response of the tissue. Basic block diagram of bio-impedance measurement is shown in Fig. 1. Tissue Under Test (TUT) is represented as bio-impedance simulator during the testing phase of bio-impedance devices.

Simulation is the artificial replication of an actual-world process with sufficient fidelity by providing immersion, feedback, reflection, and practice without the risks involved in real-time measurement. Medical domain uses simulators for education and training, testing of new drugs in pharmacology and for validating newly developed health monitoring devices. Patient safety, ethical sensitivities of subjects where people refuse to use newly developed devices on them, depleted resources such as unavailability of a fully fledged open source database, shortage of experimental animals, lack of skilled training personnels urge the use of simulators in these applications. Accuracy and safety of a bio-medical device is to be ensured before it is used for real-time measurement in human subjects [8]. Such simulators which help to test and validate the developed bio-impedance measurement devices are called bio-impedance simulators. Bio-impedance simulator-based measurement yields multitude of bio-impedance values for creation of detailed open source database.

2 Evolution of Bio-Impedance Simulators

Tissue consists of cells comprising of extracellular region, intracellular region and cellular membrane. The selectively permeable cell membrane has storage capabilities and resembles a capacitor whereas the extracellular and intracellular regions are resistive in nature due to ions and protein molecules. Frequency response of tissue is divided as α, β and γ dispersions. Since the capacitive element is present, the characteristics of tissue is fully categorised only in mid-range β dispersion (5 kHz–1 MHz). Electrical equivalent of tissue is a 2R-1C network and is the basic circuit of bio-impedance simulator [9]. Over the years, bio-impedance devices are tested by equivalent electrical circuits, in-vitro cultures of materials resembling human tissue, ex-vivo measurements of real tissue, plant tissues, software-based simulations, comparison of results with that of bio-impedance analysis device and use of experimental animals. All these methods have been proven to be reliable by fitting the resistances and reactances obtained in each to the Cole plot. Frickse-Morse model, Hayden model, etc. are simple 2R-1C networks which can be considered as the equivalent circuit for a biological tissue. Several combinations of resistors and capacitors constitute different tissue behaviours and vary according to applications [1].

Passive components are to be manually removed and inserted causing inaccuracies due to loose connections [10, 11]. In-vitro cultures involve the creation of living tissue environment by externally employing bio-chemicals but this method do not consider the real-time artifacts which can influence the measurement [6]. Ex-vivo bio-impedance measurement is done by measuring the impedance of actual organs outside the body. This method is also does not consider environment noises and done at room temperature instead of body temperature [12]. Plant tissues does not ensure the proper testing of bio-impedance devices before being used on tissues of human subjects as cells comprising these tissues consist of an extra cell wall in addition to a cell membrane. Cell walls add extra resistive and capacitive elements which are not accounted by researchers [13]. Experimental animals mimic humans due to live conditions and similar physiological events but these are not always available for testing purpose due to the stringent legal and ethical measures of their protection [14]. The specific impedance distribution and tissue behaviour on application of AC are simulated using softwares like MATLAB and COMSOL but these methods are just approximations of real-time measurement, consume a lot of CPU time and cannot be incorporated with portable devices [15, 16]. Recently developed bio-impedance simulators generate bio-impedance signals like ICG, IRG, IPG, IG, etc. These can be used only in the specific application and needs additional bio-signals, bio-parameters like heart rate, breathe rate, etc. to correlate with the bio-impedance signals. Standalone architecture for a general bio-impedance simulator can be implemented by using a micro-controller so that it caters to a basket of applications. Existing micro-controller-based bio-impedance simulators are semi-automated as they employ analog switches and passive components which affect the accuracy and speed of the measurement [17]. It is necessary to choose a component value from their respective banks automatically instead of choosing them through conventional methods of switching. Passive analog components have to be replaced by off the shelf digital similitude. Fully automated systems have a total control of components by an embedded board. Digitalised resistor bank, capacitor bank, display and all associated modules are to be controlled by serial buses and pins of the controller. This ensures reduced power consumption, increase accuracy and increase in speed of measurement.

In this paper, bio-impedance simulator setup is fully automated micro-controller based on digital components which do not require manual handling. The module implemented is noise prone, consumes less power, made with easily available components, does not involve ethical issues, user friendly and represents an accurate representation of biological tissue. In the equivalent circuit of simulator, ranges of resistance is chosen from 50 Ω to 1000 kΩ and that of capacitance is chosen from 10 to 1000 nF. Automated resistance banks and automated capacitor banks are implemented through digital potentiometer and 16 channel multiplexer, respectively. Control of simulator is done by the Arduino Uno board through the buses and digital pins. Theoretical and practical values of simulator are compared and after that consistency and accuracy of the developed simulator is proved by using it as variable load for a precise constant current source [18].

3 Proposed Bio-impedance Simulator Setup

From the behaviour of tissues with respect to frequency, Cole and Cole proposed equivalent circuit of tissue consists of resistive and reactive components. Different combinations of resistances and capacitors are suggested by researchers to represent this equivalent circuit.

3.1 Model of Tissue Under Test

Frickse-Morse model is simple, mostly used and is used for the present work too which is shown in Fig. 2. R_i is intracellular resistance offered to the current due to proteins and ions within the cell as well as resistance offered by cell membrane; R_e is the extracellular resistance due to ions outside the cell and C_i is capacitance due to membrane which lets in ions selectively just like a capacitor.

3.2 Design

Implementation of bio-impedance simulator requires the use of different values of resistances and capacitances. Different combinations of resistances and capacitors represent different types of tissues. Range of resistance is chosen from 50 Ω to 1000

Fig. 2 Frickse–Morse equivalent circuit of tissue under test

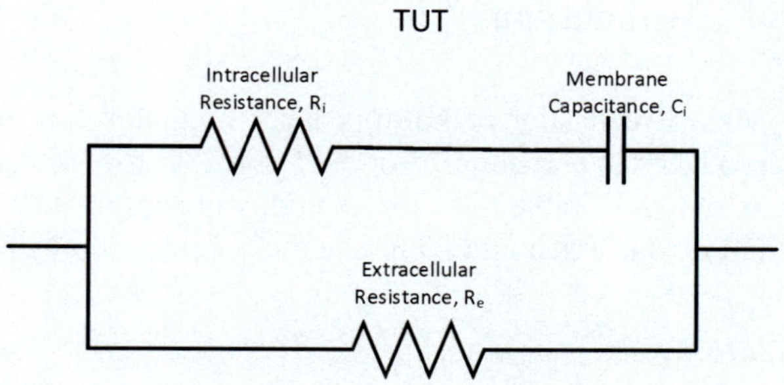

kΩ and that of capacitance is chosen from 10 to 1000 nF according to previous researches [19, 20]. Simulator circuit is implemented in manual as well as automatic way.

Manual hardware realisation of bio-impedance simulator is done initially by using a simple 2R-1C network. All the simulations and initial hardware testing are done by this methodology. Measurements of resistors and capacitors are done practically using Fluke 17 B+ series digital multimeter. Behaviour of this equivalent circuit is in concordance with the real tissue. Several values were tried for the resistors and capacitors. Fifty sets of arrangements were done with different combinations of resistors for each capacitor value as shown in Table 1. Even though the components for manual realisation are easily available around us, it requires the need for careful removal and insertion of resistors and capacitors. This is both tiresome and time-consuming.

Table 1 Result analysis of manual realisation of bio-impedance simulator

Resistance-R_i, R_e (Theo.) (Ω)	Resistance-R_i, R_e (Prac.) (Ω)	Error (%)
10	12.5	25
120	121	0.8
220	219	−0.45
470	465	−1.06
750	744	−0.8
1000	986	−1.4
Capacitor, C_i (Theo.)(nF)	Capacitor, C_i (Prac.) (nF)	Error (%)
10	12.3	23
47	49.2	4.68
110	116.09	5.54
310	320.07	3.25
410	422.28	2.99
710	729.52	2.75

4 Automation

Manual realisation of bio-impedance simulator using network of passive components produces an maximum error of 25 and 23% for resistors and capacitors, respectively, as shown in Table 1. As this is highly unacceptable in measurement and instrumentation, it is essential to improve the accuracy through automation.

4.1 Automation of Resistance Bank

The bio-impedance measurement device is intended to be portable and user friendly. So it is necessary to ensure that the testing circuit also should be automated and separated from the measuring device. Independent operation of the simulator requires a controller and here an Arduino Uno which is easily available and familiar is used. Automated implementation of simulator was done in a step-by-step manner. Initially the resistors were digitally controlled. Passive resistances were replaced by digital potentiometers. In order to vary the value of both the resistances in the suggested range fixed resistors were substituted by potentiometers. This is shown in Fig. 3. Digital Potentiometers are available in market under many makes. They basically consist of a variable resistor within it and has three categories of pin in general; i.e. power supply pins (5V), communication pins (SPI or I2C); Potentiometer pins (P_a, P_{wiper}, P_b). Potentiometers are to be chosen according to application. These are available in a resolution of 64, 128 and 256; i.e. the minimum resistance which can be measured is defined by maximum resistance divided by resolution. Maximum value of resistance for this application is 1 kΩ. So digital potentiometer by Analog Devices AD8402 is chosen [21]. Here the potentiometer pins are represented as A, W and B.

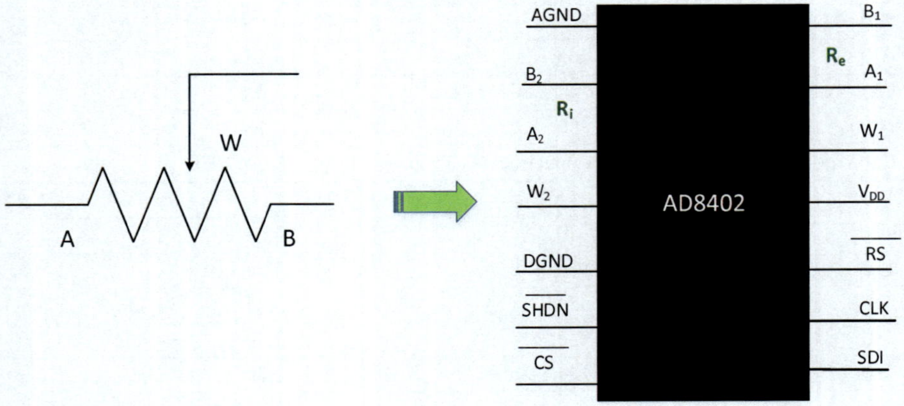

Fig. 3 Automated resistance bank using digital potentiometer AD8402

Considering a resolution of 256 bits the minimum increment is calculated as in Eq. 4.

$$\text{Resolution} = \frac{\text{Maximum Resistance}}{2^{\text{Bit Resolution}}} \tag{1}$$
$$= \frac{1000}{2^{256}} \tag{2}$$
$$= 3\Omega \tag{3}$$

Minimum resistance offered by potentiometer is 50 Ω due to the wiper resistance and it is incremented by 3 Ω if automatically varied.

$$\text{No. of resistance values} = \frac{\text{Max. Resistance} - \text{Min. Resistance}}{\text{Resolution}} \tag{4}$$
$$= \frac{1000 - 50}{3} \tag{5}$$
$$= 332\,\Omega \tag{6}$$

There can be total of 332 values for each resistance. AD8402 is controlled through Serial Peripheral Interface (SPI) by Arduino.

4.2 *Automation of Capacitance Bank*

Major challenge for the realisation of automated simulator is the non-availability of digitally tunable capacitors. Available ones are meant for RF applications, not in accordance with the present frequency range and cost-ineffective. As the manual implementation of capacitor bank was inefficient several ways of automatically controlling capacitor bank were tried. An open source Arduino-based capacitance meter capable of measuring capacitance in all ranges is used to measure the on-board capacitance in each method [22]. These methods are shown in Fig. 4. As shown in Fig. 4, in order to realise capacitor bank, switches are needed (Sw). Each switch adds different values of capacitor (Cp_n) to circuit. Mechanical push buttons used as switches created unwanted delays in adding capacitance values and required application of pressure to function properly on breadboards. MOSFETS were the second option chosen for implementing switch due to their fast operation and lesser power consumption. But the effective capacitance value dropped to $PicoFarads$ as the internal capacitance of MOSFET added up with the capacitance in the capacitor bank. The capacitance meter displays the measured on-board capacitance on the LCD display monitor. Finally a 16 channel analog multiplexer is employed to act as switches. Each capacitor is connected to each channel of the multiplexer. 16 channels enables to have 16 values of capacitances. Along with 332 values of each of R_i and R_e this makes a huge number of combinations of simulator circuit. The detailed circuit diagram of the bio-impedance simulator is as shown in Fig. 5. The signal pin

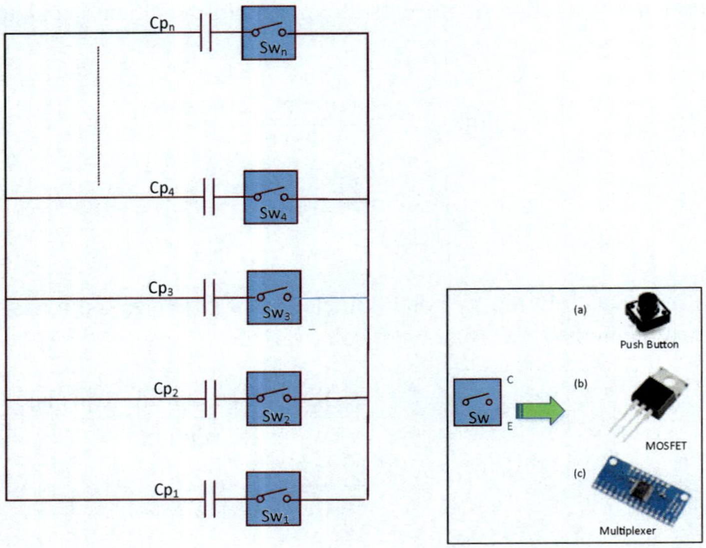

Fig. 4 Automated capacitance bank using **a** push button **b** MOSFET **c** multiplexer

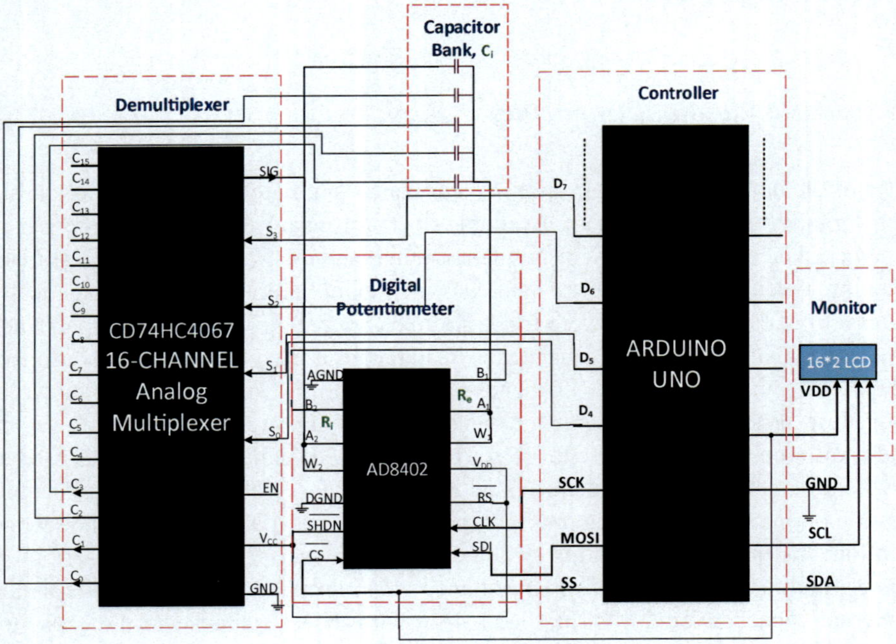

Fig. 5 Detailed circuit diagram of the proposed bio-impedance simulator

of the multiplexer is connected to the default capacitor of capacitance bank. This capacitance gets connected to corresponding capacitances depending on the status of D7–D4 pins of Arduino. These pins vary from 0000 to 1111 thereby selecting different channels of multiplexer. Each channel is a switch and is turned on for corresponding D7–D0 bits. AD8402 is controlled by the SPI pins of Arduino, namely, SCK, MOSI and SS. The live values of resistances and capacitances are displayed on a 16 by 2 LCD monitor and this display is controlled by Arduino via the Inter Integrated Circuit (I2C) bus.

5 Results and Discussions

Manual realisation TUT phantom using network of passive components produces a maximum error of 25 and 23% for resistors and capacitors, respectively, Table 1 because of which automation of bio-impedance simulator is carried out.

5.1 *Experimental Setup of Bio-impedance Simulator*

Experimental setup of the proposed bio-impedance simulator is shown in Fig. 6.

The SHDN pin of AD8402 is made high to retain the value of digital potentiometer after power off. This enables the measurement of its resistance using a multimeter. The relationship between the resistance value expected and the one measured by a multimeter is found out and plotted. It was seen that the relationship was linear as depicted in Fig. 7. A slope of 1 is obtained experimentally and proves the linear operation of resistors in the suggested simulator. Once the resistance showed accurate automated operation it is combined with the automated capacitance bank. R_i and R_e is varied from 50 Ω to 1 kΩ for each value of capacitance set by the condition of S_3-S_0 pins which are controlled by the D7-D4 pins of Arduino. While using 16 channel analog multiplexer as switch, theoretical capacitances (as marked on the ceramic capacitor) are compared with the capacitance measured by the capacitance meter and is depicted in Table 2. It shows that error in each capacitance gets accumulated as the value of capacitance increases. The error in 10 nF capacitance is 2.3 nF. Accuracy of these components can be increased by using 1% tolerance capacitances. The comparison of theoretical and practical values of capacitor (measured using Arduino-based capacitor meter) is depicted through a graph as shown in Fig. 8. Inference from the graph is that the values of practical values of capacitors are within the tolerance limit of 25% suggested by the manufactures and can be further improved if less tolerance components are used.

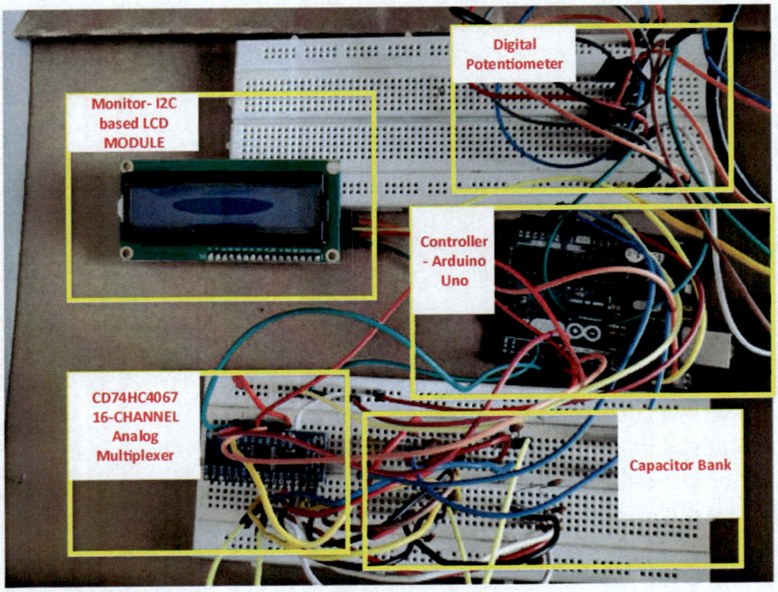

Fig. 6 Experimental setup of bio-impedance simulator

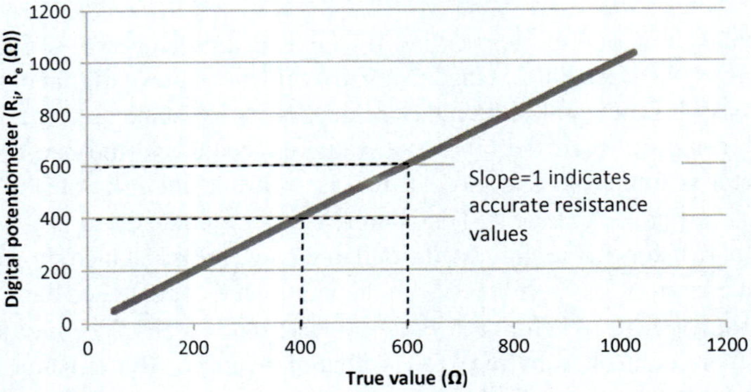

Fig. 7 Linearity in the resistance simulated through digital potentiometer AD8402

5.2 *Validation of BI Simulator*

Power consumed by implemented bio-impedance simulator is experimentally calculated by measuring current drawn and voltage supplied using Tecktronix 6.5 precision multimeter and is found to be a meagre power of 125 mW. The performance of bio-impedance simulator is assessed by using it as load for constant current source and experimental setup for the same is as shown in Fig. 9. We have already designed and implemented a precise portable constant current source in the load-in-the-feedback loop topology which offers a constant current of 159 μA is used to validate the

Table 2 Comparison of theoretical and practical capacitances while using 16-channel analog multiplexer as switch

S_3	S_2	S_1	S_0	C_i (Theo.) (nF)	C_i (Prac.) (nF)	Error (%)
Default				10	12.3	23
0	0	0	0	20	24.04	20.2
0	0	0	1	47	49.2	4.680851
0	0	1	0	67	73.12	9.134328
0	0	1	1	110	116.09	5.536364
0	1	0	0	210	217.7	3.666667
0	1	0	1	310	320.07	3.248387
0	1	1	0	410	422.28	2.995122
0	1	1	1	510	525.22	2.984314
1	0	0	0	710	729.52	2.749296

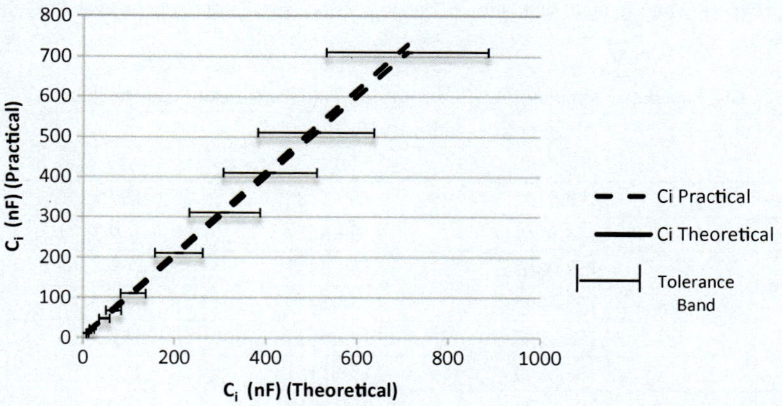

Fig. 8 Comparison of theoretical and practical values of capacitor, C_i incorporated in the simulator

bio-impedance simulator [18]. This source offers a constant current irrespective of load. Bio-impedance simulator with different combinations of resistances and capacitances act as varying load. Load currents are measured using Tektronix 6.5 precision Digital Multimeter which is connected in series to the source. R_i and R_e are varied from 50 Ω and 1 kΩ for each capacitance value. 500 samples are measured for each value of capacitance. The results obtained during the experiment are shown in Table 3. Minimum, maximum and average currents for 500 samples are analysed here. It is seen that there is only a variability of 1.69 μA which is well within the limits. Comparative analysis of implemented device over other existing techniques is elaborated in Table 4. The developed simulator consists of digital potentiometer, ceramic capacitors and Arduino Uno as controller all of which are high frequency compatible.

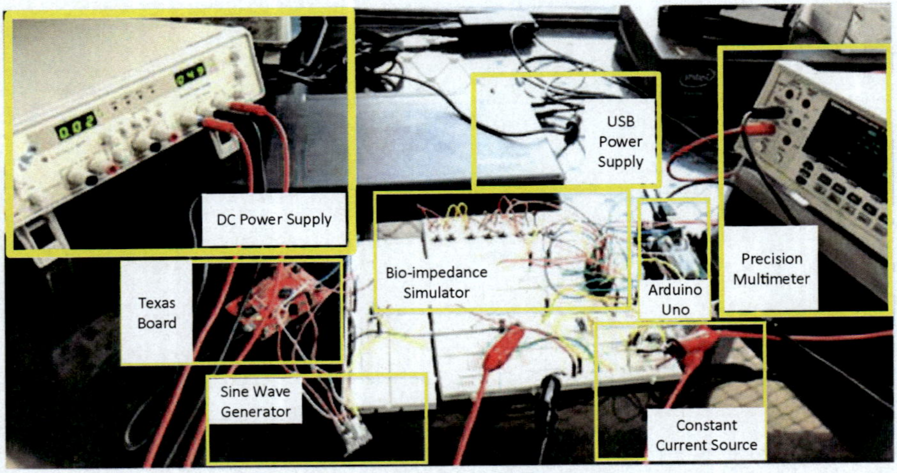

Fig. 9 Experimental setup of validation of bio-impedance simulator using precise constant current source

Table 3 Bio-impedance simulator acting as varying load for constant current source

Capacitance (nF)	Current (μA with varying R_i and R_e		
	Min	Max	Avg
4.7	158.911	159.7885	159.5256
10	158.8551	159.5258	159.3171
22	158.0882	158.5427	158.3158
47	158.1555	158.5440	158.3146
60	158.1297	158.5621	157.3325
80	157.5245	158.1151	157.3085
100	158.3373	159.0958	158.5408
1000	158.157	159.5232	158.2635

6 Conclusion and Future Scope

Bio-impedance simulator was designed and implemented to mimic biological tissue. The designed simulator can be used to test and validate bio-impedance devices developed. The known value of bio-impedance can be compared with the value measured by the developed device. It consists of accurate digital potentiometers as resistance bank and multiplexer controlled capacitor bank. Automated bio-impedance simulator with digital potentiometer yields linear operation and 100% match with theoretical values which is an improvement over manual realisation of simulator that has a percentage error of 25%. A variation of only 1.69 μA is obtained when tested as a varying load for a precise constant current source. The accuracy of the simulator can be further improved by using 1% tolerance capacitances. Power consumed by the

Table 4 Comparative analysis of bio-impedance simulators

Method	Tissue phantom	Power consumption	Frequency compatibility	Application
Ex-vivo [12]	Real organ tissue	-	100 kHz	General
LCR Meter [13]	Plant tissue	500 W	10 MHz	Nitrogen content analysis
In-vitro [6]	Bio-chemical culture	1 W	100 MHz	Uro-lithosis
Signal simulators [3]	ICG & IRG	660 mW	5 kHz–1 MHz	Cardiology and pulmonary
Analog simulators [10, 11]	2R-1C network	>5 W	100 kHz	Glottography
Software simulators [15, 16]	Impedance distribution	100–200 W	1 kHz–2 MHz	Musculo-skeletal
Experimental animals [14]	Animal tissue	500 W	5 MHz	Cancer
Our work	Automated 2R-1C network	125 mW	No frequency constraints	General

developed simulator is experimentally got as 125 mW. The system will be improved to study the behaviour of tissues at different frequencies and will be modified to suit bio-impedance spectroscopy-based disease diagnosis.

References

1. Simini F, Bertemes-Filho P (2018) Bioimpedance in biomedical applications and research. Springer International Publishing
2. Long V, Short M, Smith S, Sénéchal M, Bouchard DR (2019) J Sports Med (5). https://doi.org/10.1155/2019/7624253
3. Kim GE, Kim SY, Kim SJ, Yun SY, Jung HH, Kang YS, Koo BN (2019) Yonsei Med J 60(3):735–741. https://doi.org/10.3349/ymj.2019.60.8.735, https://doi.org/10.1007/s10470-019-01430-0
4. Sarvaiya J, Pandey P, Pandey V (2009) IETE J Res 55(3):100. https://doi.org/10.4103/0377-2063
5. Huang JJ, Hung YH, Wang JJ (2015) Measurement 78:9
6. Chung WY, Silverio AA, Tsai VF, Cheng C, Chang SY, Zhou MY, Chen S-Y, Kao CY, Rustia DA, Lo YW (2016) Microelectron J 56:142
7. Li N, Xu H, Wang W, Zhou Z, Qiao G, Li DDU (2013) Meas Sci Technol 24
8. Webster JG, Eren H, Measurement, instrumentation, and sensors handbook: electromagnetic, optical, radiation, chemical, and biomedical measurement
9. Grimnes S, Martinsen O (2008) Bioimpedance and bioelectricity basics. Academic Press
10. Simic M, Babic Z, Risojević V, Stojanovic G, Ramos A (2015) Society for design and process science SDPS conference
11. Blomqvist KH, Sepponen RE, Lundbom N, Lundbom J (2012) 2012 13th Biennial Baltic electronics conference, pp 199–202. https://doi.org/10.1109/BEC.2012.6376851

12. Dodde RE, Kruger GH, Shih AJ (2015) J Med Devices 9(2):11. https://doi.org/10.1115/1.4029706
13. Muñz-Huerta RF, Ortiz-Melendez ADJ, Guevara-Gonzalez RG, Torres-Pacheco I, Herrera-Ruiz G, Contreras-Medina LM, Prado-Olivarez J, Ocampo-Velazquez RV (2014) Sensors 14(7):11492. https://doi.org/10.3390/s140711492
14. Gonzalo M, Martínez-Beamonte R, Palacios P, Marín J, Castiella T, Surra J, Burdío F, Sousa R, Gemes A, Osada J, García-Gil A (2012) Transplantation proceedings, vol 44, no 6, p 1579. https://doi.org/10.1016/j.transproceed.2012.05.006, http://www.sciencedirect.com/science/article/pii/S004113451200437X
15. Olmo A, Yufera A (2010) pp 178–182. https://doi.org/10.13140/2.1.4991.8402
16. Anand G, Lowe A, Al-Jumaily A (2016) J Electr Bioimpedance 7:20. https://doi.org/10.5617/jeb.2657
17. Pandey VD, Pandey PC, Sarvaiya J (2008) IETE J Res 54: https://doi.org/10.1080/03772063.2008.10876200
18. Sruthi S, Dhavse R, Sarvaiya JN (2019) Analog integrated circuits and signal processing. https://doi.org/10.1007/s10470-019-01430-0
19. Bogonez-Franco P, Nescolarde L, Galvez-Monton C (2013) Physiol Meas 34(1):1. https://doi.org/10.1088/0967-3334/34/1/1
20. Sanchez B, Bandarenkab AS, Gerd Vandersteenc RB, Schoukensc J (2013) Med Eng Phys 35
21. Analog Devices, Digital Potentiometer (2002). Rev. 3
22. Badger P (2008) Capacitance meter and rc time constants. https://www.arduino.cc/en/Tutorial/CapacitanceMeter

Analysis of Memory-Based Real Fast Fourier Transform Architectures for Low-Area Applications

Rajasekhar Turaka and M. Satya Sai Ram

Abstract The Fast Fourier Transform (FFT)– processor can be described as the most important Numerical Algorithm of our lifetime. In this paper, we have presented different Memory-based Real Fast Fourier Transform (RFFT) Architectures for low-area applications with the Processing Elements (PE) like High Radix Small Butterfly (HRSB), Urdhva Tiryakbhayam Butterfly (UTB) and a PE with vedic multiplier-carry lookahead units are utilized to reduce the Area, Delay and Area-Delay-Product (ADP). In this, the FFT processor is based on Radix-2 Decimation-In-Frequency (DIF) Algorithm, and it also supports higher radix algorithms. The architectures are implemented on various Xilinx Field Programmable Gate Array (FPGA) devices and also on 180 nm Application-Specific Integrated Circuit (ASIC) to compare different parameters like Area, Delay and Power.

Keywords Fast Fourier Transform (FFT) · Real FFT · Field programmable gate arrays (FPGA) · Application-specific integrated circuit (ASIC) · Urdhava tiryakbhayam sutra

1 Introduction

The Fast Fourier Transforms (FFT) are the basic operations in most of the digital and image signal processing applications, and they are an efficient way to compute the Discrete Fourier Transform (DFT). The FFT processor plays an important role in modern digital communications such as digital video broadcasting (DVB), high-speed digital subscriber lines, medical imaging and power line communications

R. Turaka (✉)
Department of ECE, Velagapudi Ramakrishna Siddhartha Engineering College, Kanuru, Vijayawada, A.P., India
e-mail: rajasekharphd2017@gmail.com

M. S. S. Ram
Department of ECE, RVR&JC College of Engineering, Guntur, A.P., India
e-mail: msatyasairam1981@gmail.com

Z. Patel et al. (eds.), *Advances in VLSI and Embedded Systems*, Lecture Notes in Electrical Engineering 676, https://doi.org/10.1007/978-981-15-6229-7_3

(PLC) [1]. Because of its productive parallel form, the FFT is a benchmark in evaluating the performance of the digital signal processor. Area, speed, power consumption, execution time and throughput are the key parameters of an FFT processor. Different architectures, algorithms and methodologies have been extensively used in the past. This forms the foundation for new competent novel architectures. One fundamental architecture issue is the type of hardware platform. The FFT processors are commonly implemented on various evaluation boards such as digital signal processor (DSP), FPGAs and ASICs. Digital signal processors are a generalized form of microprocessors. The primary reason most engineers choose FPGA over DSP is driven by the application's millions instructions per second (MIPS) requirements and also FPGAs having inherent advantages such as reliability, flexibility and adaptability, where as ASIC stands for application-specific IC, specifically designed for one particular application. The difference between FPGA and ASIC is that in case of ASIC, the system once embedded with a function cannot be modified, whereas FPGAs are highly flexible devices.

A standard FFT processing unit is composed of butterfly computation units, memories and an address generator. Diverse architectures are designed to enhance the overall performance and complexity of the FFT hardware units. The processing element (PE) and the memory unit are the key units in the FFT processor. FFT processors can be classified based on their attributes, architectures and algorithms they apply. Based on their architecture, FFT processors are classified as memory, pipelined and parallel structures and their subdivisions, and also a combination of these architectures like parallel-pipelined architecture is possible. Pipelined architectures use many butterfly units to attain high-speed operation with a larger die area. Many butterfly units are used in a parallel manner in parallel FFT architectures [2].

1.1 Memory-Based Architectures

The framework of a memory-based FFT processor involves a butterfly computation unit, a ROM and RAM unit for storing intermediate data. The input data, intermediate results and outputs are saved via a dual-port RAM. The ROM is used to gather the twiddle factors. The address generator unit generates the address for analyzing information for butterfly operations and also for storing the output results in RAM. The control signals for every module are provoked through the sequential control unit.

1.1.1 Single Memory-Based Architecture

The simple memory-based architecture is a single memory-based design. It contains only a single memory unit as shown in Fig. 1. For each of the $\log_r M$ phases of the FFT, the data is read from the memory and written back to the same memory.

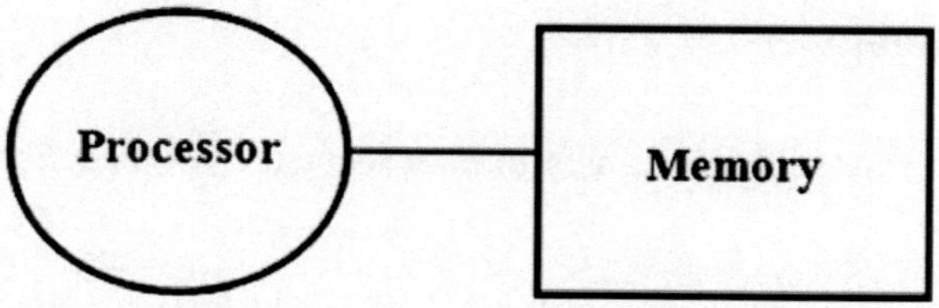

Fig. 1 Block diagram of single memory-based architecture [3]

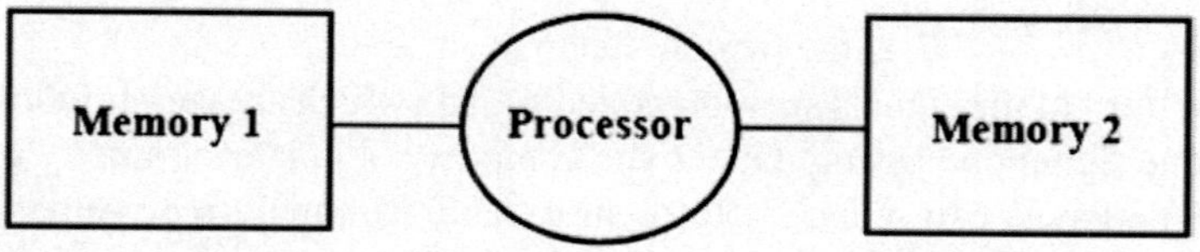

Fig. 2 Block diagram of dual memory-based architecture [3]

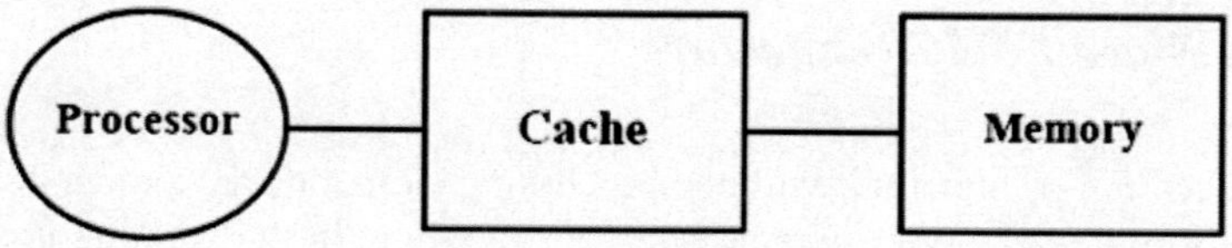

Fig. 3 Block diagram of cache memory-based architecture [3]

1.1.2 Dual Memory-Based Architecture

Dual memory-based design performs "ping-pong-like" activities to achieve read and write tasks. Every one of the two memories exchanges their features between butterfly units of input and output which is shown in Fig. 2.

1.1.3 Cache Memory-Based Architecture

This architecture is analogous to the architecture which is shown in Fig. 1 except that a small cache memory exists between the main memory and the processor unit. The aim of this architecture is to diminish the size of the main memory to reduce the power consumption for large FFT sizes also. This offers two advantages over other architectures that include higher speed and increased energy efficiency [3].

2 Methodology

The M-point discrete Fourier transform for a sequence y(n) is defined as

$$Y(k) = \sum_{n=0}^{M-1} y(n)W_M^{nk}, 0 \leq k \leq M-1$$

where $W_M = e^{-j2\pi/M}$ is the twiddle factor.

Today, most of the signals are real-valued signals which are very important in case of real-time signal processing. Due to the complex conjugate symmetry property of real-valued signals can reduce both memory and arithmetic requirements. Hence, there is no need to calculate all of the samples of FFT.

2.1 In-Place FFT Architecture

An in-place FFT architecture with one processing element (PE) shown in Fig. 4 can process 4 samples at a time, and each memory module in the architecture can store up to M/4 words of length P. There are four memory banks which are used to store the intermediate values of the butterfly computations. Each processing element is connected with four memory banks through the set of multiplexers, in which one set

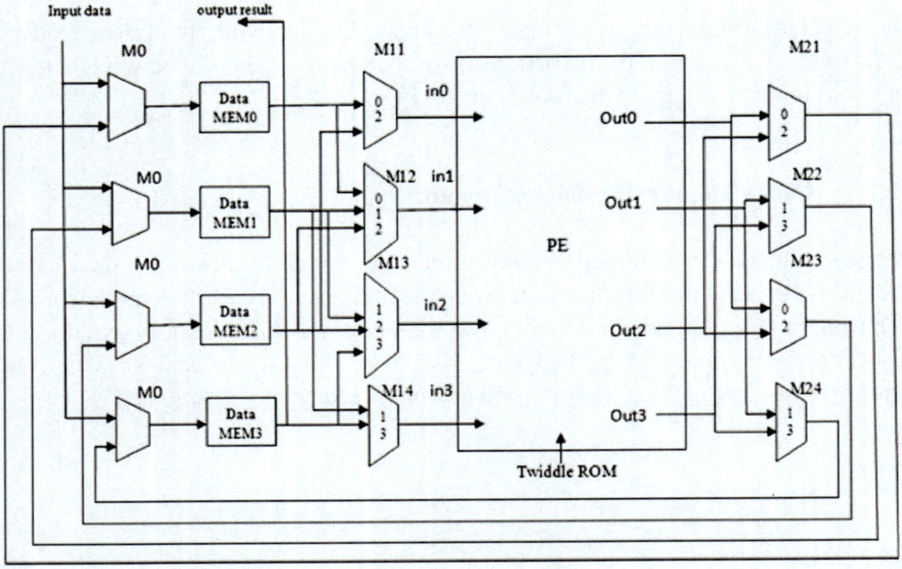

Fig. 4 In-place FFT architecture [4]

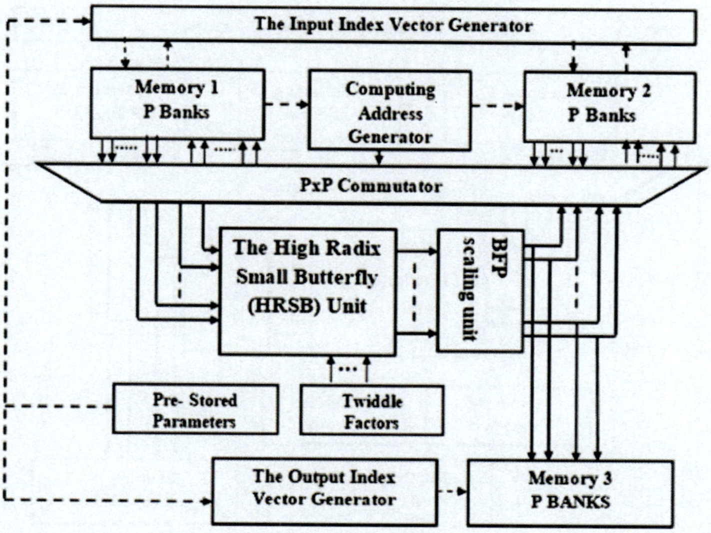

Fig. 5 Continous flow FFT architecture with HRSB [5]

decides the input sequence and another set decides which memory bank stores the intermediate values.

The processing element can read and write the data to the same memory module [4].

2.2 Continous Flow FFT Architecture with HRSB

This is the example of the index mapping technique of the prime factor algorithm and common factor algorithm. The FFT processor architecture with a high radix butterfly unit (HRSB) is shown in Fig. 5. It comprises an index vector generator unit for both input and output, three memory modules, address generator and an HRSB unit to compute the butterfly opeartion with the help of prestored twiddle factor parameters. Without any data conflicts the input data distributes to memory modules by the input index vector, and the processed data reloads into another memory module by the output index vector generator.

2.3 Continous Flow FFT Architecture with UTB

The modified conflict-free memory-based FFT architecture with urdhva tiryagbhayam butterfly (UTB) unit is shown in Fig. 6. Vedic mathematics consists of 16 sutras for computing multiplications. Among the 16 sutras, urdhava tiryakbhyam, a

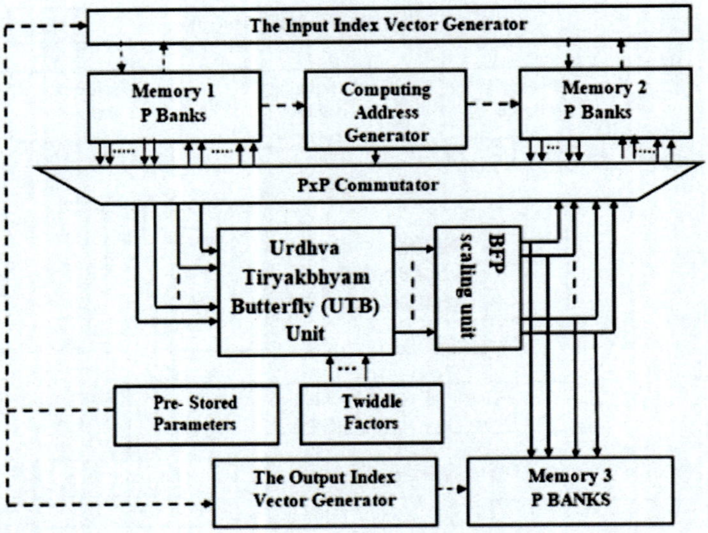

Fig. 6 Continous flow FFT architecture with urdhava tiryakbhayam [5]

crowning gem of all sutras, is the most universal and efficient method through which the product of any two natural numbers can be produced [5].

2.4 LC-CSLA-RFFT Architecture

The block diagram of low cost-carry select adder-RFFT (LC-CSLA-RFFT) architecture is shown in Fig. 7. It comprises processing element (PE), RAM I, RAM O, multiplexer (MUX), demultiplexer (DEMUX), control circuit, address generator and register bank [6].

2.5 RFFT Architecture with Dual-Port RAM and Vedic Multiplier

The proposed RFFT architecture with dual-port RAM with the processing element where the data can read from and write into the same memory is shown in Fig. 8. The PE uses a vedic multiplier and carry lookahead adder to improve the area, speed, delay and area-delay-product (ADP) and is suited for moderate and high-speed applications. It consists of a control circuit having the controlling signals, DRAM used to perform both read and write operations and a PE could perform the butterfly computations of all the stages [7].

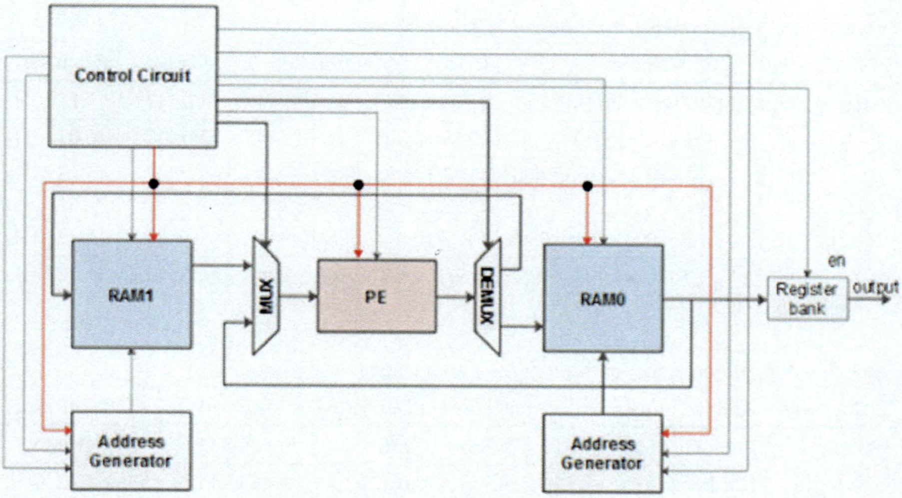

Fig. 7 Block diagram of LC-CSLA-RFFT architecture [6]

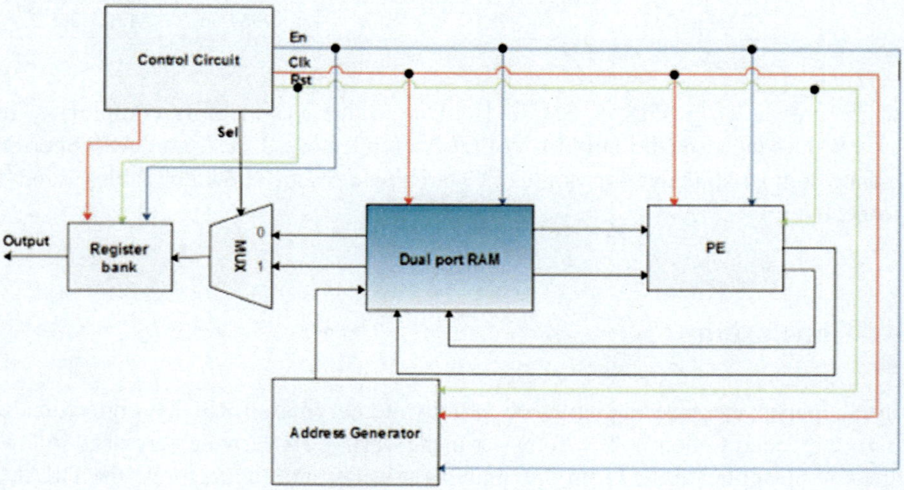

Fig. 8 Block diagram of DRAM-VM-CLA RFFT architecture [7]

3 Implementation and Performance Evaluation

In this, the FFT architectures are implemented using Verilog hardware description language (HDL) on various FPGA and ASIC platforms.

We have analyzed FPGA performance for Virtex-4 device using Xilinx ISE 14.2 tool. ASIC performance can be obtained from the Cadence Virtuoso RTL compiler tool in 180 nm technology. The DRAM-VM-CLA architecture uses only one dual-port memory bank whereas other architectures use more than two memory banks

Table 1 FPGA performance for different FFT architectures

FPGA device	Architecture	LUT	Flip flop	Slice	Frequency
Xc4vfx12 Virtex-4	Zhen-Guo [4]	1376/10,944	94/10,944	737/5472	45.13 MHz
	Kai-Feng [5]	1245/10,944	92/10,944	721/5472	46.07 MHz
	Kai-Feng_modified [5]	1198/10,944	88/10,944	687/5472	46.45 MHz
	Rajasekhar.T [6]	1177/10,944	65/10,944	630/5472	47.42 MHz
	Turaka.R [7]	609/10,944	152/10,944	361/5472	65.24 MHz

Table 2 ASIC performance for different FFT architectures

Technology	Architecture	Area (μm^2)	Delay (ps)	Power (nW)	ADP (μm^2xps)
180 nm	Zhen-Guo [4]	3,425,843	170.7	9,459,835	584,791,400
	Kai-Feng [5]	3,314,752	162.5	9,583,945	538,647,200
	Kai-Feng_modified [5]	3,225,612	162.5	9,648,692	524,161,950
	Rajasekhar [6]	3,145,451	170.7	10,366,843	534,726,670
	Turaka. R [7]	592,981	92.2	1,996,096	54,672,848.2

which are shown in Figs. 4, 5, 6 and 7. Due to this less-memory complexity, the area was reduced in the DRAM-VM-CLA architecture. FPGA and ASIC performance comparison table for various FFT architectures are shown in Tables 1 and 2, respectively.

4 Conclusion

In this paper, we have implemented various memory-based RFFT architectures in Xilinx ISE and Cadence RTL compiler using Verilog HDL, which are used in low-area VLSI applications. From the analysis of different architectures, the DRAM-VM-CLA architecture is best in terms of area, delay and ADP in case of ASIC performance, and also LUTs, slices and frequency are improved in case of FPGA performance with the use of a dual-port RAM and a vedic multiplier. This architecture can give speed up to 65.24 MHz, slices count 361 and delay of 92.2 ps.

References

1. Jo BG, Sunwoo MH (2005) New Continuous-Flow Mixed-Radix (CFMR) FFT processor using novel in-place strategy. IEEE Trans Circ Syst-1 52(5):911–919
2. Joshi SM (2015) FFT architectures: a review. Int J Comput Appl 116(7)

3. Bass BM (1999) A low-power, high-performance, 1024-point FFT processor. IEEE J Solid State Circ 34(3)
4. Ma Z-G, Yin X-B, Feng Yu (2015) A novel memory based FFT architecture for real valued signals based on a Radix-2 decimation-in-frequency algorithm. IEEE Trans Circ Syst II 62(9):876–880
5. Xia K-F, Wu B, Xiong T (2017) A memory based FFT processor design with generalized efficient conflict-free address scheme. IEEE Trans Very Large Scale Integr Syst 25(6)
6. Turaka R, Ram MSS (2018) Low area high speed LC-CSLA-RFFT architecture for Radix-2 decimation-in-frequency algorithm. J Adv Res Dyn Control Syst 10(09-Special Issue)
7. Turaka R, Ram MSS (2019) Low power VLSI implementation of real fast fourier transform with DRAM-VM-CLA. Microprocess Microsyst 69:92–100

Optimization of MEMS-Based Capacitive Sensor with High-k Dielectric for Detection of Heavy Metal Ions

Dinesh Rotake and Anand D. Darji

Abstract Nowadays, all the countries worldwide facing lots of environmental issues out of which Heavy Metal Ions (HMIs) contamination is extremely harmful and hazardous to human health. Many countries were facing enormous challenges to solve this HMIs problem. Air and Water pollution due to HMIs is a global issue and required to solve as early as possible to maintain today's air and water quality demands. A portable device made using MEMS-based sensors technology recommended to detect the multiple analytes simultaneously for environmental monitoring applications. Accordingly, our main objective is to optimize and design a capacitive MEMS-Based sensor platform utilized for sensing the low concentration of HMIs. As we know, a proposed capacitive sensor is designed using MEMS-based technology, which ultimately produces the capacitance in the femtofarads (fF) range. It is tough and costly to measure this fF range of capacitance. An individual capacitance to digital converter (CDC) circuit is required to measure this capacitance produced by MEMS-based sensors. Hence, we have used the HfO_2 as High-k dielectric layer to increase the sensor capacitance in picofarads (pF) range to use the market available CDC circuit AD7150. This mixed dielectric approach is beneficial to solve the problem of measurement in capacitive sensors for the low-cost application using the market available embedded systems. In this work, we try to optimize the microcantilever-based capacitive MEMS sensor with a different thickness of HfO_2 as a High-k material using COMSOL Multiphysics 5.3 software. The FEA analysis for an optimized capacitive sensor with HfO_2 as a High-k dielectric shows the maximum capacitance variation 3.5 fF compared to 0.3 fF without High-k for HMIs mass between 1 and 1000 μg.

Keywords BioMEMS · COMSOL multiphysics · Capacitive sensor · Finite element analysis (FEA) · SAM (Self-assembled Monolayers) · High-k · Capacitance to digital converter (CDC)

D. Rotake (✉) · A. D. Darji
Electronics Engineering Department, Sardar Vallabhbhai National Institute of Technology, Surat 395007, India
e-mail: dinesh.rotake@gmail.com

Z. Patel et al. (eds.), *Advances in VLSI and Embedded Systems*, Lecture Notes in Electrical Engineering 676, https://doi.org/10.1007/978-981-15-6229-7_4

1 Introduction

Today many environmental problems have a significant effect on the life of humans all around the globe, out of which HMIs contamination is proven to be very harmful, according to World Health Organization (WHO) and Central Ground Water Board (CGWB). According to reports, 15 million children die each year under the age of five because of diseases caused by drinking water, out of which HMIs contamination is the most hazardous and cannot be removed easily. To use any type of sensor in the Biosensor application requires to have a surface modification to capture the targeted molecules. So, we need to deposit the gold layer on the top surface of the sensor. Chromium was used below the gold surface to improve the adhesion of gold to polysilicon. The Au-layer is used to immobilize the biolinker on the top surface of the sensor to detect the required analytes. Air and Water pollution due to HMIs is a global problem that required prime attention and needs to solve as early as possible. The proposed application assumes variation in HMIs mass between 1 and 1000 μg per liter range as per the WHO data [1]. The proposed MEMS-based sensor utilized the capacitive effect to detect the targeted HMI biomolecules. This range of HMIs mass exerts a force in the range of 9.80665×10^{-9}N to 9.80665×10^{-6}N and pressure of 0.78453–784.53 Pa for a proposed design. We have also calculated the number of particles need to detect in order to cross the WHO limit depending on the molar mass of different HMIs. According to the average particle radii of HMIs, we have decided the width of the microcantilever beam to be 250 μm.

We have also investigated the previous work done in the field of HMIs detection, such as microcantilevers functionalized with metal-binding protein AgNt84-6 is demonstrated as suitable sensors for the detection of heavy metal ions like Hg^{2+} and Zn^{2+} [2]. SAMs (self-assembled monolayers) modified microcantilevers used for the detection of Ca^{2+} ions are presented in [3]. Arrays of microcantilever sensors encapsulated in fluidic wells and fluidic channels are discussed in [4] and [5], respectively. A Chitosan (CS)-graphene oxide (GO) Surface Plasmon Resonance (SPR) sensor is explained in [6], while simple microcantilever beam based detection is given in [7–9]. Since all of these methods use optical readout, they require heavy setup and costly lab equipment. Also, the optical readouts for microcantilever-based sensors have disadvantages when used with a microfluidic biosensor environment when the refractive index of liquid changes [10].

A portable system made of MEMS sensors capable of detecting multiple analytes simultaneously in air and water is highly demanded. The HMI detection in the vapor phase can be a solution for laboratory-based detection, but for the field instrument using MEMS, the temperature cannot be raised beyond a specific limit. Hence, the microfluidic detection is the only option which required high sensitivity [11, 12]. Capacitive detection has enormous potential to outperform over other sensing techniques for the applications which require to have high sensitivity, stability, and low-pressure. It is simple to take contact measurements and simple design principles. Compatible with CMOS technology and easily fabricated as compared to the other sensor types. This method does not require continuous bias current and unaffected by temperature, mechanical misalignment, etc. Among most of the electronic devices fabricated in microtechnology, the capacitors have their characteristics closest to the ideal behavior. Accordingly, our main objective is to develop a platform that can be used for sensing HMIs in air and water using a capacitive microcantilever-based sensor fabricated using MEMS technology.

The most challenging task in a microfluidic application is to avoid shorting of contacts as a complete assembly embedded in liquid. In case of a capacitive pressure sensor, the problem of shorting can be solved either by modifying the upper surface of the cantilever with an insulating layer of silicon dioxide (SiO_2) or silicon nitride (Si_3N_4) or incorporating the bilayer of air and High-k dielectric between the two plates of the capacitor. This approach solves two existing problems; first, it will avoid the shorting between two plates of the capacitor to reduce the power consumption, and second, it will increase the change in capacitance to few femtofarads in order to use the market available CDC integrated circuit AD7150 (evaluation board) having full-scale (changing) capacitance range of 13 pF and with a resolution of 1 fF [13]. The proposed capacitive MEMS sensor also reduced the additional cost of fabricating the CDC interfacing circuit for HMIs application [14].

2 Concept of High-k for Capacitive Sensor

The relation between capacitance (C) of the parallel plate capacitor and applied potential (V) is given by

$$C = \frac{Q}{V} \tag{1}$$

where C = capacitance between two plates, Q = amount of charge stored, V = applied potential.

Now, by substituting V = Ed and Q = εAE Eq. (1), we have parallel plate capacitance (C) is given by

$$C = \frac{\varepsilon A}{d} \tag{2}$$

where A = area of capacitor plate, ε = permittivity and d = distance between two plates.

Also, Capacitive sensor value with mixed dielectric theoretically given by formula with bilayer dielectric (Air + High-k) is

$$C = \frac{\varepsilon A}{d - t\left(1 - \frac{1}{ke}\right)} \tag{3}$$

where, $\varepsilon = \varepsilon_0 \varepsilon_r$, ε_0 = Vacuum permittivity, ε_r = relative permittivity, t = thickness of High-k and ke = relative permittivity of High-k.

2.1 Sensitivity

In case of BioMEMS, sensor sensitivity is a very important parameter and this value required to be maximum for sensor application. The sensitivity (Sc) for the capacitive sensor is calculated by using the following formula:

$$Sc = \frac{\Delta C / C}{\text{Pressure}(\text{Pa})} \tag{4}$$

where ΔC = Change in capacitance, C = Initial value of capacitance.

2.2 Material Properties

The materials properties used in the FEA simulation of the capacitive sensor using COMSOL Multiphysics software are listed in Table 1.

3 Design of Optimized Capacitive Sensor for BioMEMS Application

FEA analysis of polysilicon and Silicon Nitride based capacitive sensor has been performed for different lengths and thicknesses of microcantilever to optimize the dimension [11]. To improve the sensitivity of the proposed capacitive sensor, we have performed the sensitivity analysis with respect to different thicknesses of High-k material. We have used the HfO_2 as a High-k material because of its compatibility with MEMS devices and well-known foundry processes. The fabrication step of the proposed modified capacitive sensor with High-k is shown in Fig. 1.

Table 1 Material properties used in COMSOL simulation

Material	Young's modulus (GPa)	Poisson's ratio
Polysilicon	160	0.22
Silicon Nitride	250	0.23
Aluminum	70	0.35
Gold	70	0.44
Chromium	279	0.21

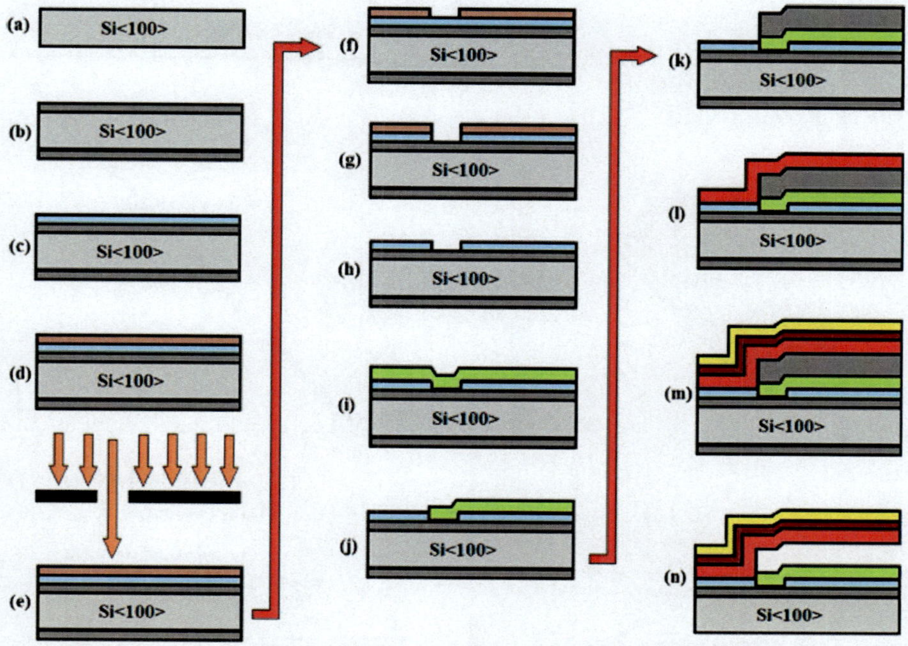

Fig. 1 Fabrication steps of proposed modified capacitive sensor with High-k: **a** Silicon Wafer <100> **b** 1 μm SiO_2 grown by thermal oxidation at 1000 °C **c** Aluminum deposition 200 nm **d** Photoresist (PR) coating 1 μm **e** UV- exposure through mask to have desired pattern **f** development of PR **g** etching of desired material **h** acetone dip for PR removal **i** deposition of HfO_2 3.2 μm **j** patterning of HfO_2 by repeating steps (**d**, **e**, **f**, **g**, **h**) **k** SiO_2 sacrificial layer deposition 0.3 μm **l** patterning of Polysilicon = 0.475 μm by repeating steps (**d**, **e**, **f**, **g**, **h**) **m** patterning of Chrome (Cr = 10 nm) and Gold (Au = 15 nm) by repeating steps (**d**, **e**, **f**, **g**, **h**) **n** SiO_2 sacrificial layer etching using Buffered Hydrofluoric Acid (BHF) to released microcantilever

3.1 FEA Analysis of Capacitive Sensor Without High-K

FEA simulation of the polysilicon-based capacitive sensor without High-k has been performed using COMSOL software. The analysis is carried out for different pressure corresponding to HMIs mass between 1 and 1000 μg as per WHO data. The maximum value of change in capacitance for finalized dimension without High-k dielectric is 0.3 fF for length $(L) = 250$ μm, width $(W) = 80$ μm and thickness $(T) = 0.5$ μm with a maximum displacement of 3.18 μm for the polysilicon-based capacitive sensor using COMSOL Multiphysics software [15], as shown in Fig. 2. The sensitivity of this capacitive sensor without High-k using Eq. (4) is found to be 8.39 μF/F/Pa (see the appendix for calculation).

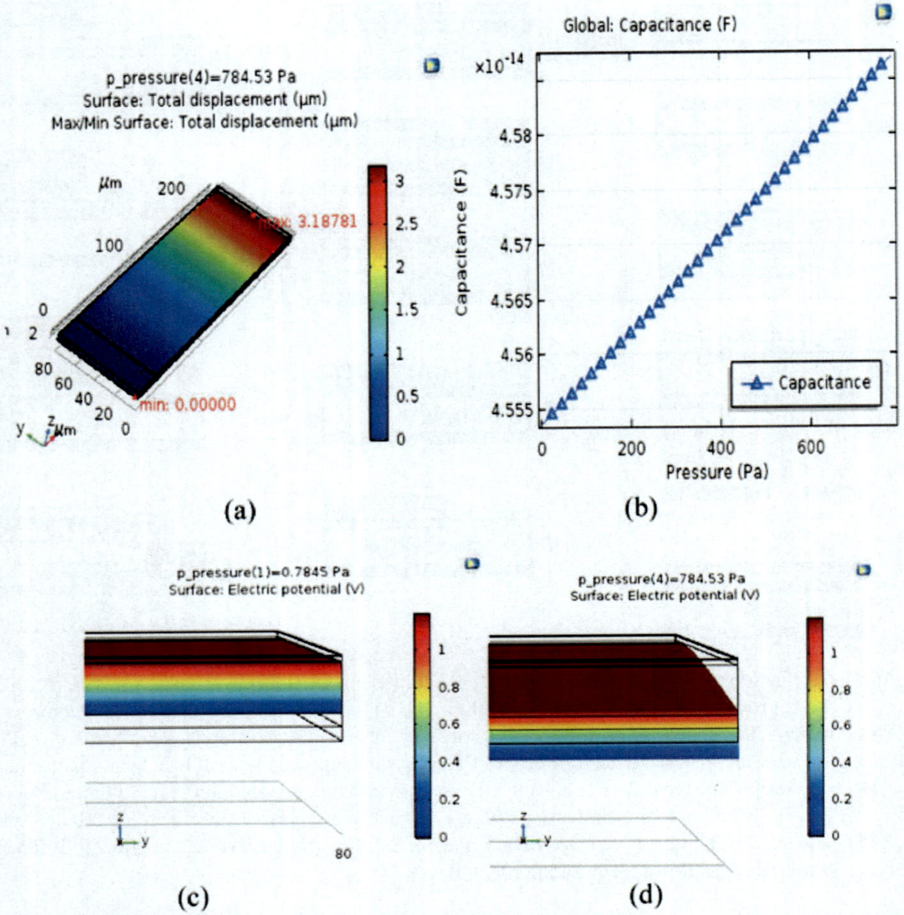

Fig. 2 **a** Maximum displacement for 784.53 Pa, **b** pressure versus capacitance graph without High-k dielectric, **c**, **d** showing variation of maximum displacement for 0.784 Pa and 784.53 Pa respectively for Applied potential (1.2 V)

3.2 FEA Analysis of Capacitive Sensor with High-k

Incorporating the bilayer of air and High-k dielectric between the two plates of capacitor solves two existing problems; first, avoid the shorting between to plates of the capacitor to reduce the power consumption and second, it will increase the capacitance few hundreds of femtofarads to use the market available CDC integrated circuit AD7150 (evaluation board with 0–0.5 pF sensor input and a resolution of 1 fF) to directly measure the capacitance in the digital domain and do the further processing.

The maximum value of change in capacitance for an optimized polysilicon-based capacitive sensor with High-k dielectric of 0.3 μm is 0.35 fF using COMSOL Multiphysics software as shown in Fig. 3.

The maximum value of change in capacitance for an optimized polysilicon-based capacitive sensor with High-k dielectric of 1.3 μm is 0.5 fF using COMSOL Multiphysics software as shown in Fig. 4.

The maximum value of change in capacitance for an optimized polysilicon-based capacitive sensor with High-k dielectric of 2.5 μm is 1.1 fF using COMSOL Multiphysics software as shown in Fig. 5.

The maximum value of change in capacitance for an optimized polysilicon-based capacitive sensor with High-k dielectric of 3.2 μm is 3.5 fF using COMSOL Multiphysics software as shown in Fig. 6.

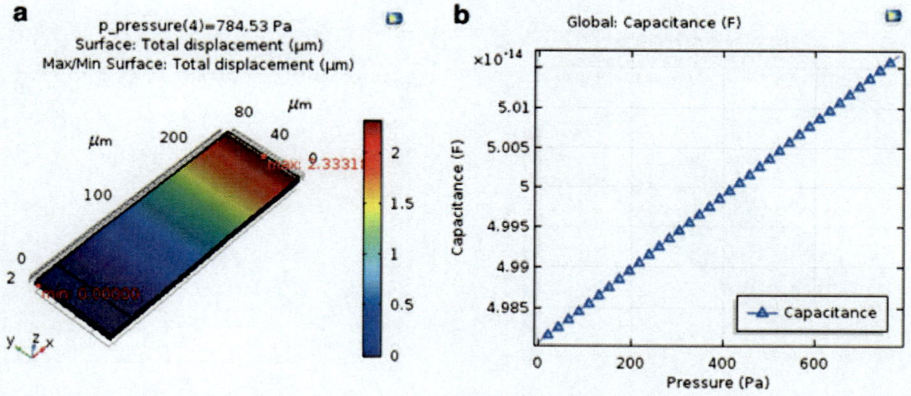

Fig. 3 **a** Maximum displacement for 784.53 Pa **b** pressure versus capacitance graph with High-k dielectric of 0.3 μm

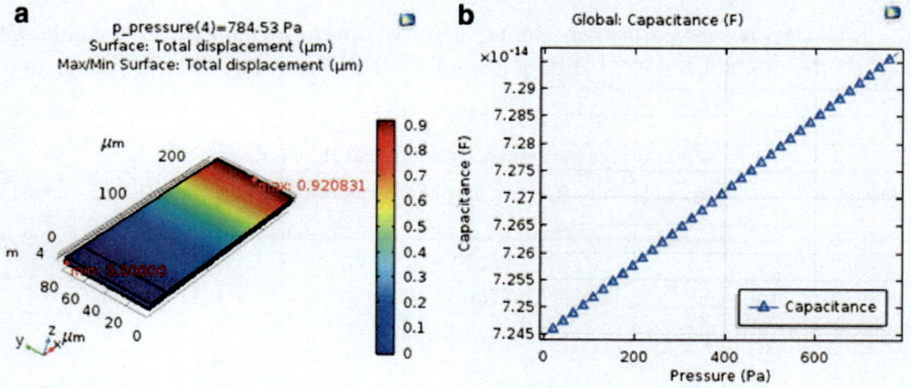

Fig. 4 **a** Maximum displacement for 784.53 Pa **b** pressure versus capacitance graph with High-k dielectric of 1.3 μm

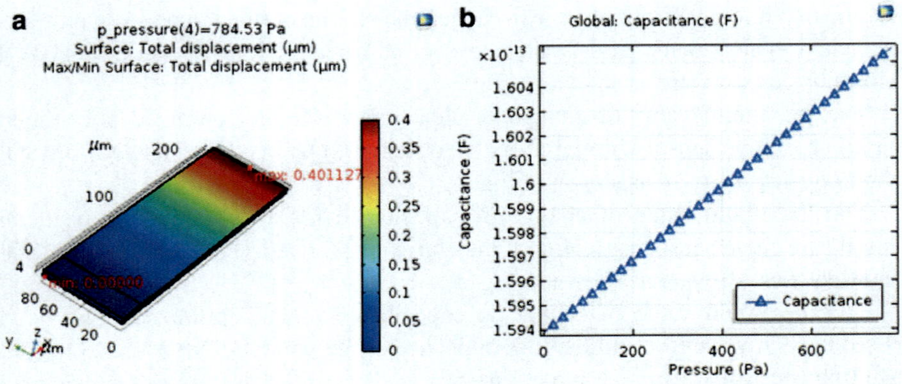

Fig. 5 **a** Maximum displacement for 784.53 Pa **b** pressure versus capacitance graph with High-k dielectric of 2.5 µm

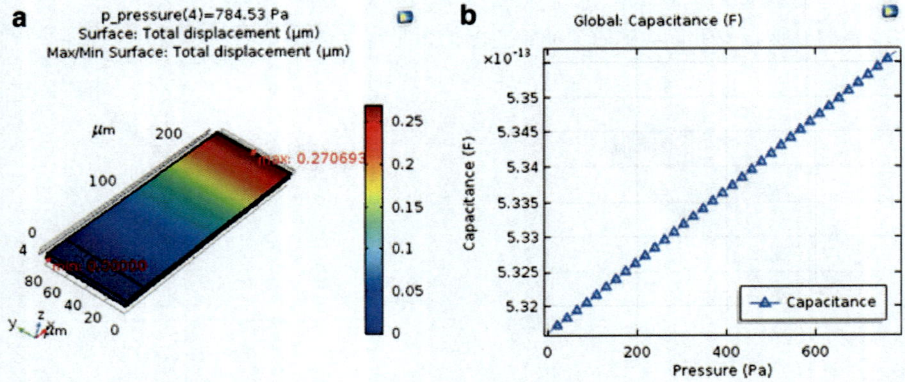

Fig. 6 **a** Maximum displacement for 784.53 Pa **b** pressure versus capacitance graph with High-k dielectric of 3.2 µm

Table 2 shows the performance comparison of optimized capacitive sensor design with different thicknesses of High-k using COMSOL Multiphysics FEA simulation.

Table 2 Capacitive sensor design performance for different High-k thickness

Design	Air dielectrics (µm)	Capacitance range (C) (fF)	Change in capacitance (ΔC) (fF)
Without High-k	3.5	45.85–45.55	0.3
High-k (0.3 µm)	3.2	50.15–49.80	0.35
High-k (1.3 µm)	2.2	72.95–72.45	0.5
High-k (2.5 µm)	1.0	160.5–159.4	1.1
High-k (3.2 µm)	0.3	535.5–532	3.5

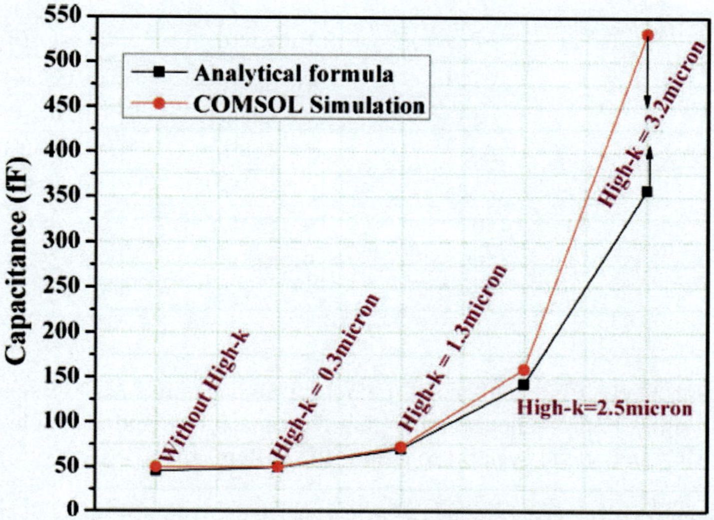

Fig. 7 Comparison for Capacitance using Analytical method and COMSOL simulation with maximum applied pressure of 784.53 Pa (see Appendix-C for theoretical calculation) is shown in Fig. 7.

The maximum value of change in capacitance for an optimized polysilicon-based capacitive sensor with High-k dielectric 0.3 μm, 1.3 μm, 2.5 μm, and 3.2 μm is 0.35 fF, 0.50 fF, 1.1 fF, and 3.5 fF, respectively, using COMSOL Multiphysics software (Fig. 7).

4 Conclusion

We have proposed an HfO_2-based MEMS capacitive sensor for HMIs detection in air and water. Many of the cantilever based design studied in the literature required the costly lab instruments for characterization and processing to detect the HMIs. Today's world requires a portable, low-cost embedded system with fast and accurate detection of contaminants. The proposed capacitive method overcomes all the above problems with HfO_2 as a High-k dielectric which ultimately increases the base capacitance of the proposed sensor to picofarads range. This approach solves two existing problems; first, avoid the shorting between two plates of the capacitor to reduce the power consumption, and second, it will increase the change in capacitance to few femtofarads in order to use the market available CDC integrated circuit AD7150 (evaluation board). The FEA analysis for an optimized capacitive sensor with HfO_2 as a High-k dielectric shows the maximum capacitance variation 3.5 fF compared to 0.3 fF without High-k for HMIs mass between 1 and 1000 μg. The maximum value of change in capacitance for an optimized polysilicon-based capacitive sensor with High-k dielectric 0.3 μm, 1.3 μm 2.5 μm, and 3.2 μm is 0.35 fF, 0.50 fF, 1.1 fF, and

Table 3 Mass to force conversion

Mass (μg)	Force (N)	Pressure (Pa)
0.001	9.80665E-12	0.000784532
0.01	9.80665E-11	0.00784532
0.1	9.80665E-10	0.0784532
1	9.80665E-9	0.784532
10	9.80665E-8	7.84532
100	9.80665E-7	78.4532
1000	9.80665E-6	784.532

3.5 fF, respectively. FEA analysis of the optimized capacitive MEMS sensor shows nearly ten times improvement in changing the capacitance and also able to use the market CDC (embedded system) to reduce the system cost.

Acknowledgments The authors would like to thank the Ministry of Electronics and Information Technology (MeitY), New Delhi for partial financial support through "Visvesvaraya Ph.D. Scheme for Electronics and IT" and SMDP-C2SD project.

Appendix

Conversion Table

There are various HMIs out of which Cadmium (Cd^{2+}) Mercury (Hg^{2+}), Lead (Pb^{2+}), Arsenic (Ar^{2+}) and Chromium (Cr^{2+}) are hazardous and WHO (World Health Organization) has provided limits 3, 1, 10, 10, 50 μg/L, respectively. Copper has the largest limit (Cu^{2+}) = 1000 μg/L so, from this we thought of mass variation in 1–1000 μg/L and conversion of mass to force shown in Table 3.

Calculation for Microcantilever Dimensions

- The molar mass of mercury (Hg) = 200.79 g/mole
- 1 mol = 6.02×10^{23} particles (Avogadro's number)
- 1 particle mass = (molar mass/Avogadro's number) = 3.33353×10^{-22} g
- WHO limit Hg = 1 μg/L
- So, No. of particles to be detected for 1 μg = (1 μg/1 particle mass) = 3×10^{15}
- No. of particles to be detected for 1 ng = 3×10^{12}
- No. of particles to be detected for 1 pg = 3×10^{9}
- No. of particles to be detected for 1 fg = 3×10^{6}
- Atomic Radii of Hg = 150 pm

- Here, we have selected area ($L = 250$ μm and width $= 80$ μm) $= 20000$ (μm × μm)
- So, No. of particles that can sit in this area of microcantilever $= \left\{\frac{20000(\mu m \times \mu m)}{\Pi \times 150 \times 150(pm \times pm)}\right\} = 2.83085 \times 10^{11}$
- Hence, approximately we can detect mercury ions in the range of 1 pg to 1 ng.

Theoretical Calculation for Capacitance

- Capacitive sensor value is verified theoretically with parallel plate capacitance formula using air ($\varepsilon_r = 1$) as a dielectric given by

$$C = \frac{\varepsilon A}{d} = \left\{\frac{9.954 \times 10^{-12} \times 1 \times 18000 \times 10^{-12}}{3.5 \times 10^{-6}}\right\} = 45.53\ \text{fF}$$

{* Effective overlap area is less by 25 × 80 = 2000 (μm × μm)} and Observed value using COMSOL5.3, (C) = 45.55 fF Fig. 2d.

- Capacitive sensor value with mixed dielectric are verified theoretically by using the formula with bilayer dielectric (Air = 3.2 μm + HfO_2 = 0.3 μm) is given by

$$C = \frac{\varepsilon A}{d - t\left(1 - \frac{1}{ke}\right)} = \left\{\frac{9.954 \times 10^{-12} \times 1 \times 18000 \times 10^{-12}}{3.5 \times 10^{-6} - 0.3 \times 10^{-6}\left(1 - \frac{1}{22}\right)}\right\} = 49.59\ \text{fF}$$

and Observed value using COMSOL5.3, (C) = 49.85 fF Fig. 3b

- Capacitive sensor value with mixed dielectric are verified theoretically by using formula with bilayer dielectric (Air = 2.2 μm + HfO_2 = 1.3 μm) is given by

$$C = \frac{\varepsilon A}{d - t\left(1 - \frac{1}{ke}\right)} = \left\{\frac{9.954 \times 10^{-12} \times 1 \times 18000 \times 10^{-12}}{3.5 \times 10^{-6} - 1.3 \times 10^{-6}\left(1 - \frac{1}{22}\right)}\right\} = 70.54\ \text{fF}$$

and Observed value using COMSOL5.3, (C) = 72.45 fF Fig. 4b.

- Capacitive sensor value with mixed dielectric are verified theoretically by using formula with bilayer dielectric (Air = 1.0 μm + HfO_2 = 2.5 μm) is given by

$$C = \frac{\varepsilon A}{d - t\left(1 - \frac{1}{ke}\right)} = \left\{\frac{9.954 \times 10^{-12} \times 1 \times 18000 \times 10^{-12}}{3.5 \times 10^{-6} - 2.5 \times 10^{-6}\left(1 - \frac{1}{22}\right)}\right\} = 143.1\ \text{fF}$$

and Observed value using COMSOL5.3, (C) = 159.4 fF Fig. 5b.

- Capacitive sensor value with mixed dielectric are verified theoretically by using the formula with bilayer dielectric (Air = 0.3 μm + HfO_2 = 3.2 μm) is given by

$$C = \frac{\varepsilon A}{d - t\left(1 - \frac{1}{ke}\right)} = \left\{\frac{9.954 \times 10^{-12} \times 1 \times 18000 \times 10^{-12}}{3.5 \times 10^{-6} - 3.2 \times 10^{-6}\left(1 - \frac{1}{22}\right)}\right\} = 357.6\ \text{fF}$$

and Observed value using COMSOL5.3, $(C) = 532$ fF Fig. 6b.

Theoretical Calculation for Sensitivity

- The sensitivity (Sc) for capacitive sensor without High-k is calculated by using a formula:

$$Sc = \frac{\Delta C/C}{\text{Pressure(Pa)}} = \frac{0.3\ \text{fF}/45.5\ \text{fF}}{784.53\ \text{(Pa)}} = 8.39\ \mu\text{F/F/Pa}$$

- The sensitivity (Sc) for capacitive sensor with High-k is calculated by using a formula

$$Sc = \frac{\Delta C/C}{\text{Pressure(Pa)}} = \frac{3.5\ \text{fF}/532\ \text{fF}}{784.53\ \text{(Pa)}} = 8.39\ \mu\text{F/F/Pa}$$

- From the calculated sensitivity of the capacitive sensor with and without High-k; it is clear that change in capacitance is improved without affecting the sensitivity of the sensor.

References

1. Thompson T, Fawell J, Kunikane S, Jackson D, Appleyard S, Callan P, Bartram J, Kingston P, Water S, Organization WH et al (2007) Chemical safety of drinking water: assessing priorities for risk management. https://apps.who.int/iris/bitstream/handle/10665/43285/9789241546768_eng.pdf
2. Cherian S, Gupta RK, Mullin BC, Thundat T (2003). Detection of heavy metal ions using protein-functionalized microcantilever sensors. Biosens Bioelectron 19(5):411–416. https://doi.org/10.1016/S0956-5663(03)00226-4
3. Ji H-F, Thundat T (2002) In situ detection of calcium ions with chemically modified microcantilevers. Biosens Bioelectron 17(4):337–343. https://doi.org/10.1016/S0956-5663(01)00270-6
4. Venstra W, Wien W, Sarro P, Van Eijk J (2012) Microcantilevers encapsulated in fluid wells for sensing in liquids. Microelectron Eng 97:247–250. https://doi.org/10.1016/j.mee.2012.03.030
5. Xiang W, Lee C (2009) Nanophotonics sensor based on microcantilever for chemical analysis. IEEE J Selected Topics Quantum Electron 15(5):1323–1326. https://doi.org/10.1109/JSTQE.2009.2016578
6. Kamaruddin NH, Jamaludin NMA, Lokman NF, Bakar AAA (2014) Effect of bi-metallic structure on the performance of chitosan graphene oxide surface plasmon resonance sensor. In: 2014 IEEE 5th international conference on photonics (ICP). IEEE, pp 185–187. https://doi.org/10.1109/icp.2014.7002350

7. Wang B, Huang F, Nguyen T, Lin Q (2010) Microcantilever-based label-free thermal characterization of biomolecular affinity binding. In: 2010 IEEE 23rd international conference on Micro Electro Mechanical Systems (MEMS). IEEE, pp 855–858. https://doi.org/10.1109/memsys.20https://doi.org/10.5442341
8. Feng JY, Ye XY, Shang YF, Wu K, Chen F (2013) Integrated dual grating inteferometric detection with polymer microbeams for biochemical sensing in liquid environment. IET Micro Nano Lett 8(10):629–632. https://doi.org/10.1049/mnl.2013.0237
9. Lang H, Huber F, Zhang J, Gerber C (2013) MEMS technologies in life sciences. In: 2013 Transducers & Eurosensors XXVII: The 17th international conference on solid-state sensors, actuators and microsystems (TRANSDUCERS & EUROSENSORS XXVII). IEEE, pp 1–4. https://doi.org/10.1109/transducers.2013.6626686
10. Thaysen J (2001) Cantilever for bio-chemical sensing integrated in a microliquid handling system. Ph.D. dissertation, Technical University of DenmarkDanmarks Tekniske Universitet, Department of Micro-and NanotechnologyInstitut for Mikro-og Nanoteknologi. https://orbit.dtu.dk/files/4681389/Jacob_Thaysen.pdf
11. Rotake D, Darji A (2018) Heavy metal ion detection in water using mems based sensor. Mater Today: Proc 5(1):1530–1536. https://doi.org/10.1016/j.matpr.2017.11.242
12. Rotake DR, Darji AD, Kale NS (2019) Design of microfluidic experimental setup for the detection of heavy metal ions using piezoresistive BioMEMS sensor. Microelectron Int. https://doi.org/10.1108/MI-05-2019-0025
13. Zargar ZH, Islam T (2019) A novel cross-capacitive sensor for noncontact microdroplet detection. IEEE Trans Ind Electron 66(6):4759–4766. https://doi.org/10.1109/TIE.2018.2863205
14. Rotake D, Darji A (2018) CMOS based capacitance to digital converter circuit for MEMS sensor. In: IOP conference series: materials science and engineering, vol 310, no 1. IOP Publishing, pp 012030. https://doi.org/10.1088/1757-899X/310/1/012030
15. COMSOL (2017) Comsol multiphysics 5.3. [Online]. https://www.comsol.co.in

Statistical Analysis of Vehicle Detection in the ITS Application for Monitoring the Traffic and Road Accident Using Internet of Things

Diya Vadhwani and Devendra Thakor

Abstract Intelligent transportation system involves the improvement in public safety using many emerging technology like wireless sensor network and internet of things. Intelligent transportation system works on major application area for development of smart cities in current scenario. One of the ITS application i.e. vehicle detection and analysis of moving vehicle in the accident notification system was implemented using Arduino Mega 2560 microcontroller and NodeMCU device. The distance of vehicle from the restriction area was calculated which results in the statistical analysis of vehicles, flow rate and analysis of traffic. In order to avoid the accident the vehicle detection, flow rate and traffic prediction are main goals of concern. Flow rate analysis results some of the traffic prediction statistics in order to avoid the accidents roadside or construction areas. This paper gives an immense and innovative idea to evaluate vehicle detection, flow rate of vehicles and traffic prediction at road side using statistical Poisson's distribution method.

Keywords Internet of things · Intelligent transportation system · Vehicle detection · Traffic · Poisson's distribution

1 Introduction

The major areas of concerns like intelligent transportation system and intelligent traffic management in smart cities are useful for economic growth and development, improvement in road transportation, use of information technology in the road infrastructure modification. This enables to achieve smart city challenges in India. The ITS application and systems are implemented with varieties of domain like wireless technology, Bluetooth, sensor-based, internet of things, wireless sensor networks, artificial intelligence, machine learning, image processing etc.

D. Vadhwani (✉) · D. Thakor
CGPIT, Uka Tarsadia University, Bardoli, India
e-mail: dnvadhwani@gmail.com

Z. Patel et al. (eds.), *Advances in VLSI and Embedded Systems*, Lecture Notes in Electrical Engineering 676, https://doi.org/10.1007/978-981-15-6229-7_5

In this paper is the area of domain used is internet of things and wireless sensor network to integrate the transportation system with the information and mobile technology. The modification in current system will be by permission of government agencies and ministry of transportation and by law and rules if applicable.

1.1 ITS Application

The intelligent transportation system had wide range of applications which are useful for public safety measures and public concern domains like smart parking or parking guidance and information system, intelligent traffic management system, integrated multi-modal transport, vehicle detection, accident notification system, traffic analysis, car navigation, automatic number plate recognition, traffic signal, toll collection, bill payment etc.

1.2 Accident Notification System

The accident notification system is one of the challenge were researcher focuses in order to save public lives. The major cause for accident in urban area or the construction area are due to uncontrolled vehicles, driver's ambiguous behavior, traffic at roadside, lack of rules, unavailability of traffic police etc. In the construction area most of time accidents occurs at day time or night. So in order to avoid the accidents the notification to the workers in advance will be given.

Accidents can be prevented if vehicle crossing the prone area is detected and its detected distance from prone area is notified to cloud service via the proposed methodology. In the proposed system the distance value is calculated using the ultrasonic sensor configured with Arduino-mega micro-controller. As soon the value fed in controller will be send to cloud service via NodeMCU device. To predict the traffic, vehicle analysis will help to give the prediction of number of vehicles crossing the prone area considered. Also the accidents can be prevented if workers working in prone area are warned in advance with calculated distance.

1.3 Vehicle Detection

The accident notification system involves the vehicle detection module in the application of intelligent transportation system considered in this paper. The vehicle detection in the accident alert module is one of the crucial tasks. In order to detect the vehicles real time the experiments had been carried out. Experiments are carried out to detect the nearest vehicle crossing the construction prone area by analyzing real time investigation.

The real time vehicle detection will give the distance calculated from prone area and flow rate of vehicles will help to evaluate the traffic and to avoid the accident. For public safety the intelligent transportation system plays important role when there are traffic issues or accident at road side or construction area. In order to develop and achieve the goals of ITS there is vast scope of improvement for system modeling and enhancing.

2 Literature Review

Literature review will give an idea to know the current trends used for intelligent transportation system and its application. The study is been carried out for vehicle detection and its analysis in the traffic evaluation and accident notification application in the field of Intelligent transportation system.

The road accidents and vehicle monitoring in the intelligent transportation system are main area of inter-est. Using VANET and connected technology the vehicle was monitored [2]. The intelligent traffic monitoring system based on cloud computing was considered as future scope [2].

Real time vehicle detection and counting of vehicles using image processing can also be achieved. Data processed from video vehicle detected and compared with the existing algorithms [3].

Parking is one of the issues to be solved in intelligent transportation system; the online parking information system was developed which also shows the vehicle detection using web application method [4].

Vehicle detection based on anisotropic magneto-resistive sensor can be used in traffic monitoring. Vehicle detected with its digital distance measurement for finding the speed of vehicle [5].

Vehicle detection in the traffic evaluation and accident alert or accident notification system is one of the tasks to be performed. The vehicle detection and speed of vehicle is evaluated using ultrasonic sensor and Arduino microcontroller [7].

A prototype for vehicle monitoring, driver condition, fuel monitoring and location finding was developed using microcontroller, GPS and cloud services [8].

An internet of things based vehicle tracking system was studied to find various challenges based on IOT based solutions [9].

The traffic congestion system developed to find the traffic density using the Intel Galileo and IR sensor. Real time monitoring of traffic is to be considered as future evaluation parameter [11].

DSRC (dedicated short range communication) help in sharing the street light, road side infrastructure with ISP's [12].

Study of different types of detectors for vehicle in traffic monitoring or accident alert or ITS application [15].

The traffic congestion, traffic management and vehicle queue management was the focused issues. The two models i.e. TDMM (traffic density monitoring module) and TMM (Traffic management module) were proposed which uses the internet of things. Real time traffic monitoring using internet of things based will gives the experimental results [17].

Traffic and real road monitoring in single lane and multiple lanes. Vehicle detection in single and multiple lanes can be monitored in the model implemented. Traffic acquisition based on wireless sensor network for single lane of road monitoring was analyzed [21].

From the detailed literature study some the parameters were found which are vehicle speed, vehicle distance, traffic condition, driver behavior, Volume of traffic, Time gap, vehicle count, parking slot, fuel detection, vehicle count, street light sharing, infrastructure sharing etc.

Above Table 1 shows the statistical analysis of vehicle detection, traffic and accident by consideration of various parameters like traffic density, vehicle count, speed, distance, time gap, accident rate, hazardous location etc. The literature study shows the statistical method implemented to find the vehicle count, number of accidents occurred and traffic conditions.

In the study of existing statistical method like Poisson distribution, Binomial distribution, Exponential distribution etc., were used for statistical analysis for different parameter evaluation. In order to avoid the accidents at construction site there should be some statistical evaluation of vehicle crossing near that prone area.

The real time traffic evaluation and real time vehicle detection and notification are the scope found from the above literature studies.

The statistical Poisson method [6, 10] is used for evaluating the random number. So in order to evaluate the vehicle count using Poisson method [6, 10] was used. The existing system implemented to detect vehicles or for traffic evaluation.

In this paper we came with new system implemented to find the distance of vehicles crossing near the prone area in order to avoid the accident or crash at construction site and save the workers and vehicle driver lives.

The main contributions of this paper are as follows:

- In the proposed method the distance is calculated from prone area using experimental set-up.
- This distance parameter is the random value of vehicle distance which is used in the statistical evaluation to find the number of vehicles crossing the prone area.
- The experimental setup is portable.
- Internet of Things technology is used for real time recording of distance value.

In order to consider the parametric evaluation of vehicle detection for accident notification system in the construction area following section explains the prototype evaluation of vehicle detection using internet of thing and cloud services.

Table 1 Literature on statistical analysis for vehicle detection, traffic congestion and accident

Sr. No	Paper title	Data sets	Statistical methods	Traffic parameters	Remarks
1.	Probability approach for ranking high-accident locations [1]	Data sets obtained from traffic department in Riyadh (Capital of Saudi Arabia)	Poisson and binomial distributions	Accident rate, hazardous location	Statistical approach estimates the probability of having a certain number of sites which can be classified as hazardous locations
2.	Distances between vehicles in traffic flow and the probability of collision with animals [6]	An automatic detector (radar) was used to give the value for distance	Poisson's Distribution of probabilities, negative-exponential model	Volume of traffic, speed density, time gap, distance	The simplified models have shown the road barrier effect with certain traffic intensity on wildlife
3.	Mathematical study for traffic flow and traffic density in kigali roads [10]	Datasets collected from the national police of Rwanda in 2012. Data sets collected from Kigali roads specifically at roundabouts from Kigali Business Center (KBC) to Prince House as our study sites	Poisson distribution	Traffic flow, density and speed	To traffic congestion can be reduce and the number of accidents on Kigali roads are analyzed depending on the existing roads and the number of vehicles travelling in a particular place
4.	A low-cost IoT application for the urban traffic of vehicles, based on wireless sensors using GSM technology [13]	Data sets collected using ARDUINO UNO R3 GSM /GPS module, laser sensor	Poisson distribution and exponential distributions w	Vehicle count, density of traffic	System predicts alternatives to reduce the traffic congestion, fuel waste, and air pollution

(continued)

Table 1 (continued)

Sr. No	Paper title	Data sets	Statistical methods	Traffic parameters	Remarks
5.	A semi-nonparametric Poisson regression model for analyzing motor vehicle crash data [19]	Datasets collected from 1443 rural highway sections in California State of USA from 1993 to 2002	Poisson regression model (Semi nonparametric (SNP) distribution)	Vehicle count, density of traffic	A random parameter model can be developed and compared with the SNP model and the traditional NB model

3 Proposed System

The proposed system for the notification of accident in the construction area is mention in [18] and its prototype is implemented using Arduino 2560 microcontroller, two ultrasonic sensors, NodeMCU wifi module. For evaluating and analyzing the distance values generated from sensors is recorded on the thingspeak cloud service.

From Fig. 1 the diagram shows the proposed system used in the current prototype implementation [18].

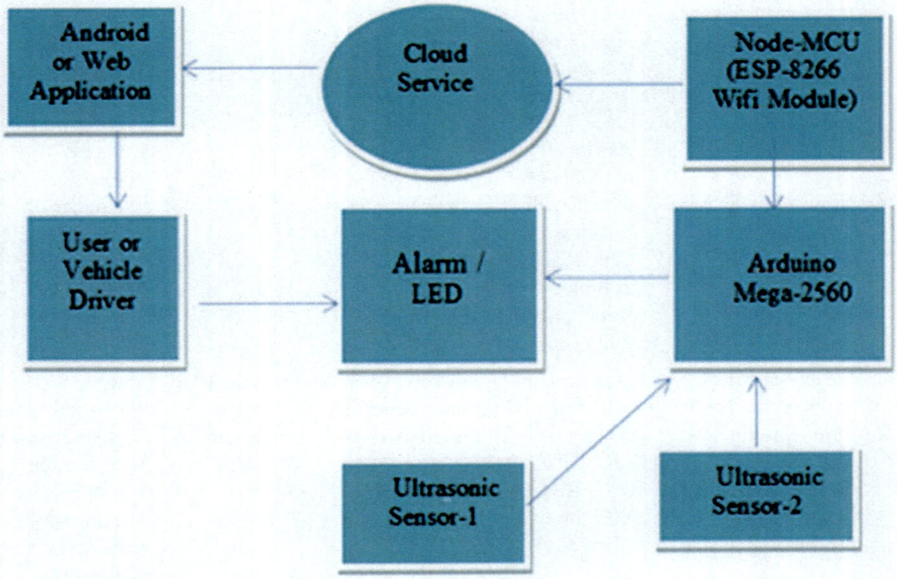

Fig. 1 Block diagram for proposed system for vehicle infrastructure, accident alert and mobile application [18]

Also from [18] my system model was proposed to alert the workers in construction area or urban area, the distance was assumed to be 4 to 400 cm. When the vehicle is detected is nearer to the prone area i.e. distance less than 4 cm than the warning will be send to workers using a buzzer, here LED is considered as output. So the LED output will be HIGH and when the distance is more than 4 cm than LED output is LOW. The proposed system [18] is implemented by considering the first scenarios as explained in next sections.

Assumption for Vehicle Detection and Analysis:

Consider the following assumption for algorithm making:

- Find the traffic zone or construction areas were the accidents are occurring frequently.
- Mount the proposed prototype or scenario in that zone.
- Install the mobile application in the user mobile or vehicle driver mobile.
- Connect through Wi-Fi the Sensors, Microcontroller, Web or Mobile application to the cloud service thingspeak.com
- Assume that model proposed may be install externally in the prone area or may be install in the vehicle itself which will be evaluated after configuration and experimentation.
- Assume distance of vehicle between 4 and 400 cm for calculation.
- Availability of internet or Wi-Fi for communication.

Algorithm Steps for System Model for Vehicle detection which can be installed externally:

(For Vehicle Detection and Analysis)

Step: 1 Find the distance of vehicle crossed in prone area.
Step: 2 When the vehicle is detected is nearer to the prone area i.e. distance less than 4 cm than the warning will be send to workers using a buzzer or LED, here LED is considered as output. So the LED output will be HIGH.
Step: 3 When the vehicle is detected is far from the prone area i.e. distance more than 4 cm than no warning will be send to workers, here LED is considered as output. So the LED output will be LOW.
Step: 4 Find the number of vehicles detected 'n' in prone area, (Vehicle Count = n/Here Random Number).
Step: 5 Send the distance of number of vehicles detected to the cloud service for analysis.
Step: 6 Find the vehicle detected per unit time(1 h/2 h/3 h).
Step: 7 Apply the Poisson distribution method on random values (Vehicle count/Vehicle detected).
Step: 8 Find E (n) = λ i.e. average arrival rate of vehicles.
Step: 9 Find P (n) = Probability of exactly 'n' vehicles arriving in given time.

$$P(n) = \lambda^n e^{-\lambda}/n! \quad (1)$$

where,

P(n) = Probability of exactly 'n' vehicles arriving in given time.
λ = Average arrival rate (vehicles / unit time).
n = Number of vehicles arriving in specific time interval.
e = 2.71828 (constant) ('e' is the base of the natural logarithm system).

As the system implemented the value for distances from the prone area was calculated for 1, 2 and 3 h. All the distances from prone area to the road were recorded on the cloud service thingspeak.com. After the measurement of the distances using sensors the live data of vehicle detection, the statistical analysis by poisons distribution method from [6, 10] and [13] was calculated on the recorded data for prediction of vehicle detection, traffic conditions and accident evaluation.

4 Implementation

The prototype proposed in Fig. 1 is implemented using Arduino-mega-2560 microcontroller, ultrasonic sensor, NodeMCU Wi-Fi module and Arduino-IDE. The distance is calculated as soon the vehicle, here simple object is considered as moving vehicle.

The NodeMCU is the cheapest IoT Device and it is easy for installation so it used for real time values calculation.

The prototype implemented was assumed to be installed on road-side in construction area. The distance of vehicle object is calculated using the sensors. The distance from the prone area was between the ranges i.e. 2–400 cm. If the distance is less than 2 cm from pone are than LED will be HIGH and distance is more than 2 cm than LED will be LOW. The system model implemented show the different calculated whenever any object crosses nearer to prone area or far from prone area. Some of the assumptions from [18] were considered in the implementation of prototype or system model.

Figure 2 shows the prototype developed as follows.

Figure 2 shows the system model implemented were the two moving object were detected and its distance was calculated using the ultrasonic sensors. The LED output is HIGH when the object comes nearer to the system installed in prone area.

Similarly in the Fig. 3 the system model implemented were the two moving object were detected and its distance was calculated using the ultrasonic sensors. The LED output is HIGH when the object comes nearer to the system installed in prone area and the LED output is LOW when the object is far from system installed in the prone area.

From Fig. 4 the output was generated and the distance value calculated from the sensors after the implementation in Arduino-IDE. The implementation of proposed model shows that when the object is nearer or far in the available prone area.

Fig. 2 Prototype for proposed system model with distances less than 4 cm (LED = HIGH)

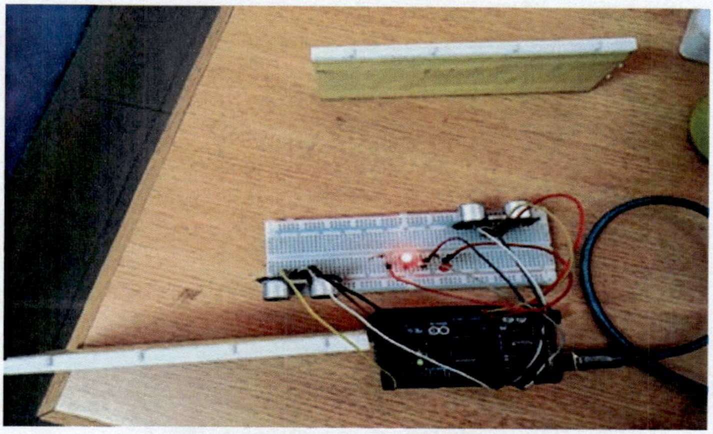

Fig. 3 Prototype for proposed system model with one side distance less than 2 cm (LED = HIGH) and another side distance is greater than 4 cm (LED = LOW)

The evaluation results show that, when the vehicle is the crossing the prone area than its distance is measured and according to the distance generated the LED output responses.

The generated distance are send to the cloud service in order to warn the vehicle drivers and the workers working in construction prone area. The distance will helps to analyze the amount of traffic in that area and the number of vehicles coming nearer to prone area. These distances will help in detection of the number of vehicles crossing the prone area and the number of accidents occurred at road side or construction area.

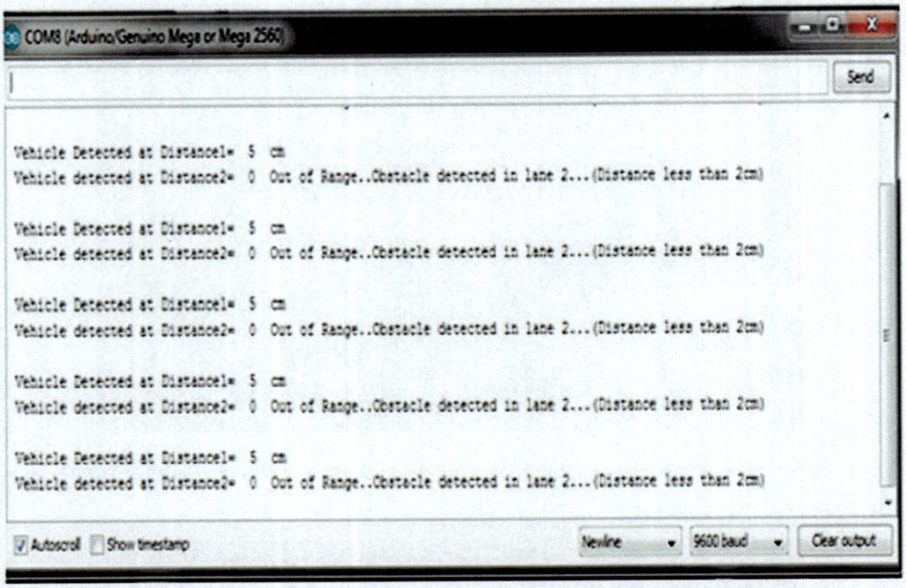

Fig. 4 Output when the vehicle or object is nearer i.e. less than 4 cm distance and when object is far i.e. distance is 5 cm

The real time data is generated from sensor and controller and is sent to cloud service for monitoring the vehicle condition. The real time value is posted and is shown in Fig. 5.

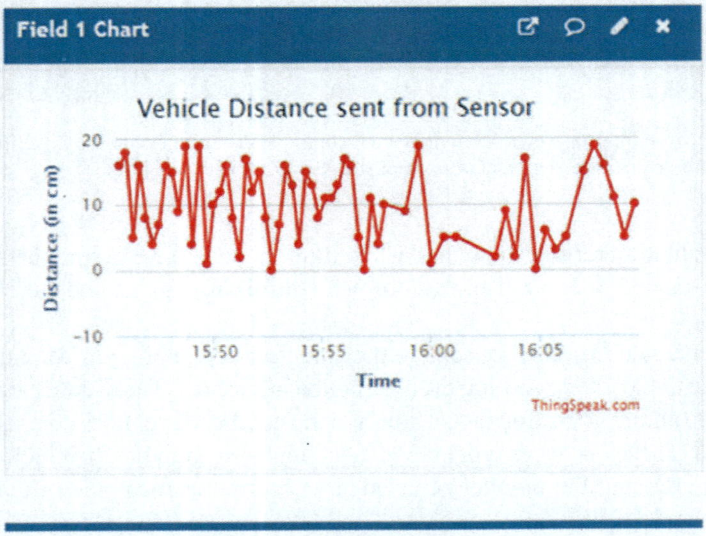

Fig. 5 Real time Distance value of vehicle from cloud service

5 Statistical Analysis for Vehicle Detection

For the accident detection at construction site, to find the vehicle crossing nearer to prone area is one of the crucial tasks. When the vehicle is detected nearer to prone area, workers can be alerted using some notification device or alarm. For the proposed system gives an immense idea to find the distance of vehicle crossing nearer to prone area. The number of vehicle were crossed nearer to prone area were computed. The number of vehicle passed in unit time was calculated using poison's distribution method [14–20].

Here prone area is the construction site where workers are working, so for safety of workers we need to evaluate traffic and vehicles crossing near that area. To find the number of vehicles crossing near the prone area, assume distance is 4 to 400 cm. If vehicle distance is less than 4 cm than there are chances of accidents, if vehicle distance is greater than 4 cm there may be no accidents.

From the real datasets collected the number of vehicle or vehicle count per hour was calculated.

Total number of vehicle arrived in one hour = 96 vehicles per hour.

Total number of vehicle arrived in two hour = 193 vehicles per hour.

Total number of vehicle arrived in three hour = 310 vehicles per hour.

For Poisson's distribution [14, 19],

$$E(n) = \lambda.$$

Where, λ = Average arrival rate,

n = Random number (Count of vehicles).

t = Time in second.

I = Intensity of Vehicles.

For one hour data collected total of 96 vehicles per hour, $\lambda 1 = I/3600 * t$ [19]

$$\lambda 1 = 96 * 60/3600 = 1.6$$

$\lambda 1 = 1.6$ vehicles per second.

For two hour data collected total of 193 vehicles per hour, $\lambda 2 = I/3600 * t$

$$\lambda 2 = 193 * 60/3600 = 3.21$$

$\lambda 2 = 3.21$ vehicles per second.

For three hour data collected total of 310 vehicles per hour, $\lambda 3 = I/3600 * t$

$$\lambda 3 = 310 * 60/3600 = 5.16$$

$\lambda 3 = 5.16$ vehicles per second.

Now lets us consider the calculation for Poisson's distribution [19],

Table 2 Probability values of vehicle arrivals calculated using Poisson distribution

N	P(n)	P(x <= n)
0	0.20189	0.20189
1	0.323	0.52489
2	0.2584	0.78329
3	0.13783	0.92112
4	0.05519	0.97631
5	0.017641	0.99392
6	0.004704	0.998624
7	0.0010752	0.999692
8	0.00215	0.999907
9	0.0003823	0.9999945

$$P(n) = \lambda^n e^{-\lambda}/n!$$

where,

P(n) = Probability of exactly 'n' vehicles arriving in given time.

λ = Average arrival rate (vehicles/unit time)

n = Number of vehicles arriving in specific time interval

e = 2.71828 (constant) ('e' is the base of the natural logarithm system.)

Probability values of the vehicles arrivals calculated using Poisson's distribution is as shown in Table 2 below.

The graph of probabilities is shown in Fig. 6.

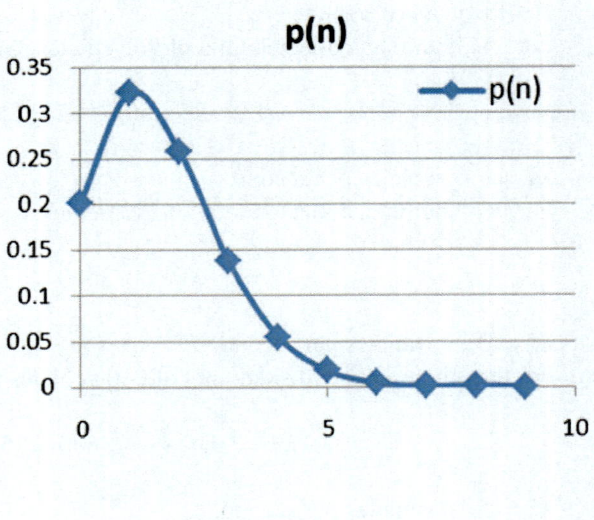

Fig. 6 Graph of Probability values of vehicle arrivals calculated using Poisson distribution

6 Results and Discussions

The minimum number of people died annually in Indian construction sector from 2008 to 2012 was 11,614. These estimates number of accidents are based on available reliable information derived from the construction sector of National Capital Territory (NCT) Delhi region using different sources [6] (Fig. 7).

The road accidents are frequent in time. So there should be model to evaluate the random number. In this paper the Poisson's distribution is proposed to compute the probability and to model the number of vehicles arrived in the available duration time. Since the Poisson's distribution is used to describe the random numbers [7, 10, 13] the probability for 'n' vehicles arrived in given time interval and the probability that certain number of vehicles arrived in given time interval is computed using Poisson's distribution.

The probability of number of vehicles arriving in unit time is computed using P (n) function by Poisson's distribution.

After the implementation the results were recorded on the cloud service things-peak.com. Prototype implementation calculates the distance using ultrasonic sensors. The two sensors are used to calculate the distance values from 2 to 20 cm and 2 to 400 cm. After the distance values recorded using the module, the cloud service shows the live data recording for 1, 2 and 3 h.

From the data collected for one hour the vehicle count is 96 vehicles per hour, so $\lambda 1 = I/3600 * t$ [19], i.e.

Value of $\lambda 1 = 1.6$ vehicles per second.

From the data collected for two hour the vehicle count is 193 vehicles per hour, so $\lambda 2 = I/3600 * t$ [19], i.e.

Value of $\lambda 2 = 3.21$ vehicles per second

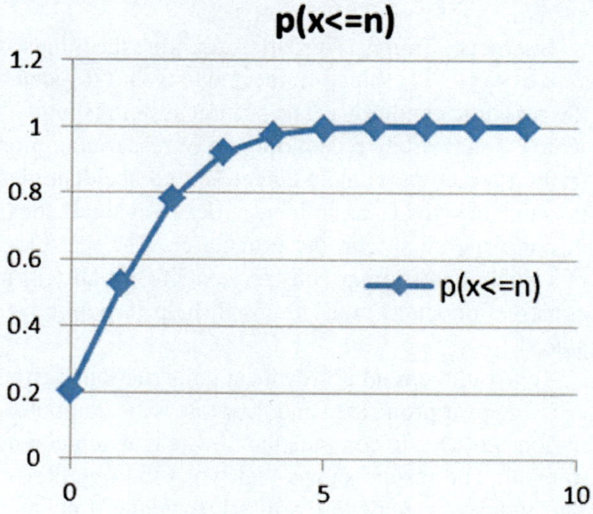

Fig. 7 The graph of cumulative probability values of vehicle arrivals calculated using Poisson distribution

From the data collected for one hour the vehicle count is 310 vehicles per hour, so $\lambda 3 = I/3600 * t$ [19], i.e.

Value of $\lambda 3 = 5.16$ vehicles per second

The hourly flow rate at construction site gives 96 vph using proposed system model and from data set available shows that out of 96 vehicles in hour has 29 vehicles are having the distances less than 4 cm so there is chance of 30% that there may be an accident occurred.

The work performed by authors in [10], shows analysis of data of accidents obtained from the national police using Poisson distribution that the probability of having more accidents on the specified roads in the study in [10] is very high.

Also the authors of [1] finds a statistical approach to estimate the probability of having a certain number of sites which can be classified as hazardous locations (a location with at least k accidents in a defined period of time) within a certain region.

Now from our results generated, the hourly rate of vehicle's crossing near the prone was calculated.

Also it gives the distance value of vehicles which are crossing nearer to prone area, which is used find the count of vehicles which may cause the hazardous environment in construction area. In our work the distance parameter was evaluated to check the number of vehicles coming nearer to prone area which may decrease the accident rate at construction site.

7 Conclusion and Future Work

The public safety on road and safety of workers on construction side in urban areas is one the major area of application in the field of intelligent transportation system. The current system consists of vehicle detection in accident alert or accident notification system.

Internet of Things is research area were the things, devices can be controlled using web of world, i.e. things connected to web. This idea helps the researcher to monitor the real time conditions. The current system is implemented using internet of things which is the widely recent domain of research to monitor the real time scenario for traffic prediction, vehicle detection and accident alert. The results generated from the statistics givens an immense idea to evaluate the distance of vehicle at road side or construction side in the prone area. The real time values will gives an analysis of vehicle detection at construction site which will help to reduce the accidents at construction site. Finally this will help in public safety and reduce deaths at road side.

To avoid the road accidents at construction site there are parameters, i.e. vehicle crossing near prone area and flow rate to be computed. In this paper Poisson's distribution was used to compute the flow rate of vehicles and vehicle arrival in given time interval. The results shows that when the number is random we can find the flow rate and can evaluate the traffic at particular time interval. Also all the vehicles have the distances values calculated using the system model implemented in the prone area. Flow rate of vehicles is used to evaluate the traffic at given time interval. The number of vehicles which are crossing nearer prone can give the prediction results of accident occurrence. As if the distance of vehicle is less than 4 cm than there is chance of accident and to avoid this, the LED output unit or an alarm gives the notification to workers at construction site to be safe.

In future the traffic prediction algorithms will be used to reduce the traffic and reduce the accident rate at construction site. Another future work will be real time data evaluation for real time traffic monitoring at roadside using some statistical analysis with comparative analysis of existing methods.

References

1. Al-Ghamdi AS (2000) Probability approach for ranking high-accident locations. In: Brebbia CA, Sucharov LJ (eds), Urban transport VI. WIT Press. www.witpress.com. ISBN 1-85312-823-6
2. Agrawal Y, Jain K, Karabasoglu O (2018) Smart vehicle monitoring and assistance using cloud computing in vehicular Ad Hoc networks. Int J Trans Sci Technol 60–73. Publishing services Elsevier B. V
3. Anandhalli M, Baligar V (2017) A novel approach in real-time vehicle detection and tracking using Raspberry Pi. Faculty of Engineering, Alexandria University, Production and hosting by Elsevier B.V
4. Bagenda D, Parulian C (2011) Online information of parking area using ultrasonic sensor through Wifi data acquisition. In: ICon-ITSD IOP Publishing IOP conf. series: earth and environmental science, vol 175
5. Daubaras A, Zilys M (2012) Vehicle detection based on magneto-resistive magnetic field sensor. Electron Elect Eng 2(118). ISSN 1392–1215
6. Martolos J, Anděl P (2013) Distances between vehicles in traffic flow and the probability of collision with animals. Transact Trans Sci 6(2)
7. Jain A, Thakrani A, Mukhija K, Anand N, Sharma D (2017) Arduino based ultrasonic radar system using Matlab. Int J Res Appl Sci Eng Technol (IJRASET) 5(4). ISSN: 2321-9653
8. Josea D, Prasad S, Sridhar V (2015) Intelligent vehicle monitoring using global positioning system and cloud computing. In: 2nd international symposium on big data and cloud computing (ISBCC'15), Elsevier B.V
9. Kannimuthu S, Somesh C, Mahendhiran P, Bhanu D, Bhuvaneshwari K (2016) Certain investigation on significance of internet of things (IoT) and big data in vehicle tracking system. Indian J Sci Technol 9(39). https://doi.org/10.17485/ijst/2016/v9i39/87427
10. Idrissa K (2017) mathematical study for traffic flow and traffic density in kigali roads. In: World Academy of Science, Engineering and Technology International Journal of Mathematical, Computational, Physical, Electrical and Computer Engineering, vol 11, no 3
11. Lakshminarasimhan GV, Parthipan V, Mohammed Irfan A (2017) Traffic density detection and signal automation using IoT. Int J Pure Appl Math 116(21):389–394
12. Ligo A, Peha J (2018) Cost-effectiveness of sharing roadside infrastructure for internet of vehicle. In: IEEE transactions on intelligent transportation systems, IEEE
13. Nugra H, Abad A, Walter F, Galárraga F, Aules H, Villacís C, Toulkeridis T (2016) A low-cost IoT application for the urban traffic of vehicles, based on wireless sensors using GSM technology. In: IEEE/ACM 20th international symposium on distributed simulation and real time applications, IEEE
14. Onoja A, Oluwadamilola A, Adewale L (2017) Embedded system based radio detection and ranging (RADAR) system using arduino and ultra-sonic sensor. Am J Embed Syst Appl 5(1):7–12

15. Parulekar G, Desai D, Gupta A (2009) Vehicle detect and monitor techniques for intelligent transportation—a survey. IOSR J Mechan Civil Eng (IOSR-JMCE) 57–59
16. Prasetyo M, Latuconsina R, Purboyo T (2018) A proposed design of traffic congestion prediction using ultrasonic sensors. Int J Appl Eng Res 13(1):434–441. ISSN 0973-4562
17. Sadhukhan P, Gazi F (2018) An IoT based intelligent traffic congestion control system for road crossings. In: International conference on communication, computing and internet of things (IC3IoT), IEEE
18. Vadhwani D, Buch S (2019) A novel approach for the ITS application to prevent accidents using wireless sensor network, IoT and VANET. In: The 2019 third IEEE international conference on electrical, computer and communication technologies IEEE ICECCT (2019)
19. Ye X, Wang K, Yajie Z, Lord D (2018) A semi-nonparametric poisson regression model for analyzing motor vehicle crash data. Plos One. https://doi.org/10.1371/journal.pone.0197338
20. Youngtae J, Jinsup C, Inbum J (2014) Traffic information acquisition system with ultrasonic sensors in wireless sensor networks. Int J Distribut Sensor Netw 961073:12. Hindawi Publishing Corporation
21. Youngtae J, Jung I (2014) Analysis of vehicle detection with WSN-based ultrasonic sensors. Sensors. www.mdpi.com/journal/sensors. ISSN 1424-8220

Qualitative and Quantitative Analysis of Parallel-Prefix Adders

Sudhanshu Janwadkar and Rasika Dhavse

Abstract Binary adders are one of the most recurrent architectures in digital VLSI design, and the choice of adder architecture can boost or bust the overall performance of the design. Parallel-Prefix Adders are preferred over conventional adders for higher wordlengths. In this paper, a comprehensive, qualitative and quantitative analysis of popular Parallel-Prefix Adders for various wordlengths (N = 4, 8, 16 and 32) is presented. The adders are implemented using VHDL coding and Vivado 2016.2 HLx platform targeted for Basys3 board and compared on the basis of Device Utilization, Speed and Power Consumption. Results indicate that Kogge–Stone adder is the fastest adder with F_{max} = 104.93 MHz but is most area-power inefficient consuming 133 LUTs and 23.756 W power at 10 GHz. Sklansky adder is most power efficient consuming 22.857 W power at 10 GHz. Brent–Kung adder is area optimum consuming 62 LUTs.

Keywords Binary adders · Brent–Kung adder · Carry computation · High-speed adders · Kogge–Stone adder · Parallel-prefix adders · Power-efficient adders · VLSI

1 Introduction

Recent advances in complicated information processing systems have dictated the need for hardware implementation of complex algorithms in the field of signal, image and video processing, speech processing, cryptography, etc., each involving complex mathematical procedures. A critical challenge in these hardware implementations is processing of enormous captured data in relatively smaller amount of time [1, 2]. Real-time processing of data has dictated the need for high-speed VLSI architectures. Furthermore, the advent of a plethora of portable battery-operated devices performing complex information processing has posed limits on energy consumption. Often, it

S. Janwadkar (✉) · R. Dhavse
Electronics Engineering Department, Sardar Vallabhbhai National Institute of Technology, Surat, India
e-mail: mr.sudhanshu.janwadkar@gmail.com

Z. Patel et al. (eds.), *Advances in VLSI and Embedded Systems*, Lecture Notes in Electrical Engineering 676, https://doi.org/10.1007/978-981-15-6229-7_6

becomes customary for a VLSI designer to choose an architecture which has high speed or low power or both.

Digital hardware implementation of any signal processing algorithm can be mapped to three arithmetic computational blocks, viz., Adders, Multipliers and Shifters. Of these, adders are the most frequently occurring sub-circuit elements in processors, ALU units, multipliers, dividers, etc. Beyond all doubts, adder architecture can boost or bust the overall performance of the system. Performance of traditional adders such as Ripple Carry adders, Carry Propagate adders, etc., deteriorates with increase in wordlength [3]. Carry Look Ahead adders are viewed as best alternative for adders which offer good speed and consume less power beyond 32-bit wordlength input [4]. Variations of Carry Look Ahead adders, collectively known as Parallel-Prefix Adders, are potential candidates for the abovementioned scenario. A VLSI designer may fail to choose an apt adder if all these architectures are not properly evaluated for the application under consideration.

In apropos, we present an exhaustive comparative analysis of various binary adders. The most popular Parallel-Prefix Adders are Sklansky Adder (SA), Kogge–Stone Adder (KSA), Brent–Kung Adder (BKA), Han–Carlson Adder (HCA), Ladner–Fischer Adder (LFA) and Knowles Adder (KA). We have thoroughly reviewed architectures of the aforementioned adders [5–11]. Moreover, we have compared them on a single platform for various wordlengths so that a consensus can be reached on the choice of adder architecture. This paper will aid any designer to choose the adder architecture based on constraints in the design.

Section 2 deals with a brief literature review of the Parallel-Prefix Adders, followed by analysis of Parallel-Prefix Adders with Boolean algebra in Sect. 3. The experimental results obtained using Xilinx Vivado 2016.2 HLx tool for Basys3 board are represented in Sect. 4. Section 5 covers detailed analysis and discussion based on the results.

2 Literature Review

Prior to the advent of Parallel-Prefix Adders, conventionally Carry Propagate adders were known for high-speed adders. Sklansky [10] proposed conditional sum adder whose computation speed was superior to the conventional adders and hence it was proposed to be used in parallel high-speed addition in digital computers. The design consisted of numerous identical AND-OR-NOT subnetworks leading to a modular structure and was considered a suitable candidate for VLSI implementation. Speed was further enhanced in prefix adder design by Kogge and Stone [7], who invested in recursive doubling technique hardware approach. Recursive doubling resulted in a highly regular adder structure with minimum logic depth and is considered the most time-efficient adder [12, 13]. However, recursive doubling causes increased gate count and explosion of wires. Area complexity of KSA is $O(N^2)$ while time complexity is improved to $O(log_2 N)$, for N-bit adders.

The gate-count explosion problem is solved in adder structure by Brent and Kung [5], who proposed another regular adder structure for the computation of prefixes. Brent–Kung approach increases logic depth of the adder and avoids long wires. While still maintaining the computation time complexity at $O(2 1og_2 N - 1)$, Brent Kung approach achieves area complexity as $O(N log_2 N)$. Han and Carlson [9] combined BKA and KSA structures and proposed an intermediate adder structure whose depth grows in proportion to $2kN + N^2/2k$, where N is the number of bits and k is the increase in depth of the hybrid adder structure. The area complexity of HCA is $O(N log_2 N)$.

Ladner and Fischer [8] conceptualized that adders could be represented by directed acyclic oriented graph tree. Each node of the graph tree would represent the Boolean product ($y_n = x_1 x_2 \ldots x_n$, $y_{n-1} = x_1 x_2 x_{n-1}, \ldots y_1 = x_1$; $n \in D$) at the node n. The edges in graph reflect the wiring on the integrated circuit adder. The computation time for parallel carry computation depends on the depth of the graph, while the amount of hardware required to implement is dictated by the number of nodes in such graph. Ladner and Fischer [8] adders have a depth proportional to $2(log_2 N)$ and size $< 4N$ for N-bit binary addition.

Knowles [6] established that KSA and LFA are two extreme ends of prefix adders having minimal depth and proposed a family of prefix adders which serve as a trade-off in terms of speed versus power between the two extremes. Dimitrakopoulus and Nikolos [4] proposed an adder architecture which offers reduced delay in comparison with KA and also saves one-logic level of implementation. Das and Khatri [14] proposed an architecture which is significantly faster than the BKA, but with some area overhead. At the same time, it offers marginally high speed in comparison to the KSA, but with much reduced area.

Roy et al. [15] proved that prefix-adder architecture is most efficient when bit-width N is a power of 2 (size optimality criteria) and logic depth level (L) is $log_2 N$. Detailed analysis of impact of voltage scaling and process variation on efficiency of 32-bit Parallel-Prefix Adders is investigated by Bahadori et al. [16].

3 Parallel-Prefix Adders

Consider two input numbers to be added, each of wordlength N, denoted as $A = a_{N-1} a_{N-2} a_{N-3} \ldots a_2 a_1 a_0$ and $B = b_{N-1} b_{N-2} b_{N-3} \ldots b_2 b_1 b_0$ and the carry from previous stage denoted by C_{in}. Their addition results in sum, of equal wordlength N, denoted by $S = s_{N-1} s_{N-2} s_{N-3} \ldots s_2 s_1 s_0$ and carry to next stage C_{out}.

Parallel-Prefix Adders compute the sum S and output carry C_{out} in three stages, preprocessing stage, Prefix Tree carry computation stage and Post-processing stage [4, 14] The scheme is represented in Fig. 1.

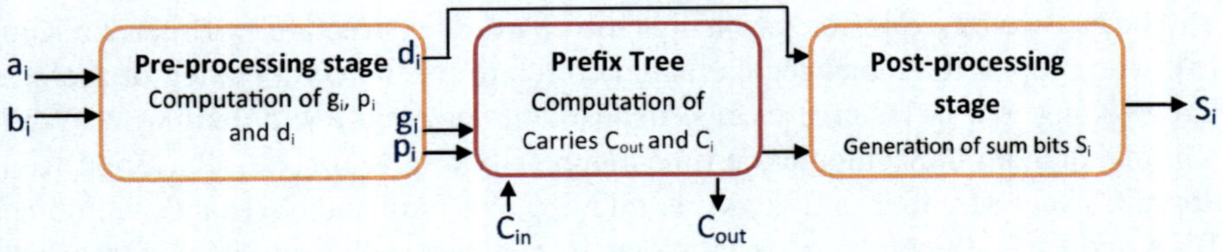

Fig. 1 Representation of parallel-prefix adders [17]

3.1 Preprocessing Stage

The initial preprocessing stage computes the generate (g_i) and propagate signals (p_i) and half-sum (d_i) , $i \in (0, N)$ defined as

$$g_i = a_i.b_i \tag{1}$$

$$p_i = a_i + b_i \tag{2}$$

$$d_i = a_i \wedge b_i \tag{3}$$

where '.' denotes AND operation, '+' indicates OR operation and '∧' denotes XOR operation.

3.2 Parallel-Prefix Carry Computation Stage

This stage computes the carry signals c_i using the outputs of the preprocessing steps g_i and p_i. The parallel-prefix circuit with input of wordlength N and inputs $u_1, u_2 \ldots u_N$ computes outputs $v_1, v_2 \ldots v_N$ determined using an associative operator $\circ$ such that $v_1 = u_1, v_2 = u_2 \circ u_1 \ldots v_N = u_N \circ u_{N-1} \circ \ldots u_2 \circ u_1$ [8].

Carry Computation stage is modeled as a prefix problem by Han and Carlson [9] and defines the association between the pairs of generate and propagate bits as

$$(g_i, p_i) \circ (g_{i-1}, p_{i-1}) = (g_i + p_i.g_{i-1}, p_i.p_{i-1}) \tag{4}$$

where $i \in (0, N)$ and '.' denotes AND operation and '+' denotes OR operation.

Extending Eq. 4 results in a series of consecutive generate and propagate pairs (g, p) produced from bits k, k−1, k−2..., j, defined as

$$(G_{k:j}, P_{k:j}) = (g_k, p_k) \circ (g_{k-1}, p_{k-1}) \circ \ldots \circ (g_{j+1}, p_{j+1}) \circ (g_j, p_j) \tag{5}$$

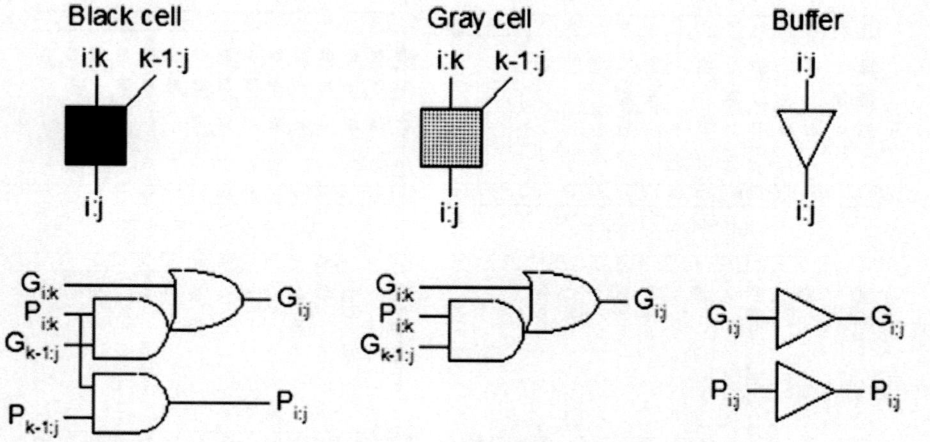

Fig. 2 Black cell and gray cell representation in parallel-prefix adders

The idempotency property of $\circ$ operator allows group term $(G_{i:j}, P_{i:j})$ to be derived from the association of two overlapping terms $(G_{i:k}, P_{i:k})$ and $(G_{m:j}, P_{m:j})$ where $i > m > j$, defined as

$$(G_{i:j}, P_{i:j}) = (G_{i:k}, P_{i:k}) \circ (G_{m:j}, P_{m:j}) \tag{6}$$

The operator $\circ$ can be represented as a node in graph and the pairs $(G_{i:j}, P_{i:j})$ form the edges of the graph.

Equation 6 corresponds to a set of (G,P) pairs computed using combinational circuits, known as Black cell and Gray cell. The representation is shown in Fig. 2. The buffers indicated by inverted triangular notation are used for data synchronization at the outputs. At the end of this stage, the carry of the stage is available as

$$c_i = G_{i:0} \tag{7}$$

3.3 *Post-processing Stage*

The various Parallel-Prefix Adders achieve high speed of operation through variation of the prefix-tree stage. In essence, the number of gray and black cells and their arrangement (i.e., depth of the graph and the interconnections between the cells) dictate the speed of the design. The characteristic of Parallel-Prefix Adders is that the carry-computation time doesn't increase linearly with increase in Wordlength N and the carry bits are computed in parallel in the prefix-tree stage, hence justifying the name. At the end of the prefix carry-computation stage, the carry bits are available at the output of the prefix tree. The carries are fed to the post-processing stage.

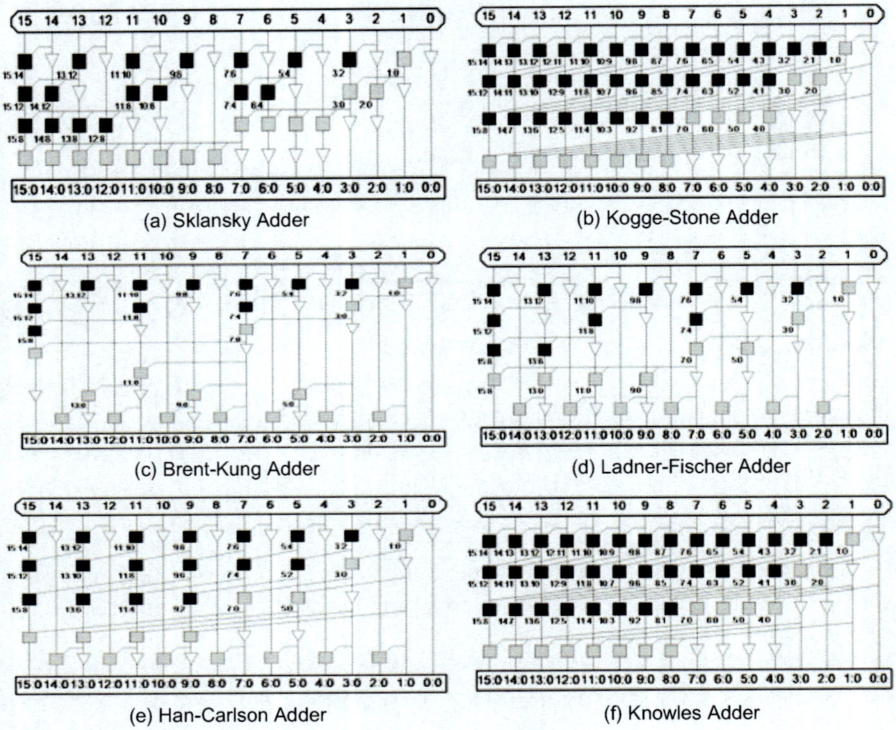

Fig. 3 16-bit parallel-prefix adder representations

In the post-processing stage, the carry bits are XORed with the half-digit bits produced in the preprocessing stage to produce the final sum bits. This stage computes the sum bits $S = s_{N-1}s_{N-2}s_{N-3}\ldots s_2s_1s_0$ as [18]

$$s_i = d_i \wedge c_{i-1} \tag{8}$$

The various Parallel-Prefix Adders viz., SA [10], KSA [7], BKA [5], HCA [9], LFA [8] and KA [6] differ in the architecture used in this stage, i.e., depth of the adder and the amount of routing [12].

Various literature sources [4, 17, 19–21] describe the architecture of the Parallel-Prefix Adders using the gray and black cells represented in Fig. 2. The 16-bit representations of these adders are shown in Fig. 3.

4 Experimentation and Results

To have a quantitative analysis with apt justification, of the Parallel-Prefix Adder architectures, the designs were modeled using hardware description language VHDL (standard 2000). Each of the designs was targeted for Artix-7 family-based Basys3 board (Xilinx part number *xc7a35tcpg236-1*), by Digilent, Inc. Artix-7 family Field Programmable Gate Array (FPGA) devices are based on 28 nm high-K metal gate (HKMG) process technology [22]. The particular FPGA has 20,800 LUT elements, 41600 Flip-flops, 106 IOBs, 50 Block RAMs and 90 DSP units among other peripherals, and FPGA core operates at 1 V supply [23]. Functional Verification (using test-bench) Synthesis and Implementation were performed using 'Xilinx Vivado 2016.2 HLx Edition'.

Each architecture has been compared for the parameters, device utilization (reflects the area), timing (reflects the maximum frequency of operation of device) and power consumption. Analysis is carried out post implementation stage after completion of routing. The tool at this stage can report exact number of logic resources, routing resources, routing delays and exact activity of the internal nodes. Power analysis at this level provides the most accurate power estimation [24].

4.1 Device Utilization

The device utilization has been captured in terms of the Number of LUTs required to implement logic, Number of Slices, Number of Slice Registers and Number of IOBs. Of these, Number of logic LUTs and Number of slices are the critical parameters to compare the designs. The device utilization summary, as reported by the tool, is reported in Table 1.

4.2 Timing

To compare the speed of the architectures, the comparison has been made on the total delay of the critical path. Also, logic delay versus routing delay contribution to the total delay is studied in essence to the fact that the architectures bring about speed by either varying the depth of the adder or increasing the logic elements performing the operation leading to wire congestion. The tool reports the paths with maximum delay after implementation stage. The logic delay, routing delay and the total delay of the path are reported [25]. Details are summarized in Table 2.

The other parameter of interest is the speed at which an architecture can work under given constraints. Constraints were set for minimum clock period as 10 ns and maximum rise time delay at each of the inputs and outputs as 0.1 ns. Under these constraints, the tool reports the Worst Negative Slack (WNS) and Worst Hold Slack (WHS) [25]. WNS is the negative of all the slacks, and hence is slack of the critical path [26].

$$\text{WNS} = \min\{\text{slack}(\tau)\} \tag{9}$$

where τ is the set of timing endpoints, e.g., primary inputs and outputs to the flip-flops.

Table 1 Device utilization summary of parallel-prefix adders

	SA				KSA				BKA				LFA				HCA				KA			
Wordlength (in bits)	4	8	16	32	4	8	16	32	4	8	16	32	4	8	16	32	4	8	16	32	4	8	16	32
Number of LUTs (logic)	4	11	28	64	4	11	45	133	4	10	30	62	4	10	27	70	4	11	30	92	4	11	51	131
Number of slices	7	18	32	100	7	17	28	106	7	16	33	87	7	16	31	99	7	16	27	89	7	17	31	113
Number of slice registers	14	26	50	98	14	26	50	98	14	26	50	98	14	26	50	98	14	26	50	98	14	26	50	98
Number of bonded IOBs	15	27	51	99	15	27	51	99	15	27	51	99	15	27	51	99	15	27	51	99	15	27	51	99

Table 2 Timing summary of parallel-prefix adders

	SA				KSA				BKA			
Wordlength (in bits)	4	8	16	32	4	8	16	32	4	8	16	32
Total delay (in ns)	5.063	5.219	6.587	7.847	5.063	5.244	8.220	7.856	5.063	5.227	7.844	9.541
Logic delay (in ns)	3.203	1.093	3.067	1.338	3.203	3.086	3.067	3.053	3.203	3.086	3.067	1.632
Route delay (in ns)	1.860	4.126	3.520	6.509	1.860	2.158	5.153	4.803	1.860	2.141	4.777	7.909
% Logic delay	63.26	20.94	46.56	17.05	63.26	58.85	37.31	38.86	63.26	59.04	39.09	17.11
% Route delay	36.73	79.05	53.43	82.95	36.73	41.15	62.69	61.14	36.74	40.96	60.90	82.89
	LFA				HCA				KA			
Total delay (in ns)	5.063	5.227	6.587	7.660	5.063	5.262	8.316	8.020	5.063	5.244	6.713	7.937
Logic delay (in ns)	3.203	3.086	3.067	1.233	3.203	3.177	3.067	3.065	3.203	3.086	3.067	3.123
Route delay (in ns)	1.860	2.141	3.52	6.427	1.860	2.085	5.249	4.955	1.860	2.158	3.646	4.814
% Logic delay	63.26	59.03	46.56	16.09	63.26	60.37	36.88	38.22	63.26	58.84	45.69	39.34
% Route delay	36.74	40.96	53.44	83.90	36.74	39.62	63.12	61.78	36.74	41.15	54.31	60.65

The maximum frequency at which the architecture can be operated under given constraints is then given by [27]

$$F_{\max} = \frac{1}{\text{Actual Clock Period} - \text{WNS}} \tag{10}$$

WHS corresponds to the worst slack of all the timing paths for min delay analysis [25]. A positive or zero value of slack indicates that the design works properly at the specified frequency. Negative value of slack is undesirable, implying constraints are violated [28]. The WNS and WHS slacks are summarized in Table 3.

4.3 Power Consumption

FPGA consists of internal Phase Locked Loops (PLLs) that can generate various clock frequencies. Power Analysis has been carried out considering 10 GHz clock frequency for switching activity computations. The static switching probability of 0.5 and toggle rate 12.5% is chosen [24]. The tool reports power consumption as an estimate of static power consumption (due to leakages), dynamic power consumption (captured in terms of signal power (depends on switching activity across signals in design) and logic power (depends on switching activity in the logic)) and IO power [24, 29]. These values are reported in Table 4. The values in Table 4 are as reported by the tool, at 10 GHz clock frequency. Under the constraints on clock (minimum clock period as 10 ns and maximum rise time delay at each of the inputs and outputs as 0.1 ns), the prefix adders work at maximum frequency 100–105 MHz. However, Vivado 2016.2 tool reports power consumption at 10 GHz clock frequency, by default. Therefore, this frequency has been chosen for power consumption analysis.

5 Discussion

5.1 Device Utilization

For a given adder of wordlength N (N = 4, 8, 16 or 32), the number of IOBs and number of slice registers are fixed irrespective of the architecture style. It is evident from Table 1 that irrespective of the adder architecture, the number of logic LUTs and slices remain same for 4-bit and 8-bit adders, implying that at small wordlengths, effect of choice of prefix-adder architecture is immaterial. For any given architecture, the number of LUTs and slices increase with increase in wordlength. This is because the number of gray cells and black cells required in implementing the design increase, implying increase in gate count. However, at higher wordlengths, logic LUTs required to implement KSA (45 and 133 logic LUTS for 16-bit and 32-bit

Table 3 Maximum frequency of operation for parallel-prefix adders

	SA				KSA				BKA			
Wordlength (in bits)	4	8	16	32	4	8	16	32	4	8	16	32
WNS (in ns)	0.377	0.367	0.357	0.377	0.377	0.365	0.481	0.47	0.377	0.365	0.375	0.314
WHS (in ns)	0.23	0.211	0.29	0.429	0.23	0.085	0.302	0.318	0.23	0.19	0.239	0.19
F_{max} (in MHz)	103.918	103.81	103.702	103.919	103.918	103.788	105.053	104.932	103.918	103.788	103.896	103.242
	LFA				HCA				KA			
WNS (in ns)	0.377	0.365	0.471	0.46	0.377	0.363	0.352	0.295	0.377	0.365	0.481	0.457
WHS (in ns)	0.230	0.190	0.331	0.367	0.230	0.204	0.379	0.265	0.230	0.085	0.212	0.206
F_{max} (in MHz)	103.918	103.788	104.942	104.822	103.918	103.766	103.648	103.040	103.918	103.788	105.053	104.789

Table 4 Power consumption of parallel-prefix adders

		SA				KSA				BKA			
Wordlength (in bits)		4	8	16	32	4	8	16	32	4	8	16	32
Static power (W)		0.078	0.092	0.157	0.485	0.078	0.089	0.162	0.485	0.078	0.092	0.158	0.485
Dynamic power (W)	Signal power	0.088	0.21	0.483	1.475	0.086	0.182	0.575	1.886	0.088	0.185	0.516	1.553
	Logic power	0.035	0.085	0.186	0.405	0.036	0.082	0.298	0.786	0.035	0.076	0.205	0.433
IO power (W)		2.686	5.226	10.298	20.492	2.686	4.716	10.358	20.599	2.686	5.238	10.295	20.436
Total power (W)		2.887	5.613	11.124	22.857	2.886	5.069	11.393	23.756	2.887	5.591	11.174	22.907
Wordlength (in bits)		LFA				HCA				KA			
		4	8	16	32	4	8	16	32	4	8	16	32
Static power (W)		0.078	0.092	0.158	0.485	0.078	0.092	0.161	0.485	0.078	0.089	0.162	0.485
Dynamic power (W)	Signal power	0.088	0.185	0.494	1.589	0.088	0.212	0.536	1.738	0.086	0.182	0.572	1.852
	Logic power	0.035	0.076	0.19	0.472	0.035	0.094	0.225	0.579	0.036	0.082	0.32	0.805
IO power (W)		2.686	5.238	10.309	20.484	2.686	5.242	10.387	20.512	2.686	4.716	10.35	20.608
Total power (W)		2.887	5.591	11.151	23.03	2.887	5.64	11.309	23.314	2.886	5.069	11.404	23.75

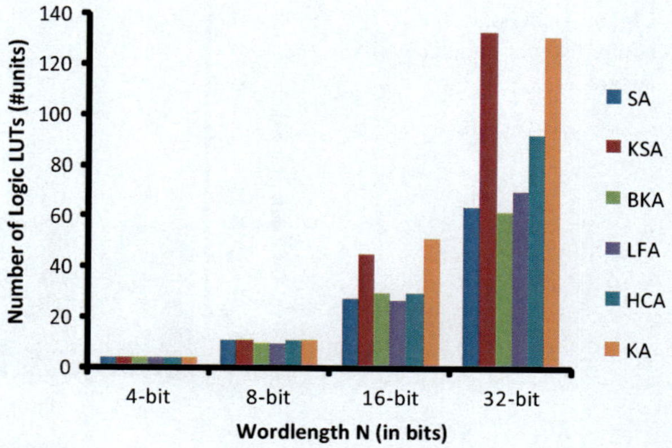

Fig. 4 Number of logic LUTs required for implementing parallel-prefix adders

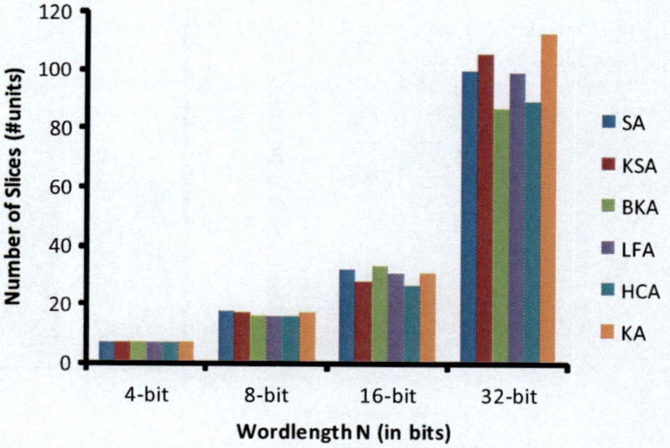

Fig. 5 Number of slices required for implementing parallel-prefix adders

implementations, respectively) and KA (51 and 131 logic LUTs for 16-bit and 32 bit implementations) shoot up and they prove to be the priciest. Explosion in gate count with recursive doubling hardware approach is clearly visible. BKA (30 and 62 logic LUTs for 16-bit and 32-bit implementations, respectively) is most economic in terms of device utilization at high wordlengths, which is less than half required for KSA and KA. BKA utilizes 89 slices against 106 and 113 required for KSA and KA implementations, respectively. Figures 4 and 5 graphically represent FPGA device utilization in terms of number of logic LUTs and number of slices, respectively.

5.2 *Timing*

From Table 2, it is clearly evident that the total delay doesn't linearly increase with increase in wordlength, as the case with conventional adders such as Ripple Carry adder and Carry Propagate adder. It is also be interpreted that, irrespective of the architecture style, at higher wordlengths routing delay supersedes the logic

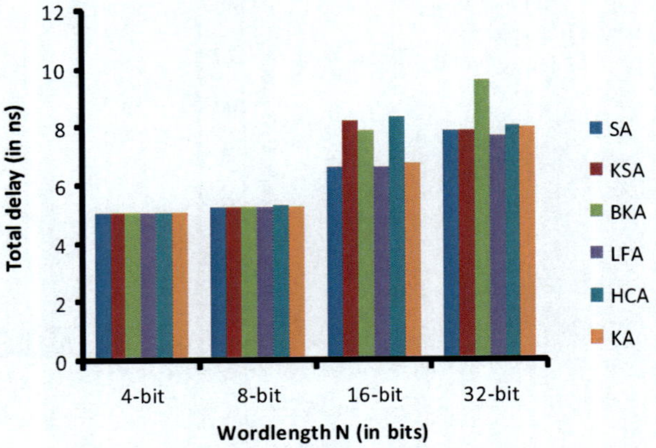

Fig. 6 Total delay comparison of parallel-prefix adders

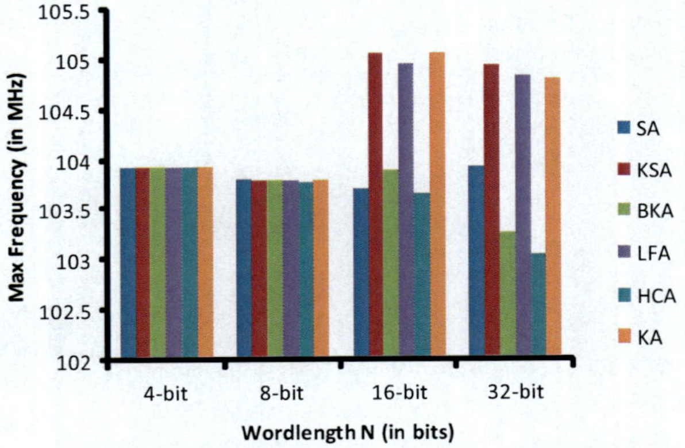

Fig. 7 Maximum frequency of parallel-prefix adders

delay. For 32-bit adders, net delay accounts for about 60–80% of the total delay. At lower wordlengths ($N = 4$ and $N = 8$), the total delay for each of the architecture is reported to be the same. However, for higher wordlengths ($N = 16$ and $N = 32$), the total delay of BKA shoots up (9.541 ns for 32-bit adder). The total delay is graphically represented in Fig. 6.

Considering the maximum frequency at which the Parallel-Prefix Adders can be operated, under the constraint of clock period as 10 ns, KSA and KA (104.932 MHz and 104.789, respectively, for 32-bit adder implementations) are obvious best choices, closely followed by LFA (104.822 MHz). The results are graphically represented in Fig. 7.

5.3 *Power Consumption*

From Table 4, it is concurred that irrespective of the architecture, the static power consumption remains fixed for a given wordlength. This is justified as static power

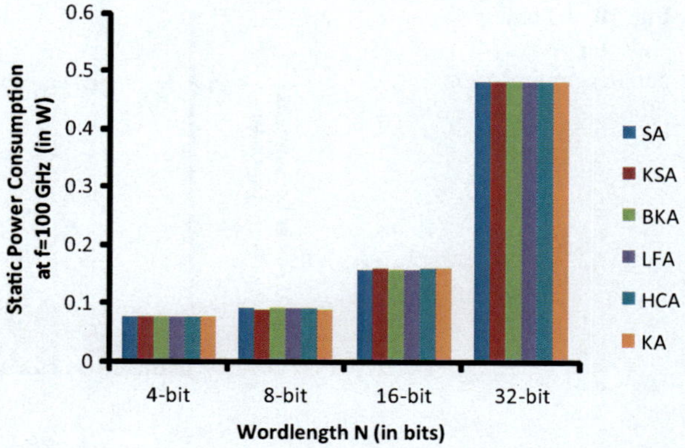

Fig. 8 Static power consumption of parallel-prefix adders

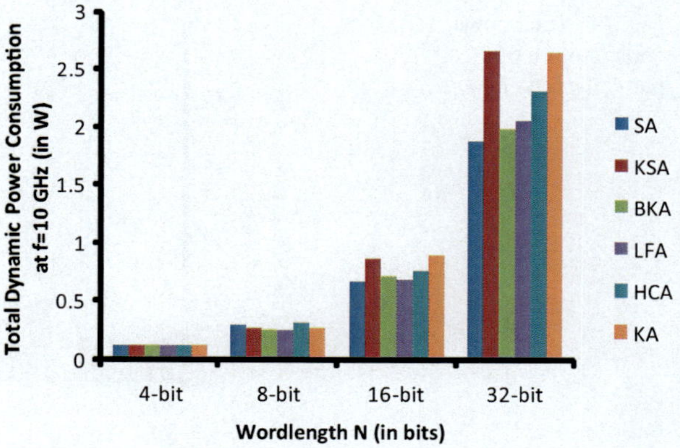

Fig. 9 Total dynamic power consumption of parallel-prefix adders

depends on leakages and for given wordlength N, leakages are similar. Figure 8 represents Static Power Dissipation.

However, for various Parallel-Prefix Adder architectures, dynamic power consumption varies. Dynamic Power dissipation consists of signal power and logic power components. Depending on the logic used in the architecture and the amount of switching activity based on wiring, the dynamic power varies. Comparisons are made on the basis of dynamic power consumption. KSA (Dynamic Power Consumption = Signal Power (1.886 W) + Logic Power (0.786 W) = 2.672 W) and KA (Dynamic Power Consumption = Signal Power (1.852 W) + Logic Power (0.805 W) = 2.657 W) are power-hungry architectures at 32-bit wordlength. SA is the most power-efficient (Dynamic Power Consumption = Signal Power (0.483 W) + Logic Power (0.186 W) = 0.669 W for 16-bit and Dynamic Power Consumption = Signal Power (1.475 W) + Logic Power (0.405 W) = 1.88 W for 32-bit implementations) adder architecture. Results are depicted pictographically in Fig. 9.

IO power doesn't vary much with the architecture style. The IO power consumption by various architectures is represented in Fig. 10.

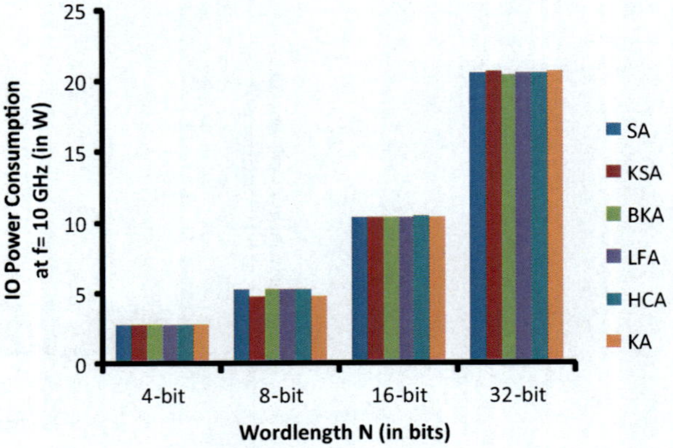

Fig. 10 IO power consumption of parallel-prefix adders

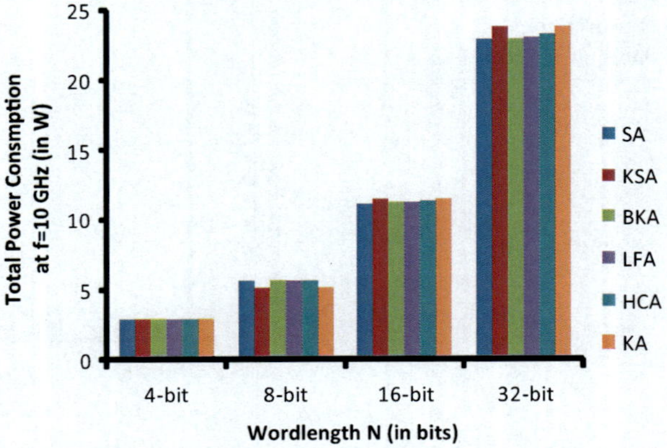

Fig. 11 Total power consumption of parallel-prefix adders

In essence, the total power consumption increases with increase in wordlength. Total Power consumed by KSA (11.124 W and 22.857 W for 16-bit and 32-bit adder implementations, respectively) and KA (11.404 W and 23.750 W for 16-bit and 32-bit adder implementations, respectively) makes them power-hungry architectures. The total power consumption by various Parallel-Prefix Adders is represented in Fig. 11.

6 Conclusion

Choosing an adder architecture in the VLSI design depends on the constraints of the application. General convention is that, for real-time processing of information, high-speed architectures are preferred and for the architectures which will be packed off in battery-operated hand-held devices, compact and power-efficient architectures are preferred. In this paper, we have attempted to compare and analyze the popular Parallel-Prefix Adder architectures, so as to present a comprehensive, qualitative and quantitative analysis. We have implemented the architectures for various wordlengths (4-bit, 8-bit,16-bit and 32-bit) targeted for Artix-7 based Basys3 board and analyzed Speed, Power Consumption and Hardware Utilization. Based on the results, we conclude that the Kogge–Stone adder and Knowles adder pose to be the best solution for high-speed adder design, but pose as giants in terms of device utilization and power consumption. However, the Brent–Kung adder is most efficient in terms of device utilization closely followed by the Sklansky adder. On the other hand, Sklansky adder is most power consumption efficient even at higher wordlengths, closely followed by the Brent–Kung adder. Therefore, the two adders are suitable for compact power-efficient designs.

Acknowledgements The authors express their gratitude to Special Manpower Development Program for Chips to System Design (SMDP-C2SD) project under Ministry of Electronics and Information Technology, Government of India, for providing access to tools required to conduct this research work.

References

1. Ashrafi A, Strollo A, Gustafsson O (2013) Hardware implementation of digital signal processing algorithms. J Elect Comput Eng 2013(782575):2
2. Woods R, Mcallister J, Turner R, Yi Y, Lightbody G (2008) FPGA-based implementation of signal processing systems, 1st edn. Wiley
3. Uma R, Vijayan V, Mohanapriya M, Pau S (2012) Area, delay and power comparison of adder topologies. Int J VLSI Des Commun Syst 3(1):153–168
4. Dimitrakopoulos G, Nikolos D (2005) High-speed parallel-prefix vlsi ling adders. IEEE Trans Comput 54(2):225–231
5. Brent RP, Kung HT (1982) A regular layout for parallel adders. IEEE Trans Comput 31(3):260–264. https://doi.org/10.1109/TC.1982.1675982
6. Knowles S (2001) A family of adders. In: Proceedings 15th IEEE symposium on computer arithmetic, ARITH-15 2001, pp 277–281
7. Kogge PM, Stone HS (1973) A parallel algorithm for the efficient solution of a general class of recurrence equations. IEEE Trans Comput C-22(8):786–793
8. Ladner RE, Fischer MJ (1980) Parallel prefix computation. J. ACM 27(4): 831–838. https://doi.org/10.1145/322217.322232
9. Han T, Carlson DA (1987) Fast area-efficient vlsi adders. In: 1987 IEEE 8th symposium on computer arithmetic (ARITH), pp 49–56
10. Sklansky J (1960) Conditional-sum addition logic. IRE Trans Electron Comput EC-9(2):226–231
11. Mittal A, Nandi A, Yadav D (2017) Comparative study of 16-order fir filter design using different multiplication techniques. IET Circ Devices Syst 11(3):196–200
12. Poornima N, Bhaaskaran VSK (2015) Area efficient hybrid parallel prefix adders. Procedia Mater Sci 10:371–380. http://www.sciencedirect.com/science/article/pii/S2211812815003077
13. A comprehensive review on the vlsi design performance of different parallel prefix adders. Mater Today 11:1001–1009
14. Das S, Khatri SP (2008) A novel hybrid parallel-prefix adder architecture with efficient timing-area characteristic. IEEE Trans Very Large Scale Integr (VLSI) Syst 16(3):326–331
15. Roy S, Choudhury M, Puri R, Pan DZ (2013) Towards optimal performance-area trade-off in adders by synthesis of parallel prefix structures. In: 2013 50th ACM/EDAC/IEEE design automation conference (DAC), pp 1–8

16. Bahadori M, Kamal M, Afzali-Kusha A, Pedram M (2016) A comparative study on performance and reliability of 32-bit binary adders. Integration 53:54–67. http://www.sciencedirect.com/science/article/pii/S0167926015001571
17. Pudi V, Sridharan K, Lombardi F (2017) Majority logic formulations for parallel adder designs at reduced delay and circuit complexity. IEEE Trans Comput 66(10):1824–1830
18. Rabaey JM, Chandrakasan A, Nikolic B (2008) Digital integrated circuits, 3rd edn. Prentice Hall Press, Upper Saddle River, NJ, USA
19. Ghosh S, Roy K (2011) Novel low overhead post-silicon self-correction technique for parallel prefix adders using selective redundancy and adaptive clocking. IEEE Trans Very Large Scale Integr (VLSI) Syst 19(8):1504–1507
20. Esposito D, De Caro D, Strollo AGM (2016) Variable latency speculative parallel prefix adders for unsigned and signed operands. IEEE Trans Circ Syst I 63(8):1200–1209
21. Moghaddam M, Ghaznavi-Ghoushchi MB (2011) A new low-power, low-area, parallel prefix sklansky adder with reduced inter-stage connections complexity. In: 2011 IEEE EUROCON—international conference on computer as a tool, pp 1–4
22. Xilinx (2019) 7 series FPGAs data sheet: overview. Available https://www.xilinx.com/support/documentation/data_sheets/ds180_7Series_Overview.pdf. Accessed 29 Nov 2019
23. Digilent (2019) Digilent Basys 3 Artix-7 FPGA Board. Available https://www.xilinx.com/products/boards-and-kits/1-54wqge.html. Accessed 25 Sep 2019
24. Xilinx (2019) Vivado design suite user guide: power analysis and optimization. Available https://www.xilinx.com/support/documentation/sw_manuals/xilinx2017_1/ug907-vivado-power-analysis-optimization.pdf. Accessed 25 Sep 2019
25. Xilinx1 (2019) Vivado design suite user guide: design analysis and closure techniques. Available https://www.xilinx.com/support/documentation/sw_manuals/xilinx2017_3/ug906-vivado-design-analysis.pdf. Accessed 25 Sep 2019
26. Kahng AB, Lienig J, Markov IL, Hu J (2011) VLSI physical design: from graph partitioning to timing closure, 1st edn. Springer Publishing Company, Incorporated, Berlin
27. Xilinx (2019) Vivado timing—where can I find the Fmax in the timing report? Available https://www.xilinx.com/support/answers/57304.html. Accessed 25 Sep 2019
28. Xilinx F (2019) Total negative slack vs worst negative slack. Available https://forums.xilinx.com/t5/Welcome-Join/Total-Negative-Slack-vs-Worst-Negative-Slack/td-p/308077. Accessed 30 Nov 2019
29. Altera (2012) Reducing power consumption and increasing bandwidth on 28-nm FPGAs, 2012. Available https://www.intel.com/content/dam/www/programmable/us/en/pdfs/literature/wp/wp-01148-stxv-power-consumption.pdf. Accessed 25 Sep 2019

A Novel Method of Multiplication with Ekanyunena Purvena

M. A. Sayyad and D. N. Kyatanavar

Abstract In this paper a novel method of multiplication is developed using Ekanyunena Purvena of Indian vedic mathematics. The multiplier designed using present method is applied for BCD(Binary Coded Decial) as well as binary number system. A 4×4, 8×8 and 16×16 bit multiplier is designed using this method and architecture is developed for BCD as well as binary number systems. The performance of this method is compared for area delay and power with array and vedic UT (UrdhvaTiryakbyham) multiplier. The proposed multiplier is implemented with VHDL on spartan3 FPGA board using Xilinx ISE 14.7 and its power dissipation is calculated using XPower analyzer. The performance of proposed multiplier is compared with conventional array multiplier.The simulation results demonstrates the improvement in processing speed as well as power consumption.

Keywords Array multiplier · Vedic multiplier · Ekanyunena purvena

1 Introduction

In Indian culture the meaning of the word Vedas is the books of wisdom. It has many aspects of life explained with education and engineering. Vedic mathematics is also a part of Vedas. This explores the methods used in mathematics for different applications in the form of vedic sutras and upasutras. There are 16 sutras (Formulae) and 13 subsutras (derived formulae) in vedic mathematics as explored by Bharati Krishna Teertha Maharaj [1]. In applications such as amplifier, increasing brightness of an image, amplifying the speech etc. the operation performed is the scaling by a number.

M. A. Sayyad (✉) · D. N. Kyatanavar
Electronics and Telecommunication Engineering Department, SRES's Sanjivani COE, Savitribai Phule Pune University, Kopargaon, Pune, India
e-mail: sayyadma@yahoo.com

D. N. Kyatanavar
e-mail: kyatanavar@gmail.com

Z. Patel et al. (eds.), *Advances in VLSI and Embedded Systems*, Lecture Notes in Electrical Engineering 676, https://doi.org/10.1007/978-981-15-6229-7_7

The EDA tools such as Xilinx and Altera Quartus II are powerful tools which give analysis of the design for different parameters such as area (in terms of number of slice or LUT), delay and power dissipation. Most of researchers have implemented the multiplier on FPGA with VHDL or Verilog. The technique for implementation of general N $\times$ N multiplication is proposed in [2], in this general method applicable for 4,8,16 and 32 bit multiplication is given. The vedic mathematics can be used for the decimal number system [3] and binary number system [4] with ASIC implementations for improvement in the performance of multiplier. The vedic multiplier is successfully implemented for applications discrete wavelet transform [5], block convolution [6], cryptographic application [7], squaring and cubing [8] with improved propagation delay to get the quick results of the multiplications. There is very less literature on the Ekanyunena Purvena sutra of vedic multiplier, as it is special case of multiplications. The Multiplier using Ekanyunena Purvena sutra has the integer multiplicand and multiplier is always 9 or array of 9 (for example 9,99,999 etc.) [9].

This paper presents BCD and binary multiplier as use of BCD numbers in displaying kinds of application gives great simplification by treating BCD numbers as separate digit. This separate digit matches with series of separate identical 7-segment displays. If the numeric quantity were stored and manipulated as pure binary, interfacing to such a display would require complex circuitry. Therefore, in cases where the calculations are relatively simple working throughout with BCD can lead to a simpler overall system than converting to binary.

2 Background of Ekanyunena Purvena

In Sanskrit word 'Veda' stands for 'knowledge'. Vedic mathematics is believed to be reconstructed from Vedas. Its methods reduce the complex calculations into simpler ones because they are based on methods similar to working of human mind thereby making them easier. Out of all these 16 sutras most of researchers used Urdhva-tiryakbyham-Vertically and crosswise method applicable to all types of multiplications and very few had used Nikhilam Navatashcaramam Dashatah-All from 9 and the last from 10 as it is efficient only in the case if multiplier and multiplicand are close to base value. The Ekanyunena Puravena method is best suitable for multiplication by maximum number.

2.1 The Present Method (Ekanyunena Purvena)

The present multiplier is based on an algorithm Ekanyunena Purvena of ancient Indian Vedic Mathematics. The present multiplier must have an integer multiplicand and multiplier must be an array of 9 (for example 9,99,999 etc.). The multiplication is performed in the following steps for decimal numbers.

Example 1. $4 \times 9 =$	Example 2. $34 \times 99 =$
Step 1: RHS of result 4–1 = 3	Step 1: RHS of result 34–1 = 33
Step 2: LHS of result 9–3 = 6	Step 2: LHS of result 99–33 = 66
Step 3: Combined result $5 \times 9 = 36$	Step 3: Combined result $25 \times 99 = 3366$

Similar multiplications steps are followed for binary coded decimal (BCD) number and binary number in which the maximum digit is 1001(in decimal 9) and 1111(in decimal 15) respectively.

Example 1. $0101 \times 1001 =$	Example 2. $0101 \times 1111 =$
Step 1: RHS of result 0101−0001 = 0100	Step 1: RHS of result 0101−0001 = 0100
Step 2: LHS of result 1001−0100 = 0101	Step 2: LHS of result 1111−0100 = 1011
Step 3: Combined result $0101 \times 1001 =$ 01000101	Step 3: Combined result $0101 \times 1001 =$ 01001011
Here the result is in BCD format and read as 45 decimal number system	Here the multiplication of decimal number 5 and 15 represented in binary with 4 bits. The result of multiplication is 01001011 in binary and its decimal equivalent is 75

3 Implementation Proposed System of Multiplication

The multiplieris designed with the ekanyunena purvena method for BCD as well as binary number system. The designed multiplier is implemented in Xilinx ISE with VHDL.

3.1 BCD Multiplication Using Ekanyunena Purvena

Let the multiplier is 1001 (decimal equivalent = 9) and multiplicand may be any integer consider as x the multiplication is represented as

$$\begin{aligned} x \times 1001 &= x(10^1 - 1) \\ &= 10x - x \end{aligned} \tag{1}$$

The multiplication can be done with one multiplication with 10 and subtracting the multiplicand from the result.

In the proposed method following approach is used for implementation

$$\begin{aligned} x \times 1001 &= 10x - 10 + 10 - x \\ x \times 1001 &= 10(x - 1) + (10 - x) \end{aligned} \tag{2}$$

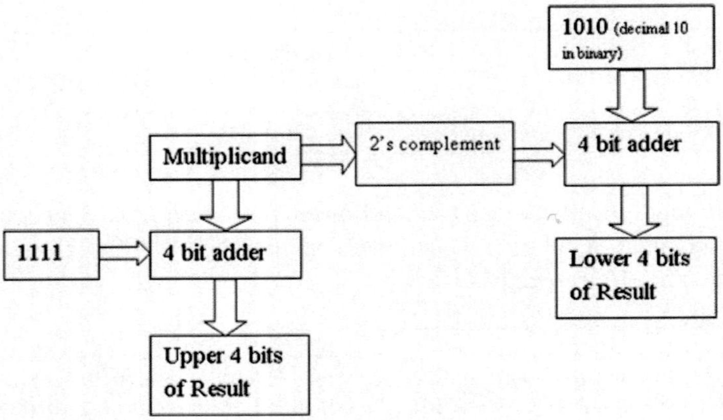

Fig. 1 4 × 4 bit Ekanyunena Purvena multiplier for BCD multiplications

The term on LHS is the upper digit of the result and the RHS term is the lower digit of the result. The result can be generated with the help of 2 subtractions only. For upper digit subtract 1 form multiplicand, which is implemented as adding 1111 to the multiplicand and neglect the carry. To get the lower digit use binary subtraction. The scheme for implementations is shown in Fig. 1.

3.2 *Proposed System for Binary Numbers Multiplication Using Ekanyunena Purvena*

Let the multiplier is 1111 (decimal equivalent = 15) and multiplicand may be any integer consider as x the multiplication is represented as

$$\begin{aligned} x \times 1111 &= x(2^4 - 1) \\ &= 2^4 x - x \end{aligned} \tag{3}$$

The multiplication can be done with one multiplication with 2^4 and subtracting the multiplicand from the result. In the proposed method following approach is used for implementation

$$\begin{aligned} x \times 1111 &= 2^4 x - 2^4 + 2^4 - x \\ x \times 1111 &= 2^4 (x - 1) + (2^4 - x) \end{aligned} \tag{4}$$

The term on LHS represents the upper 4 bits of the result and the term on RHS represents lower 4 bits of the result. The result can be generated with the help of 2 subtractions only. For upper 4 bits subtract 1 form multiplicand, which is implemented as adding 1111 to the multiplicand and neglect the carry. To get the lower

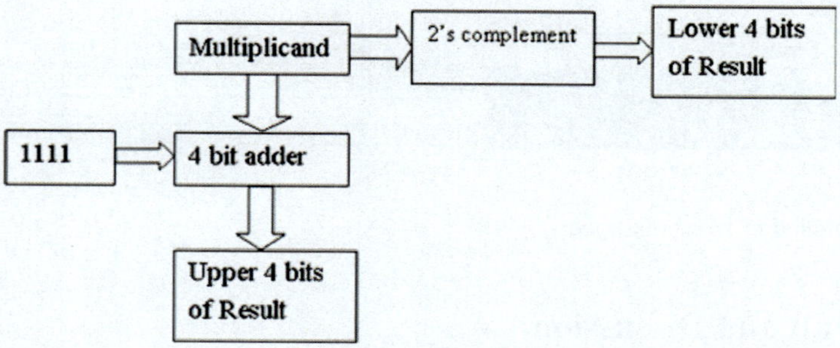

Fig. 2 4 × 4 bit Ekanyunena Purvena multiplier for binary multiplications

4 bits use binary subtraction. The scheme for implementations is shown in Fig. 2. For 4 × 4 bit multiplier the multiplication is with one 4 bit adder and one 2′ complement circuit. The 4 bit adder circuit uses 1 half adder and 3 full adder whereas 2′ complement circuit consist of 4 half adders.

3.3 *Synthesis and Simulation of Proposed Multiplier*

The proposed multiplier is implemented with Xilinx ISE in VHDL and synthesized using Xilinx XST. The synthesized circuit for 16 bit multiplier is shown in Fig. 3a, b for proposed multiplier for BCD numbers and Proposed multiplier for binary numbers respectively. All the designed multipliers are simulated using Isim simulator the simulation results 16 bit multiplier is shown in Fig. 4 for Proposed Ekanyunena Purvena multiplier with binary numbers.

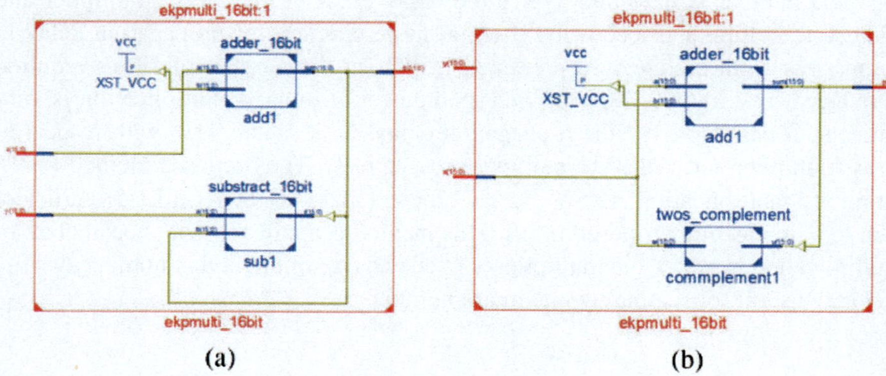

Fig. 3 16 × 16 bit Ekanyunena Purvena multiplier BCD numbers and binary numbers

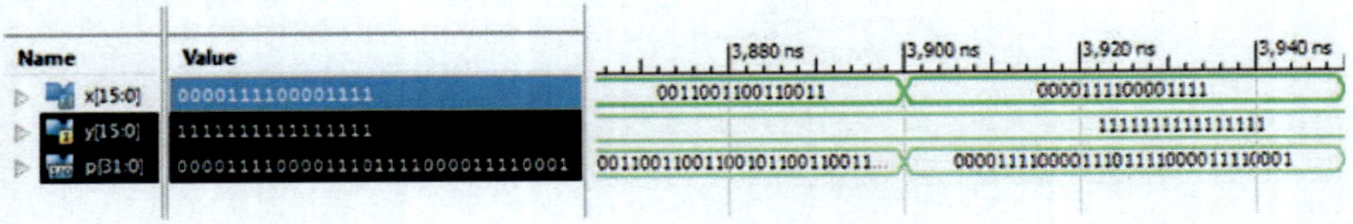

Fig. 4 Simulation 16 bit multiplier

4 Result and Discussion

Synthesis of the proposed Ekanyunena Purvena multiplier was done using Xilinx ISE 14.7 and the report related to Number of Slices, Number of 4 input LUTs and delay is presented in Table 1. For calculation of power dissipation XPower analyzer tool is used, the switching activity data can be written at the time of simulation after place and route. This switching activity data file is used by XPower analyzer to calculate total power. The power calculations will help to optimize the power during design. In our design the simulation input are kept same for array and vedic multiplier so that there should be same effect of dynamic power on the circuits. The total power dissipation of each multiplier is shown in Table 1.

Note: Table 1 contain the results of 16 bit array multiplier and UT vedic multiplier form [10] which is our own work.

5 Conclusion

This paper presents the design approach and architecture for BCD as well as binary number system multiplications. The proposed architecture for multiplication of 16 bit BCD number is 62% and 47% more delay efficient than array multiplier and vedic UT multiplier respectively. It is not more efficient for propagation delay in comparison with architecture presented in [9] but the number of LUTs required is reduced by 11.86%. The proposed method for 16 binary multiplication is very efficient, it has improved the propagation delay by 80% and 71% with respect to array multiplier and vedic UT multiplier respectively. The proposed method for 16 bit binary multiplication is 45.9% delay efficient and 30% area (in LUTs) efficient than the architecture proposed in [9]. This method is applicable for special class of multiplication in which the multiplier is fixed and maximum in that number system, and very useful for scaling by maximum number.

Table 1 Comparison of multipliers with array and UT multiplier

Implementation method	Slices	LUTs	Delay (ns) total	Delay (ns) logic	Delay (ns) route	Power (W)
4 bit multiplier						
Array multiplier 4 bits	17	30	18.301	9.456	8.845	0.069
UT vedic multiplier 4 bits	19	34	16.479	8.977	7.502	0.067
Architecture used in [9]	5	10	9.683	6.896	2.787	0.063
Ekanyunena purvena multiplier 4 bits BCD	3	6	7.926	6.103	1.823	0.063
Ekanyunena purvena multiplier 4 bits binary	4	7	7.999	6.103	1.896	0.063
8 bit multiplier						
Array multiplier 8 bits	72	126	35.075	15.204	19.817	0.083
UT vedic multiplier 8 bits	99	174	29.247	13.288	15.959	0.082
Architecture used in [9]	14	27	15.192	8.812	6.380	0.067
Ekanyunena purvena multiplier 8 bits BCD	13	23	15.128	8.812	6.316	0.072
Ekanyunena purvena multiplier 8 bits binary	9	17	10.712	7.061	3.651	0.067
16 bit multiplier						
Array multiplier 16 bits [10]	293	510	70.424	31.404	39.020	0.106
UT vedic multiplier 16 bits [10]	395	691	50.400	23.726	26.674	0.100
Architecture used in [9]	31	59	26.260	12.644	13.616	0.079
Ekanyunena purvena multiplier 16 bits BCD	29	52	26.637	12.644	13.993	0.082
Ekanyunena purvena multiplier 16 bits binary	23	41	14.194	8.019	6.175	0.079

References

1. Swami Bharati Krsna Tirtha (1965) Vedic mathematics. Motilal Banarsidass publishers, Delhi
2. Akhter S (2007) VHDL implementation of fast nxn multiplier based on vedic mathematic. In: 18th European Conference, IEEE circuit theory and design ECCTD 2007, pp 472–475
3. Saha P, Banerjee A, Dandapat A, Bhattacharyya P (2012) Design of high speed vedic multiplier for decimal number system. In: Proceedings of 16th international symposium, VDAT 2012 Shibpur, India, pp 79–88
4. Jagannatha KB, Lakshmisagar HS, Bhaskar GR (2014) FPGA and ASIC implementation of 16-bit vedic multiplier using Urdhva Triyakbhyam Sutra. In: Proceedings of International Conference, ICERECT 2012, Lecture Notes in Electrical Engineering, vol 248, Springer, India, pp 31–38. https://doi.org/10.1007/978-81-322-1157-0_4
5. Tripathi S, Singh AK (2015) High speed and area efficient discrete wavelet transform using vedic multiplier. In: International conference on computational intelligence and communication networks (ICCICN 2015), pp 363–367
6. Hanumantharaju MC, Jayalaxmi H, Renuka RK, Ravishankar M (2007) A high speed block convolution using ancient indian vedic mathematics, In: International conference on computational intelligence and multimedia applications (ICCIMA 2007), Sivakasi, Tamil Nadu, pp 169–173. https://doi.org/10.1109/ICCIMA.2007.332
7. Leonard Gibson Moses S, Thilagar M (2011) VHDL implementation of high performance RC6 algorithm using ancient indian vedic mathematics. In: 3rd international conference IEEE electronics computer technology (ICECT), vol 4, pp 140–143
8. Ramalatha M, Thanushkodi K, Dayalan D, Dharani P (2009) A novel time and energy efficient cubing circuit using vedic mathematics for finite field arithmetic. In: 2009 International conference on advances in recent technologies in communication and computing, Kottayam, Kerala, pp 873–875. https://doi.org/10.1109/ARTCom.2009.227
9. Khan A, Das R (2015) Novel approach of multiplier design using ancient vedic mathematics. Adv Intell Syst Comput. https://doi.org/10.1007/978-81-322-2247-7_28
10. Sayyad MA, Kyatanavar DN (2017) Optimization for addition of partial product in vedic multiplier. In: 2017 International conference on computing, communication, control and automation (ICCUBEA), Pune, pp 1–4

FPGA Design of SAR Type ADC Based Analog Input Module for Industrial Applications

G. Dhanabalan and T. Murugan

Abstract Programmable logic controller (PLC) is connected with analog and digital input and output modules to process physical variables which are to be maintained at the desired values. Each module has a processor to process analog or digital signals. This paper has designed analog input module (AIM) using field programmable gate array (FPGA). It has used a digital to analog converter (DAC), comparator and FPGA for a single channel analog to digital converter (ADC). It becomes an analog input module (AIM) when a DAC and comparator are added to FPGA for every addition of a channel. Conversion time of a processor based AIM is $n*t_c$ where 'n', 't_c' are the number of channels and conversion time of ADC respectively. Conversion time of 'n' channel FPGA based AIM is 't_c' as all the analog signals are processed concurrently. The design was experimented using Multisim software and the conversion time of eight channels ADC was identified 0.13 ms.

Keywords Programmable logic controller · ADC · Microprocessor · FPGA

1 Introduction

The concept of Industrial automation emerged from pneumatic system. After the invention of electric system, relay logic enhanced industrial automation. Nowadays, modern industries use programmable logic controller (PLC) that reduces the complexity of relay logic. PLC architecture has a processor, memory and input/output interface. Program and data are stored in separate memory. The function of the PLC is to execute the program available in memory. Ladder diagram (LD) representation is the familiar way of programming the PLC. A typical LD will have many rungs. PLC reads the rungs from left to right and top to bottom. Output of one rung can be

G. Dhanabalan (✉)
Sri Vidya College of Engineering and Technology, Virudhunagar, India
e-mail: dhanabalan1972@gmail.com

T. Murugan
Sethu Institute of Technology, Kariapatti, India
e-mail: muruganvlsi@gmail.com

Z. Patel et al. (eds.), *Advances in VLSI and Embedded Systems*, Lecture Notes in Electrical Engineering 676, https://doi.org/10.1007/978-981-15-6229-7_8

also used as input(s) in other rung(s). Also, many rungs shall have the same input(s). Each rung generates output based on the satisfactory input conditions. Inputs to the rungs are accessed from the input modules which are attached with the PLC. Input modules acquire the field signals and store in memory. Whenever the PLC executes the rungs, it refers the memory of the input module to access the necessary input signals [1, 2]. PLC executes the LD rung by rung and hence it is sequential. PLC stores the results of the execution of LD into the memory of output modules. A processor in the output module transfers these outputs signals to the field. Thus the PLC communicates to analog input/output or digital input/output modules to read or write data from or to the external world [3]. Number of rungs in the LD, number of inputs and outputs used in the rung impacts the scan time of the PLC. The scan time of a PLC is the cumulative time of analog and digital signals scanning, execution of ladder logic program and storage of outputs in memory [4]. The scan time varies according to the number of inputs and outputs connected with the PLC [5].

The efficiency of an automation system established through PLC is based on the ability of the PLC to respond to the change(s) in the process that happens in lesser period of time. Scan time of PLC dictates the term "lesser period of time". Therefore it is necessary to optimize the scan time so that the PLC is able to respond even when the changes happen at the level of micro second. Reducing the number of rungs is one way of reducing scan time. The total number of rungs in a ladder diagram (LD) depends upon the logic. Logic optimization can help to reduce the number of rungs. Reducing the number of digital and analog signals processed by the PLC is another way of reducing the scan time. PLC processes digital signals through digital input module (DIM). Processing speed of DIM is at satisfactory level as the processor in the DIM acquires the digital signals as such. Therefore, modification in DIM does not impact the reduction in scan time. Analog signals are processed by analog input module (AIM). AIM acquires analog signals of physical parameters like pressure, level, and temperature [6] and outputs them as digital signals. Normally, eight to sixteen analog signals are connected with an AIM. It has a processor in its architecture. Since the processor cannot read analog signal, it must be converted into digital using analog to digital converter (ADC).

Processor in the AIM activates a multiplexer to choose a channel, enables start conversion pin of analog to digital converter (ADC) and stores the digital output in memory after identifying the end of conversion signal from ADC. It repeats the process for the remaining channels. If 1 μs is assumed as the conversion time to convert one analog signal into digital, the conversion time for eight analog signals will be 8 μs. This indicates that the conversion time of an AIM depends on the total number of analog signals to be converted into digital. Also, the processor consumes time to store the ADC output in memory. If 1 μs is the storage time for one analog signal, the total time required by the AIM to process eight analog signals is 16 μs. This indicates that the conversion time increases along with the number of analog signals. This greatly affects the scan time of PLC. Increase in scan time can be otherwise viewed as a drop in the quality of a product. Therefore, reduction in conversion time of AIM will reduce the burden of PLC. Also, PLC is actually connected with more than one AIMs and the time taken by the PLC to retrieve data from all the AIMs brings down its efficiency.

AIM can be otherwise viewed as a multichannel ADC as it processes eight to sixteen analog signals. An elaborate survey reveals that much attention has been given in the design and applications of multichannel ADC. Use of multichannel ADC in the gamma-ray spectroscopic measurement is discussed in [7]. It proved that multichannel ADC performs better in the gamma-ray spectroscopic measurement by comparing its performance with the performance of single channel ADC. Nonlinearity, energy resolution and timing resolution were considered as the parameters for performance analysis. Design of multichannel ADC for medical applications is discussed in [8]. It has used field programmable gate array (FPGA) to establish a data acquisition system. FPGA based motor controller has used ADC in its design [9]. FPGA reads ADC output and initiates control action to ensure that the motor speed is always at the desired value. Algorithm for PI controller was implemented in FPGA so that it acts as a PI controller. A case study on neural network based electronic nose also has been discussed in [9]. It has used LabVIEW to establish an user interface. Design of an educational laboratory has used multichannel ADC to acquire signals from various sensors [10]. Development of a prototype model for multichannel neuronal recording has used multichannel ADC [11]. It has also used FPGA to transfer data to the host computer. One of the major observations from this survey is that the input range of the ADC is well below 5v almost in all the works. Actually, an industrial standard transmitter, that generates output for a physical parameter like pressure, temperature etc., is 0–5v. The survey has also indicated that multichannel ADC has used microprocessor to perform the conversion.

This work has developed FPGA based AIM that has reduced the total conversion time of AIM. FPGA based solutions for industrial problems are tremendous [9, 12]. FPGA performs simultaneous conversion for all the analog signals and stores the data in FPGA RAM in a single clock cycle [13]. Thus the conversion time for 'n' channel is the conversion time for a single channel. This FPGA based AIM can be used as a standalone system for an application. Otherwise, it shall be used along with the PLC.

2 Architecture of Conventional Analog Input Module

Conventional AIM architecture has multiplexer type ADC controlled by the processor as shown in Fig. 1. Whenever an analog signal is converted into digital, the processor reads ADC output and stores it in random access memory (RAM) [14]. The method by which the analog signal must be converted into digital depends on the type of the application. Many industrial ADCs employ successive approximation register (SAR) type ADC. This work has developed FPGA based AIM based on the working principle of SAR type ADC. In SAR type ADC [15], the most significant bit (MSB) of SAR register is set and the remaining bits are reset. Digital to analog converter (DAC) converts the SAR content into analog signal and compares it with the analog signal

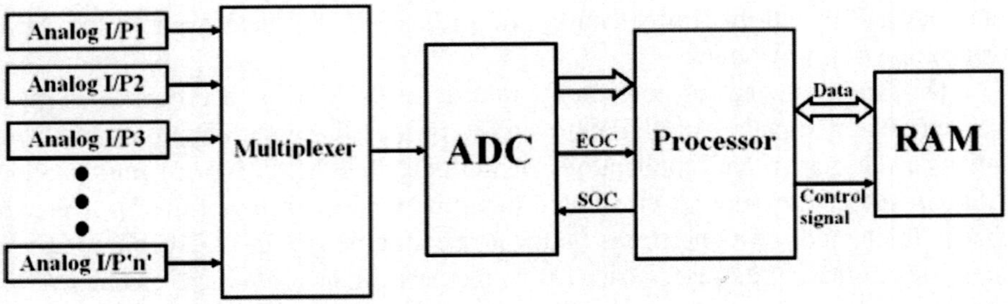

Fig. 1 Processor based conventional analog input module

that is to be converted into digital. DAC output is connected to the non inverting terminal of comparator and the analog input signal is connected to the inverting terminal of the comparator. If the output of DAC is less than the value of analog input signal then it resets the MSB bit of SAR register. Otherwise, SAR content is not disturbed. Now, the next bit towards least significant bit (LSB) is set and the above process is repeated until all the SAR bits are set and compared. An 8-bit ADC consumes eight clock pulses to complete the conversion. The content of SAR register at the eighth pulse is treated as the equivalent digital value of analog input signal. The conversion time of a single channel SAR type ADC is given in Eq. (1) as

$$t_c = t_{clk} \times n \tag{1}$$

where t_c is the conversion time, t_{clk} is the time duration of clock pulse applied to ADC and 'n' is the number of bits of ADC output. Thus, the conversion time of 'N' channels ADC is

$$t_{cN} = N \times t_{clk} \times n \tag{2}$$

where t_{cN} is the conversion time of 'N' channels. It is clear that the conversion time of a single channel gets multiplied with the total number of channels. For example, the conversion time of 8, 16 channels ADC will be 8 μs and 16 μs respectively, if the conversion time of single channel ADC will be 1 μs.

The processor can deal the ADC output in the following way.

- Processor reads the ADC output and initiates conversion for the next channel. (or)
- Processor reads the buffer register which has the digital output for the earlier sample and the ADC performs conversion for the present sample. This leads to the pipelining concept, so that the process of both reading and conversion happens simultaneously.

ADC in AIM slows down the processor execution speed, as its conversion time is high when compared to the processor clock time. Effective utilization of the processor is possible only when the conversion time of ADC matches the processor execution time. The period by which a particular signal is accessed by a processor depends on the following time constraints.

- The time required by the processor to select an analog signal using multiplexer.
- Conversion time of ADC.
- The time required to store ADC output in RAM.

Under these constraints, the time required to sample *n*th signal will be the time that meets out the above constraints multiplied by '*n*'. This makes the sampled signals as a dead signal instead of a live signal. Reduction in conversion time will reduce the sampling time which in turn increases the number of samples per second. If the conversion time can be reduced to the processor clock pulse, then the sampling time will be the conversion time plus the time required to store ADC output in memory. The sampling time can be further reduced, if the ADC output is directly stored in the RAM.

3 FPGA Design of AIM

Operating speed of a processor will be always less than that of its clock speed. Multitasking can be achieved either by sharing processor's clock time equally to all the tasks or assign one processor to each task. In time-sharing concept, tasks will be assigned on the basis of priority. This is not a multitasking as only one task is executed is at a time. Assigning one processor to each task is a real multitasking which is not possible in practical scenario. FPGA is a reconfigurable device that has configurable logic blocks (CLB) in it [16], whereas the CLB is a combinational logic circuit that executes a set of inputs applied to it. A portion of the output is fed to the CLB as input so that a logic for sequential circuit can be realized. Digital circuit designed using hardware description language (HDL) can be simulated, verified and converted into bit file so that it can be downloaded into FPGA using synthesize software. Concurrent execution of any combinational or sequential circuit by the FPGA is confirmed by assigning individual FPGA hardwares to it. Thus the delay time of a circuit is actually the propagation delay time of FPGA. This concurrent execution nature of FPGA [5] is effectively utilized in the design of AIM.

Block RAM and distributed RAM are one of the distinct feature of FPGA. These memories are available in the chip itself. Hence, the data transfer happens at a faster rate. Also, it provides the feature of storing or accessing more than one data at a time. Memory capacity of a block RAM of SPARTAN-3 generation FPGA chip XC3S500E is 360 Kb. This block RAM capacity has been divided as 20 block RAMs. Hence, the capacity of each block RAM will be 18 Kb. All these 20 block RAMs can be accessed in a single clock pulse as the memory operation is synchronous. Latest FPGA technology is flexible enough that either the number of block RAMs or the memory size of block RAM is reconfigurable. This work has proposed the design of ADC by considering the following aspects.

- Let the working principle of ADC be successive approximation register type. This helps to design concurrent conversion of more than one analog signals into digital when combined with FPGA.
- The design must provide better performance even with the lower configuration of the devices.
- Consider the DAC and the comparator that are already available in the market.
- Design the comparator using operational amplifier.

3.1 *Single Channel Analog to Digital Converter Using FPGA*

The processor and the RAM in conventional AIM are replaced by an FPGA in the proposed single channel ADC as in Fig. 2. Control logic implemented in the FPGA sets MSB bit of the SAR register. DAC converter converts the SAR content into analog signal. This analog signal is treated as the reference signal for the purpose of comparison. A comparator designed using operational amplifier compares the reference signal with the analog signal which is to be converted into digital. The reference signal and the analog signal are connected to the inverting and non-inverting input of the comparator respectively. FPGA reads the comparator output and decides whether to reset or retain the MSB bit of SAR register. MSB bit must be retained when the value of reference voltage is less than that of analog signal. Therefore, the FPGA retains the MSB bit when the comparator output is high and resets when it is low.

If the input connections to the comparator are reversed, then the condition for retaining or resetting the MSB is also reversed. Thus FPGA is flexible, as the condition of reversal can be modified using the hardware description language (HDL) coding. Now, the FPGA sets the immediate MSB bit and the above process is repeated till the status of the LSB bit is decided. The content of SAR register is equivalent to the digital output for the analog signal applied at the non-inverting input of the comparator.

Latest FPGAs provide storage facility in the form of block or distributed RAM. The design was experimented for 8-bit ADC. As the SAR register will be referred

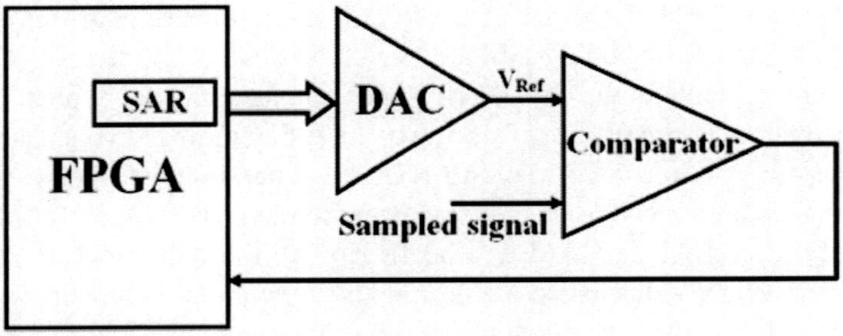

Fig. 2 Block diagram of single channel analog to digital converter

eight times during conversion, the design requires eight clock pulses to complete the conversion and the data in SAR at eighth clock pulse is the digital output. SAR register will be reset in the ninth pulse and the conversion starts again from the next pulse. The conversion time for the single channel ADC is calculated from Eq. (3) as

$$t_c = n \times \left(t_{set} + t_r + t_{pd}\right) \tag{3}$$

where t_c is the conversion time, 'n' is the number of bits of ADC output, t_{set} is the settling time of DAC, t_r is the response time of comparator and t_{pd} is the propagation delay time by FPGA logic. Delay time of t_{set} will be greater than t_r & t_{pd} as the settling of operational amplifier used in DAC is high.

Field programmable gate array analog input module–FPGA AIM. Industrial automation system deals with more number of analog signals. Circuit complexity increases when the PLC has to acquire digital data directly from many ADCs. Total number of modules that can be attached with the PLC is based on the process condition. PLC communicates to these AIMs through a common communication bus system. In AIM, a set of analog signals are grouped together to perform conversion for multiple analog signals. The digital data is transferred to PLC through suitable data transfer technique. However, this work has focused on the conversion process only.

The concept of FPGA based AIM has been developed using multiple DACs and comparators as shown in Fig. 3. Every channel is assigned a SAR register, DAC and a comparator. This confirms that all the analog signals are simultaneously converted into digital. Thus the total number of DACs and comparators used in the design is equivalent to the total number of analog signals to be converted into digital. However, only one FPGA handles the conversion for all the analog signals. Since the FPGA can directly communicate with all the DACs and comparators simultaneously, there is no need for a multiplexer.

As a set of DAC and comparator is used for each channel, the conversion time for '*n*' channel is the conversion time of a single channel only.

4 Simulation of Single and Multichannel ADC

The entire circuit was designed, simulated and verified using Multisim software. General purpose 8-bit DAC (DIG2ANACON8) was used to generate analog signal for the applied 8-bit digital input. LTC6240CS8 [17], complementary metal oxide semiconductor CMOS operational amplifier was used to design comparator. Analog signal which is to be converted into digital was connected to the non-inverting input of the comparator. The output of DAC was connected to the inverting terminal of the comparator. The operating frequency of FPGA was set as 60 kHz. Normal procedure to download a digital design into an FPGA is to write HDL description equivalent to the hardware circuit, simulate the program, verify that the HDL description generates

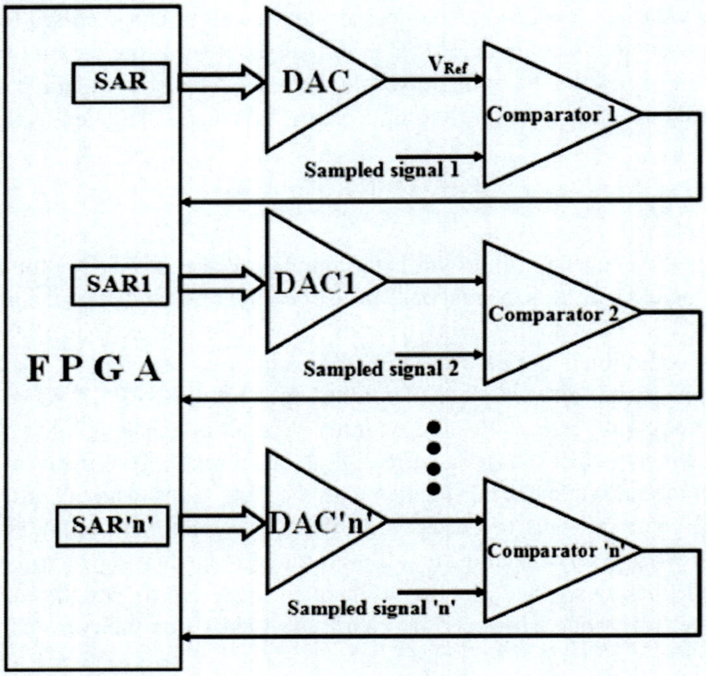

Fig. 3 FPGA based multi channel analog to digital converter

expected output, synthesize the program to convert HDL description into binary file and download it into FPGA. In case of Multisim, HDL description for single channel ADC is developed as hardware logic itself. Multisim has an option to convert this hardware logic into Very high speed hardware integrated circuit HDL (VHDL) description. FPGA was connected with other necessary components to realize single channel ADC as in Fig. 4.

FPGA logic to perform ADC conversion developed using Multisim is shown in Fig. 5. It has used 4-bit synchronous counter, 3 to 8 decoder, 2:1 Multiplexers and D-Latches and their respective identifications are also shown in the Figure. are U1, U4, U6, U8. Since the design is for 8-bit ADC, eight multiplexers and eight D-Latches were used in the design.

Eight bits of the decoder output were connected as selector lines to the eight multiplexer. D-Latches receive their inputs from the multiplexers. Outputs of the D-Latches were connected to the SAR register. Multiplexer plays a key role in setting or resetting a specific bit of the SAR register. The two inputs to the multiplexer are the comparator output and the logic '1'. Multiplexer 1 allows Logic '1' to set the MSB bit of SAR register when the counter value is one. In the next count value, multiplexer 1 allows the comparator output and the multiplexer 2 allows Logic '1' to the SAR register.

This process will continue for eight clock pulses. The counter is reset once the count value reaches eight. Actually, It is enough to have a 3-bit synchronous counter

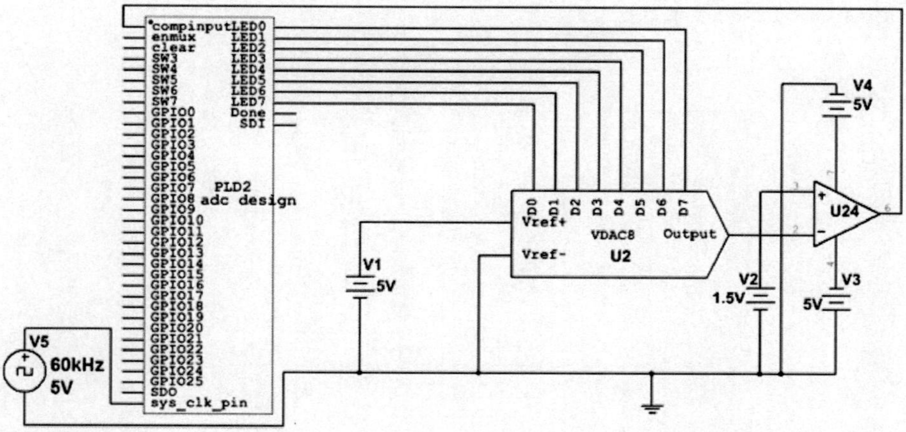

Fig. 4 Circuit diagram of single channel analog to digital converter

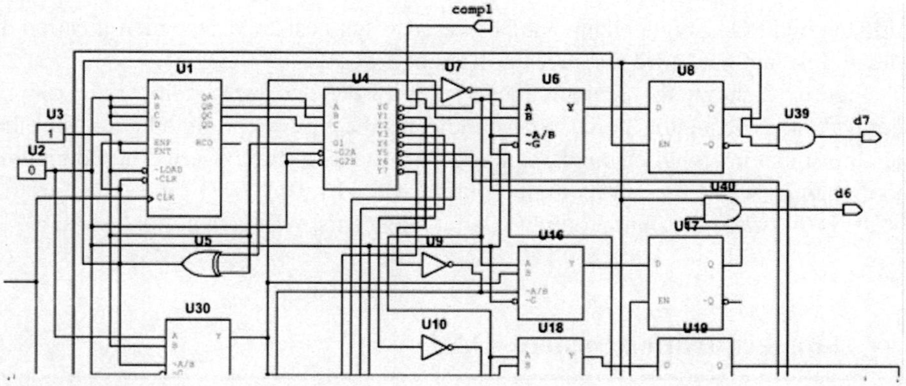

Fig. 5 Portion of the FPGA logic to perform conversion

as the specification of ADC is 8-bit. This design has used 4-bit counter to identify the end of eighth clock pulse. MSB bit of the counter is XORed with the logic '1' to reset the counter. The complete circuit was simulated and verified that the expected output is reached. VHDL description equivalent to the hardware circuit designed in FPGA will be automatically generated when the "Generate and save VHDL files" shown in Fig. 6 is selected from the option "Export to PLD" in the menu.

The aim of this work is to develop an AIM which has the capacity to convert eight analog signals into digital. 1 FPGA, 8 DACs and 8 comparators were used to design eight channel ADC. FPGA hardware logic was repeated eight times so that the design for eight channel ADC is complete. This ensures concurrent conversion for all the eight channels. Figure 6 shows FPGA logic for single channel which was converted into VHDL code and synthesized to indentify the delay time and hardware resources

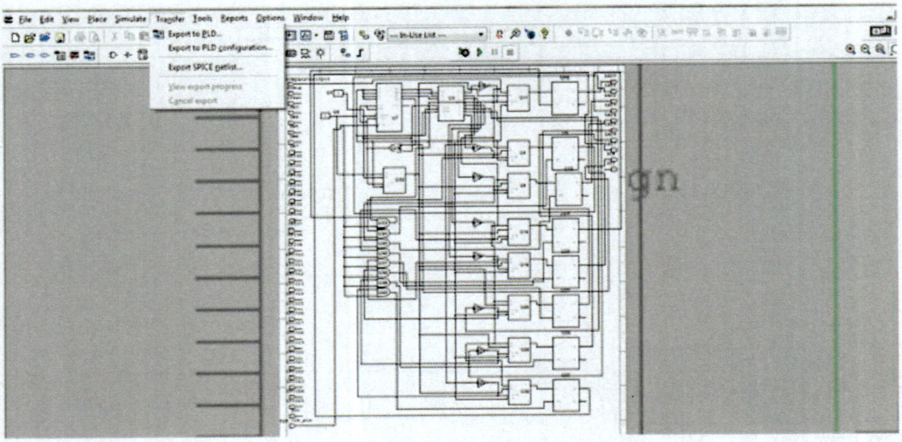

Fig. 6 FPGA hardware logic for single channel analog to digital converter

utilized by FPGA. Eight channel ADC was also synthesised for the identification of delay time and hardware resources utilized by FPGA.

Figure 7 shows the changes made by the FPGA to bring comparator output equivalent to the analog signal. It also indicates the equivalent digital output for the analog signal in eighth channel. Analog signal that is to be converted into digital in that channel was 4.0v and its digital output was 110, 011, 011. Input range to the ADC is 0–5v and the digital output is 8-bit. Hence the ADC resolution is 5/256.

5 Data Acquisition System–DAS

AIM designed using FPGA can be used as a stand alone system in small scale industries. However, it is necessary to have a data acquisition system (DAS) to monitor the field parameters [18]. Since the monitoring process happens at layman level, it is mandatory that the data acquisition software must be graphical and user friendly. Data acquisition system software installed in a computer monitors field parameters by communicating to PLC which acquires field parameters through AIM and digital input module (DIM). Precision in monitoring time can be relaxed as the software does not produce any control action. Serial data communication is adaptable to transfer data from FPGA to computer [19]. FPGA has the capacity to transfer data to computer and PLC simultaneously.

This work has developed DAS using LabVIEW software. LabVIEW is basically utilized to support the concept of virtual instrumentation. This helps to monitor the analog signals that are processed by AIM. A tool kit named "Multisim connectivity tool kit" has been used since the work is done at the level of simulation. It provides a link between Multisim and LabVIEW. LabVIEW reads the specific parameters of a circuit that has been developed in Multisim. Figure 8 shows the details for two

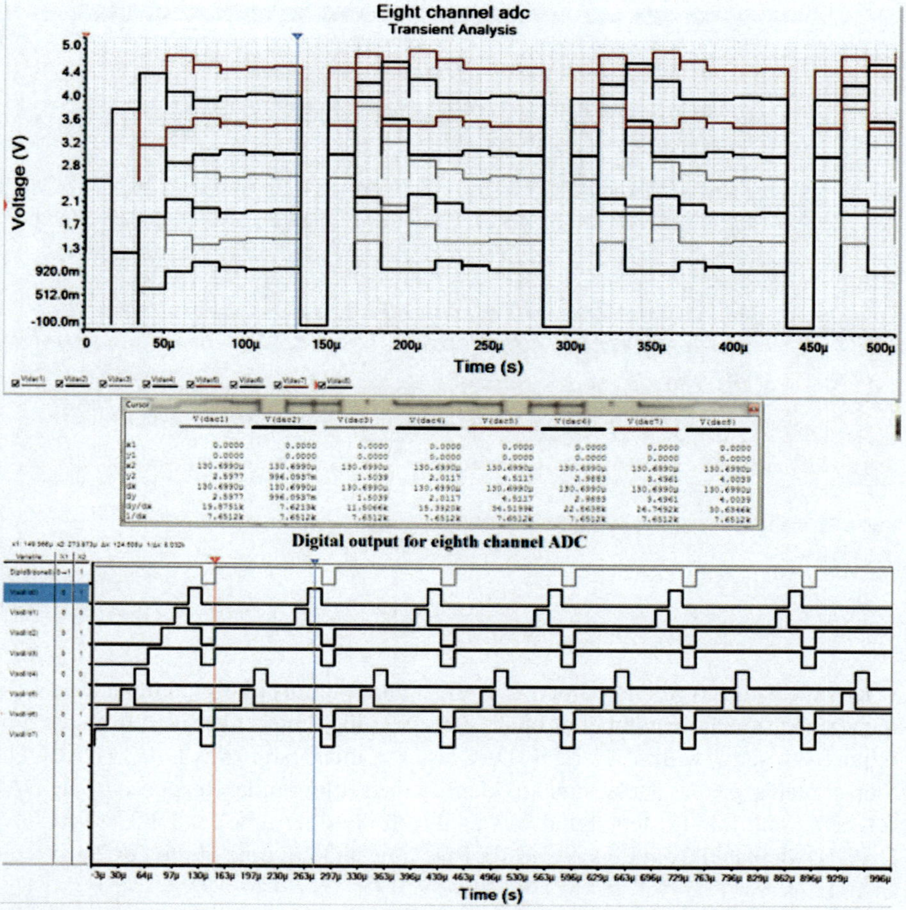

Fig. 7 Simulation of comparator output, digital output for multichannel ADC

analog signals. DAC output was changed by FPGA for every clock pulse to bring its output that is equivalent to the analog signal. Considering the analog value at the eighth clock pulse, it can be used for the purpose of monitoring or establishing control actions.

6 Result and Discussions

Efficiency of an AIM is mainly dependent on the conversion time of ADC, that is based on the settling time of DAC and Comparator. Since the comparator is designed using operational amplifier, settling time of operational amplifier is the primary factor. Operational amplifier is used in open loop configuration for it to function as a comparator. Now, conversion time for DAC is given in Eq. (4) as

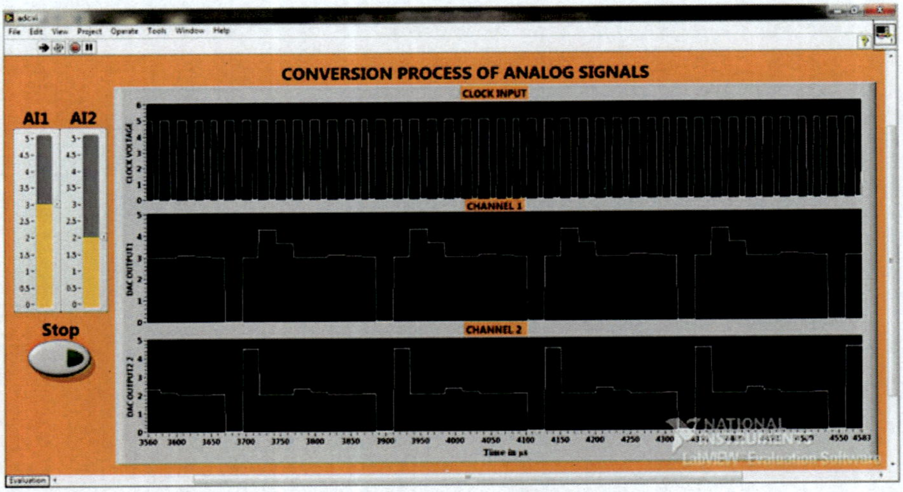

Fig. 8 LabVIEW display of multi channel analog to digital converter

$$t_{con} = n \times t_{set} \tag{4}$$

where t_{con} is the conversion time, 'n' is the number of bits of digital inputs and t_{set} is the settling time. The number of bits is 8 bit. Settling time is identified from the time required to settle within 1/2 LSB. DAC used in this design was DIG2ANACON8. Step signal is used as a test signal to identify the settling time. Response of the DAC for rising and falling step signal was analyzed. Settling time for DIG2ANACON8 DAC is calculated from Figs. 9 and 10. Rise time and fall time of the DAC was fixed as 10 ns.

Highest settling time is considered to calculate the conversion time. From Figs. 9 and 10, the highest settling time is 7.03 ns. Transient analysis of comparator was

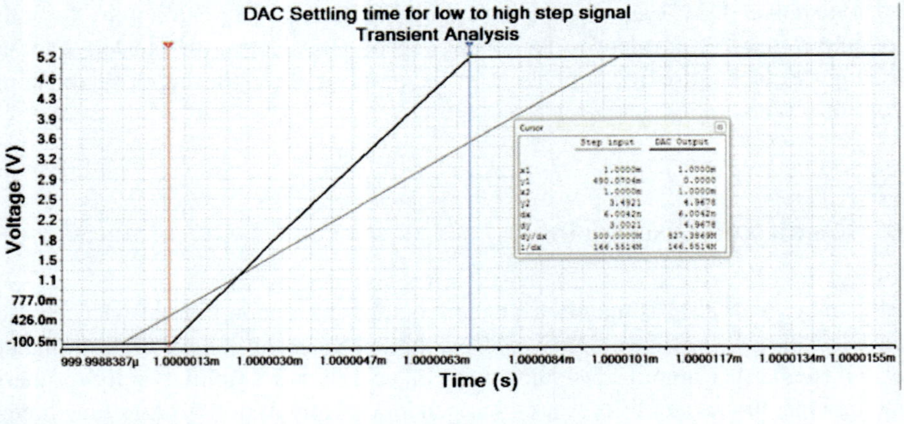

Fig. 9 DAC response for low to high input transition

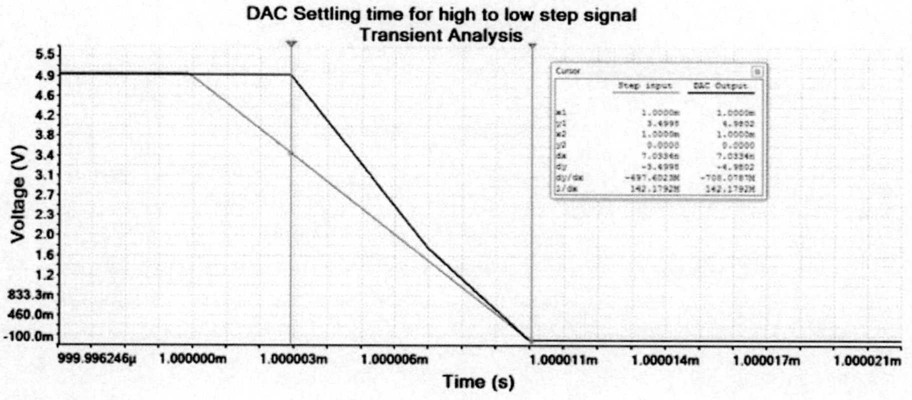

Fig. 10 DAC response for high to low input transition

done to identify its maximum response time to generate the output. The response was evaluated under two situations. In the first situation, comparator output changes from low to high while in the other situation, change in the output was reversed. Figure 11 shows the transient response of the comparator under both the conditions. Highest delay time response by the comparator was identified as 893.6 ns.

The delay time was also calculated by simulating the complete circuit. From the simulation result, it was found that the circuit produced the expected output for 60 kHz. Hardware resources utilized by single channel and multichannel ADC are shown in Table 1. It was found that logic utilization was less than 10% except for LUT-FF pair in the device XC6SLX100T-3FGG484. Since less than 10% of FPGA hardware is accommodated by the FPGA logic, remaining hardware shall be used to include other useful logics like controllers.

Delay time incurred by the FPGA to perform the conversion for eight channel ADCs were also analyzed from the synthesis result. It was found that the delay time by the corresponding FPGAs in Table 2 had proved that higher version FPGA has lesser delay time.

The analysis based on the Table 2 indicates that there is a difference of 1 ns in the delay time when comparing the delay time of single channel ADC with eight channel ADC. The basic intention of this design is to reduce conversion time which in turn helps to reduce PLC scan time. It means that the efficiency of PLC is increased. The proposed design has met the industrial requirements. Field transmitters for various physical parameters like pressure, flow, temperature, level etc., generate output as 4–20 ma signal.

Current signal is converted into voltage whose range is 0–5 v and applied to AIM. Input range in the proposed design has been also set as 0–5 v. Hence, the design is compared with the industrial AIM as in Table 3.

Conversion time of an 8-bit SAR type ADC is the sum of settling time of DAC, comparator and the delay time of FPGA. It was found that the eight ADC is able to produce the digital output at an operating frequency of 60 kHz. Therefore the

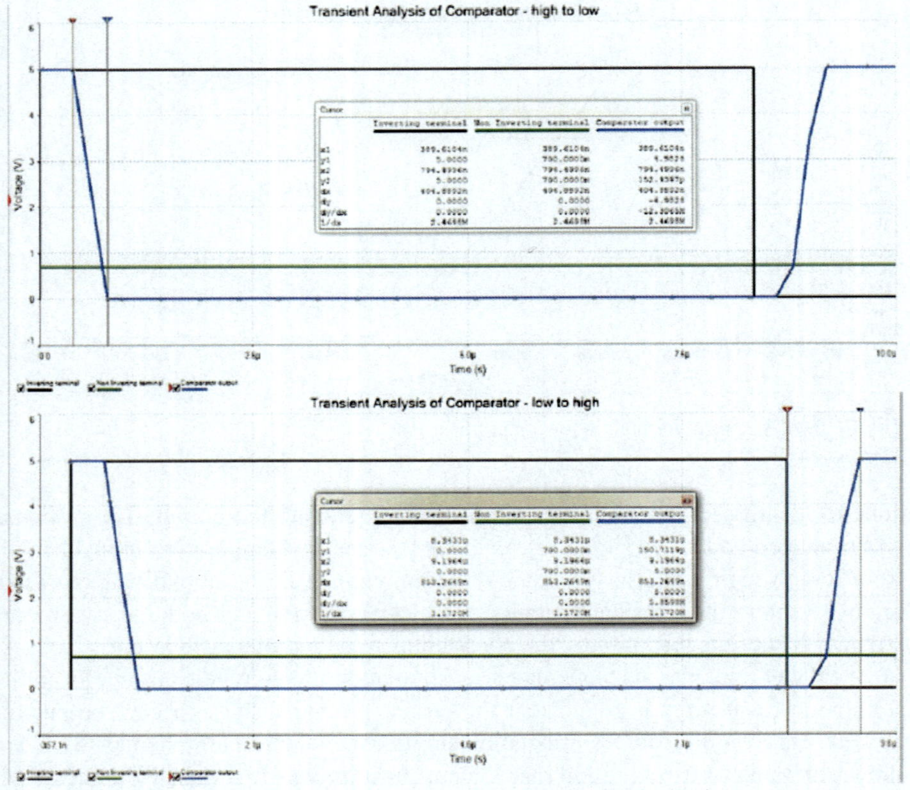

Fig. 11 Transient response of comparator

conversion of eight channel AIM is 0.13 ms and it remains same for 'n' channel ADC also. Conversion time of the proposed design was compared with the AIMs of GE Fanuc, Yokogawa and Allen Bradley AIMs [20–22]. From Table 3, it is clear that the proposed design has lesser conversion time than the other AIMs. In conventional AIMs, it depends upon the number of channels. For example, the conversion time for single channel in the model F3AD08-4R is 50 μs. Hence for 16 channels, it is 16 * 50 μs which is equivalent to 0.8 ms.

It is to be noted that this design has considered the DAC whose settling time is higher eventhough lesser settling time based DAC is available. The comparator was developed using operational amplifier instead of a dedicated comparator. Table 3 is the proof of better performance of the proposed design even under the lowest configurations of DAC and comparator.

Table 1 Logic utilization of FPGA for single and multichannel ADC

Target device	Logic utilization	Utilization	
		Single channel (%)	Multi channel (%)
XC3S500E-5CP132	Number of slices	0.6	6
	Number of slice flip flops	0.13	1.8
	Number of 4 input LUTs	0.42	4
	Number of bonded IOBs	12	79
	Number of GCLKs	4	4
XC6SLX 100 T-3FGG484	Number of slice registers	0	0
	Number of slice LUTs	0	0.4
	Number of fully used LUT-FF pairs	5	72
	Number of bonded IOBs	3.8	24.7
	Number of BUFG/BUFGCTRLs	6	6
XC5VLX 330 T-2FF1738	Number of slice Registers	0.01	0.08
	Number of slice LUTs	0.02	0.1
	Number of fully used LUT-FF pairs	0.9	22
	Number of bonded IOBs	1	7.6
	Number of BUFG/BUFGCTRLs	3	3

Table 2 Delay time by SPARTAN3E, SPARTAN6 FPGA single and Eight channel ADC

Target device	Delay time in ns	
	Single channel	Multichannel
XC3S500E-5CP132	4.326	5.152
XC6SLX100T-3FGG484	3.874	4.433
XC5VLX 330 T-2FF1738	2.872	3.537

7 Conclusions

Conversion time of SAR type ADC depends on the settling time of the DAC and the response time of the comparator. It can be reduced further if a dedicated comparator circuit is used instead of using operational amplifier as a comparator. This work has

Table 3 Comparison of proposed design with Industrial AIMs

Analog input module	Model	Number of channels	Conversion time in ms
GE Fanuc	35x/36x series	16	0.978
Yokogawa	F3AD08-4R	16	0.8
Allen Bradley	1794-IE8XOE4	08	8.0
Proposed FPGA analog input module	–	01	0.13
	–	08	0.13
	–	'n'	0.13

not focused on the establishment of compatibility between the PLC and the FPGA. Solution for FPGA networking has not been also discussed. However, these works will be done in the future.

References

1. Hennig C, Kneupner K, Kinna D (2012) Connecting programmable logic controllers (PLC) to control and data acquisition a comparison of the JET and Wendelstein 7-X approach. Fusion Eng Des 87:1972–1976
2. Milik A (2016) On hardware synthesis and implementation of PLC programs in FPGAs. Microprocess Microsyst 44:2–16
3. Lakshmi Sangeetha A, Naveenkumar B, Balaji Ganesh A, Bharathi N (2012) Experimental validation of PID based cascade control system through SCADA–PLC–OPC and internet architectures. Measurement 45:643–649
4. Bel Mokadem H, Beandrard B, Gourcuff V, De Smet O, Roussel J (2010) Verification of a timed multitask system with Uppaal. IEEE Trans Autom Sci Eng 7:921–932
5. Dhanabalan G, Tamil Selvi S (2015) Design of parallel conversion multichannel analog to digital converter for scan time reduction of programmable logic controller using FPGA. Comput Stand Interfaces 39:12–21
6. Benghanem M (2009) Measurement of meteorological data based on wireless data acquisition system monitoring. Appl Energy 86:2651–2660
7. Tan H, Hennig W, Walby MD, Breus D, Harris J (2013) Evaluation of multi-channel ADCs for Gamma-ray spectroscopy. IEEE Trans Nucl Sci 60(2):599–605
8. Abdallah M, Elkeelany O, Alouani A (2011) A low-cost stand-alone multichannel data acquisition, monitoring, and archival system with on-chip signal preprocessing. IEEE Trans Instrum Meas 60:2813–2827
9. Monmasson E, Idkhajine L, Cirstea M, Bahri I, Tisan A, Naouar M (2011) FPGAs in industrial control applications. IEEE Trans Ind Inf 7:224–243
10. Costas-Perez L, Lago D, Farina J, Rodriguez-Andina JJ (2008) Optimization of an industrial sensor and data acquisition laboratory through time sharing and remote access. IEEE Trans Ind Electron 55(6):2397–2404
11. Perelman Y, Ginosar R (2007) An integrated system for multichannel neuronal recording with Spike/LFP separation, integrated A/D conversion and threshold detection. IEEE Trans Biomed Eng 54(1):130–137
12. Dhanabalan G, Tamil Selvi S (2013) FPGA implementation of 8-bit multiplier with reduced delay time. Int J Comput Commun Eng 2(6):665–668

13. Tessier R, Betz V, Neto D, Egier A, Gopalsamy T (2007) Power-Efficient RAM mapping algorithms for FPGA embedded memory blocks. IEEE Trans Comput Aided Design Integr Circ Syst 26:278–289
14. Roth J, Darr M (2011) Data acquisition system for soil-tire interface stress measurement. Comput Electron Agric 78:162–166
15. Chang H, Huang H (2013) Adaptive successive approximation ADC for biomedical acquisition system. Microelectron J 44:729–735
16. Khedkar AA, Khade RH (2017) High speed FPGA-based data acquisition system. Microprocess Microsyst 49:87–94
17. Linear Technology (2005) LTC6240 / LTC6241 / LTC6242 CMOS Op Amps. Available https://www.linear.com
18. Zheng W, Liu R (2014) Design of FPGA based high-speed data acquisition and real-time data processing system on J-TEXT tokamak. Fusion Eng Des 89:698–701
19. An Y, Chung K, Na D, Hwang YS (2013) Control and data acquisition system for versatile experiment spherical torus at SNU. Fusion Eng Des 88:1204–1208
20. GE Fanuc automation (2002) Series 90TM–30/20/Micro PLC CPU instruction set reference manual. Available https://support.ge-ip.com
21. Yokogawa (2012) General specifications, FA-M3 analog input modules. Available https://www.yokogawa.com
22. Rockwell automation (2013) Flex I/O input, output and input/output analog modules. Available https://www.rockwellautomation.com

Need for Predictive Data Analytics in Cold Chain Management

Swati D. Kale and Shailaja C. Patil

Abstract Predictive analytics is an advanced technique that uses historical data to predict future outcomes with the help of machine learning algorithms. Cold chain logistics companies greatly depend on accuracy, timeliness, and efficiency to meet the requirements. A smart cold chain management system includes the proper monitoring and controlling of parameters such as temperature, vibrations, light, and humidity of the perishable food during the whole cold chain. In this work, predictive data analytics is proposed to make real-time predictions about time–temperature relationship considering various internal and external factors to avoid temperature abuse during transportation and thus predict the quality of food.

Keywords CCM · Predictive data analytics · Perishable food

1 Introduction

The cold chain industry is one of the fast-growing business sectors in India. The cold chain facilities will play an important role in India because of the issue of food shortage and food security. The cold chain is a logistical chain of activities involving packaging, storage, and distribution of perishable food products like fruits and vegetables, milk, meat and poultry, flowers, and vaccines from production to consumption, where the inventory is maintained in predetermined environmental parameters [1]. Refrigeration forms an important and significant part of the food and beverage retail market. It ensures the optimal preservation of perishable food. Domestic refrigeration and commercial refrigeration are important elements of the cold chain [2].

A smart cold chain management system monitors and controls temperature, vibration, light, and humidity of the perishable food during the cold chain. It also refers

---

S. D. Kale (✉) · S. C. Patil
Department of Electronics and Telecommunication, JSPM's Rajarshi Shahu College of Engineering, Tathawade, Pune, Maharashtra, India
e-mail: swadip.06@gmail.com

Savitribai Phule Pune University, Pune, Maharashtra, India

Z. Patel et al. (eds.), *Advances in VLSI and Embedded Systems*, Lecture Notes in Electrical Engineering 676, https://doi.org/10.1007/978-981-15-6229-7_9

to an information system to analyze the environmental data for reporting abnormal status to producers and consumers.

A cold chain is time-critical dynamic and complicated. It possesses various characteristics such as perishability of food, the shelf life for raw materials, long production throughput time, and need of refrigerated transportation. The safety of the food handled in the cold chain is sensitive to freshness and can easily affect human health and market prices. Therefore, it is a major challenge to maintain the quality of perishable food at all steps of the cold chain by maintaining the temperature in the desired range. Thus, to ensure the reliability of the cold chain for temperature-sensitive food required proper packaging, temperature protection, and continuous monitoring [11].

The most important aspect of building an IoT solution for cold chain applications is data analytics. The growing use of a Machine to Machine (M2M) communications allow sensors to send real-time data via mobile and wireless communications networks and will generate lots of data which might be continuous readings of temperature or other critical parameters such as carbon dioxide production, respiratory behavior, and Ethylene production that need to be monitored. One of the most challenging aspects of managing a cold chain is predicting the quality and traceability of food at every stage of the cold chain [19].

This paper is organized as follows: related work is discussed in Sect. 2, details of cold chain management are presented in Sects. 3 and 4 followed by need of data analytics, in Sect. 5 methodology of the proposed system is presented, results are discussed in Sect. 6, and conclusions are made in Sect. 7.

2 Related Work

NABARD report [1] presents the existing structure of Cold Chain Management (CCM) and future requirements. It explains various components of CCM, benefits of well connected CCM and challenges in reefer transportation for in-transit temperature monitoring. Nodali Ndrahaa et al. [2] have focused on the quality and safety of food in cold chain and also described temperature abuse in food cold chains and provided various cold chain solutions. The challenges and future research in food cold chains are also discussed. The temperature abuse is frequently occurred in the food cold chain because of poor practices of the food cold chain operators, weak design of refrigeration equipment, and the locations of food packages in the storage container. It has been observed that recent technology applied in monitoring and controlling of temperature provides a noteworthy contribution to food cold chain but appropriate data is not generated for predictive analytics. The authors Fotis Stergiou et al. [11] have presented a review on effective management of the cold chain through establishing control systems, such as intelligent packaging which includes Time Temperature Indicators (TTI). Samuel Mercier et al. [3] have applied time–temperature conditions at each stage of the cold chain. The major weaknesses in the modern cold chain have been discussed as precooling, various operations during transportation, storage at retail and in domestic refrigerators.

Riccardo et al. [4] have discussed goals and strategies for the design of an IoT architecture of the Food Supply Chain (FSC) operations and illustrated the potential benefits and opportunities for direct combination of physical food systems with virtual computer-aided control systems. Christian C. et al. [6] have developed predictive neural network models based on information that is collected using RFID temperature sensors. Spatial and temporal in-transit time–temperature monitoring is performed using neural networks. But other critical parameters are not considered for monitoring. Sjaak Wolfert et al. [7] have developed a conceptual framework that influenced the entire food supply chain where big data is used to provide predictive insights in farming operations, derived real-time operational decisions, and redesigned business processes. Abel Avitesh et al. [9] have developed a system consisting of Arduino wireless sensor networks and Xively sensor cloud which was used to monitor temperature and humidity in cold chain logistics. Khanuja et al. [12] have proposed a framework for cold chain management using IoT combined with Cloud Computing, Machine Learning, and Big Data Analytics to revolutionize the cold transport industry to monitor, visualize, track, and control various platform-dependent parameters with assured freshness and palpability (sensitivity). Halima et al. [13] have used machine learning algorithms such as neural network, SVM, Decision Tree, Linear Regression, Random Forest in the planning and transportation activities of the supply chain. The same concept can be implemented in cold chain management.

Hongmin Sun et al. [15] have designed the real-time monitoring system on the raw milk transport process based on the GPS, GPRS, and RFID technologies and the system fulfilled the whole-process monitoring on the raw milk transport and temperature from livestock farm to a processing plant, thus improving the quality safety of dairy products and the work efficiency of food administration and supervision department. Similarly, Ronak et al. [5] have addressed the problem of analyzing data collected by the dairy products to optimize the supply chain management and maximize profit in the manufacturing of milk and other dairy items. M. Subburaja et al. [16] have discussed various issues required in improving the operational efficiency of the dairy supply chain in TamilNadu, India. Elisabeth Ilie-Zudor et al. [17] have examined the challenges and potential of big data in heterogeneous business networks and relate those to an implemented logistics solution.

C. N. Verdouw et al. [18] have implemented a system based on the concept of virtual food supply chain on the IoT platform. On that basis, N.Indumathi et al. [19] have proposed an IoT based system to determine the quality of milk by measuring its pH value. The system has used the routing technique to govern the nearest milk booth. The system also traces various activities throughout the supply chain in food processing environments.

Ning Wang et al. [20] have presented an overview of the recent development of wireless sensor technologies in the agriculture and food industry. The paper also discussed the significance of wireless sensors to achieve market growth. Atefe Zakeri et al. [24] have developed a system to detect the level of microorganisms in milk. The system also gives the earlier determination of milking cycle events using a proactive approach with high accuracy. Sichao Lu et al. [25] have proposed a system based on artificial intelligence, cloud computing, and IoT to detect the temperature violation and route deviation accurately. Chaug-Ing Hsu et al. [26] have considered various features of perishable food for distribution and thus determined the vehicle count for delivering perishable food, loads and departure times of vehicles, and the shortest routing path. Maarten L et al. [27] have developed a real-time system considering shelf life inventories to monitor supply chain conditions at different nodes. Bastian Lange et al. [28] have presented challenges in existing structures of cold chains and suggest possible solutions to improve cold chain management.

From the study of previous work, it is found that the quality of food is affected mostly during transportation and many critical internal and external factors are responsible for this. Therefore, some mechanism is required to automate this process for predicting the status of food during transportation.

3 Cold Chain Management

Perishable food products such as vegetables, fruits, or generic pharmaceutical products often demand more than just refrigerated air on their long journey before they reach the end consumer. Many products need fixed levels of Temperature, Humidity, Oxygen, Carbon dioxide, Light, Gas content, and other factors to ensure that they arrive in a condition those consumers will still purchase and enjoy. There can be various causes of food losses. Since food loss reduction plays a significant role in sustainably improving food security, it is important to have economically suitable, ecologically and socially acceptable solutions. In such a situation, a well-integrated cold chain is a perfect fit to handle the problem. Based on type and perishability, the food is handled at different temperatures in the cold chain.

Based on storage temperature, the broad classification of the cold chain is provided in Table 1.

Typically, the cold chain starts from the farm level and ends at the consumer level [17]. Sorting, grading, and precooling of perishable food occurred at primary stage of the cold chain, then the food passes through various processing and storage phases. Processes involved in the integrated cold chain are

Table 1 Categories of the cold chain [1]

Category	Temperature	Examples
Frozen	<–18 °C	Frozen ingredients, processed vegetables & fruits, frozen meats, Ice cream, etc
Chilled	0 °C to 10 °C	Fresh meats, milk, butter, fresh fruits & vegetables, etc
Mild Chilled	10 °C to 20 °C	Chocolates & seeds, some milk products
Normal	>20 °C	Whole Onion, pickle, jams, oils & extracts

- Packing and cooling fresh food products (It is immediately after harvest or collection).
- Food processing (i.e., ripening, chilling, or freezing of foods).
- Refrigerated or cold transportation.
- Cold storage (short or long term warehousing of chilled foods).
- Retail markets and foodservice outlets.

The following parameters are considered for the design of an efficient and effective cold chain to provide the best conditions for inhibiting any undesirable changes.

- **Temperature**: Achieving and maintaining the perishable food at a specific lowest suitable temperature (lowest safe temperature). If safe temperature levels for food are not adhered to, this will result in chilling or freezing injury and increased food losses.
- **Humidity**: Providing relative humidity in the storage environment that prevents water loss along with a corresponding loss in weight and quality of food and also avoids excess humidity. Excess humidity can also result from temperature fluctuations, causing water to condensate on product surfaces (sweating). Free water on the product surface can lead to favorable growing conditions for bacteria and, therefore, promote deterioration. Humidity is primarily influenced by the cooling applied technology, temperatures inside the storage space, and airflows.
- **Atmospheric composition**: The composition of the atmosphere within a storage space influences the rate at which metabolic processes, such as the ripening of fruit progress. The metabolization depends on the oxygen. Therefore, a simple way to exert influence on atmospheric composition within storage is the control of airflow (i.e., limiting or increasing the influx of outside air). More advanced methods include modified atmosphere packaging [24].
- **Coefficient of Performance (CoP)**: It is a measure of the energy efficiency of a refrigerating system and is defined as the ratio between the refrigerating capacity and the power consumed by the system. It is mainly dependant on the working cycle, the evaporating/condensing temperature levels, and type of the refrigerant.
- **Depth of Discharge (DoD)**: A measure used to describe the degree to which a battery is discharged. DoD is defined as the percentage of battery capacity discharged expressed as a percentage of a battery's maximum capacity [27].
- **Global Warming Potential (GWP)**: An index comparing the climate impact of a greenhouse gas relative to emitting the same amount of carbon dioxide. The GWP of carbon dioxide is standardized to 1. GWP includes the radiative efficiency, i.e., infrared-absorbing ability of the gas as well as the rate at which it decays from the atmosphere.
- **Refrigerant**: It is a type of fluid used in a refrigerating system for heat transfer. Refrigerant absorbs the heat at a low temperature and pressure and rejects it at a higher temperature and pressure of the fluid. Usually it involves change of the phase of the fluid [28].

4 Need for Data Analytics

Analytics algorithms are required to ensure cold chain integrity for converting raw data into actionable recommendations and warnings that can improve cold storage processes, guide business decisions, and prevent cold chain failures before they occur. This includes descriptive and predictive analytics [16].

- **Descriptive**: Descriptive analytics derive properties of the cold storage system that is to be monitored. These properties are deciding the typical range of temperature of specific perishable food, determining the set point of the system's thermostat, the duty cycle of the system's compressor, etc. Because of this information the user came to know whether their storage unit is properly configured to store a particular food or not. In maximum consumer refrigerators, thermostats are available with only basic high and low settings that are not corresponding to a specific temperature. Due to this, it becomes very difficult to determine the setting temperature of a simple refrigerator's thermostat.
- **Predictive**: Predictive analytics is increasingly important to cold chain management which makes the process more accurate, reliable at a reduced cost. For keeping the cold chain safe, it is very important to do the predictions of the critical parameters in the cold chain such as temperature, humidity, coefficient of performance, i.e., energy consumption, depth of battery discharge and thus adjusting the temperature in transit or by diverting to the nearest cold storage, i.e., finding the shortest path of cold storage. Predictive analytics is used to detect problems in the cold chain before temperature abuse has occurred [29].

  Following problems generally occur

- Poor Equipment Configuration:

The predictive model will determine the basic properties of a cold storage system using information obtained from descriptive analytics. If the system's configuration is wrong then predictive analytics algorithms will indicate these configuration problems and alert the system's users about it.

- Human Error:

Simple human errors are common causes of many cold chain infractions. During transportation, vendors and drivers frequently forget to shut refrigerator or freezer doors or to ensure that a good seal is established. Therefore, it is necessary to determine when a cold storage system door will be left open alerting the system's users via a web interface, text message, or audible alert from temperature trend data before temperature bounds are violated.

- Equipment Failure:

Sometimes cold chain violations are occurred because of compressor breakdown and power failure. If a compressor failure occurs at night or on a holiday then it will be more destructive. In that case, predictive analytics can be used to detect power or compressor failure and thus alert the users through message or an automated phone call. The predictive system can also calculate the estimated time and the value of energy consumed before the cold storage system reaches on the verge of a risky temperature. Thus the system can prevent spoilage of perishable food.

Predictive analytics and machine learning are closely associated. Therefore, predictive analytics is effectively used to evaluate and control cold chain risks. Generally, two types of predictive models are used. First is the classification model that predicts class membership while second is the regression model that predicts a number. These models are used to perform data mining and statistical analysis. Built-in algorithms are available in predictive analytics software solutions [29].

Losses of perishable foods are most important in developing countries. These losses represent more than 400 million tons per year. The large post-harvest losses affect food quality. The typical cold chain involves three steps of harvesting at a warehouse, transportation, and distribution. As the change in temperature is occurred mostly during transportation because of some internal critical and environmental parameters. There is no any provision found in previous work to predict these losses in transit. Therefore, some mechanism is required to automate this process for predicting the status of food with time–temperature management. Predictive analytics can be used to avoid the wastage of perishable items.

The loss of fresh food storage in developing countries according to the study of Kitinoja [22] is shown in the following Table 2.

Table 2 Durability of perishable food

Type of food	At optimum cold temp. 0 °C	At optimum temp. + 10 °C	At optimum temp. + 20 °C	At optimum temp. + 30 °C
Fresh fish	10 days at 0 °C	4–5 days at 10 °C	1–2 days at 20 °C	A few hours at 30 °C
Milk	2 Weeks at 0 °C	7 days at 10 °C	2–3 days at 20 °C	A few hours at 30 °C
Vegetables	1 month 0 °C	2 weeks at 10 °C	1 week at 20 °C	Less than 2 days at 30 °C
Meat	2 days at 1 °C to 3 °C	1 day at 10 °C	12 h at 20 °C	A few hours at 30 °C
Mangoes	2–3 weeks at 13 °C	1 week at 23 °C	4 days at 33 °C	2 days at 43 °C
Apple	3–6 months at –1 °C	2 months at 10 °C	1 month at 20 °C	A few weeks at 30 °C

5 Methodology

The quality of food is affected during transportation due to the change in temperature of food, its pH value, and CO_2 emission.

There is no any provision found in previous work to predict these losses in transit. Therefore, some mechanism is required to automate this process for predicting the status of food with time–temperature management. Predictive analytics can be used to avoid the wastage of perishable items. The proposed system will give a prediction about the quality of food using predictive machine learning algorithms [30].

Following supervised M. L. algorithms are used for the prediction of food quality:

(i) Regression Analysis
(ii) Decision Tree
(iii) Support Vector Machines
(iv) Random Forest

Some critical factors shown in Fig. 1 are responsible for these changes.

The main aim of the proposed system is to maintain cold storage conditions during transport. Transport vehicles are nothing but cold stores on wheels, and it is

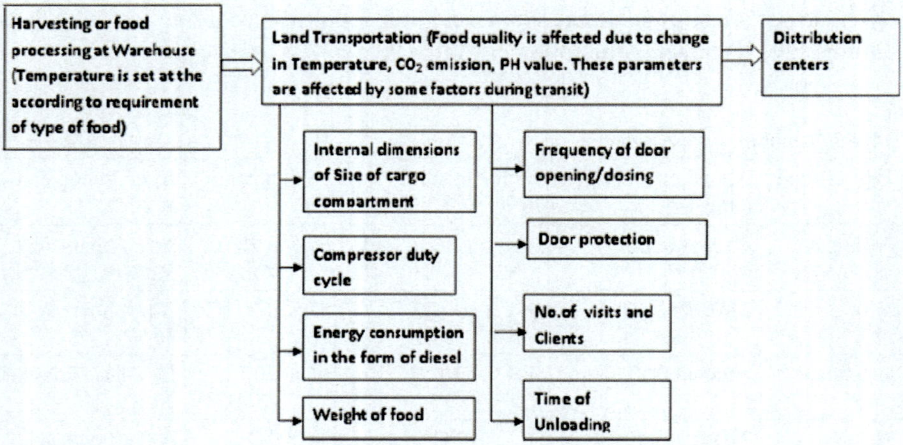

Fig. 1 Process in cold chain management

Fig. 2 Case 1

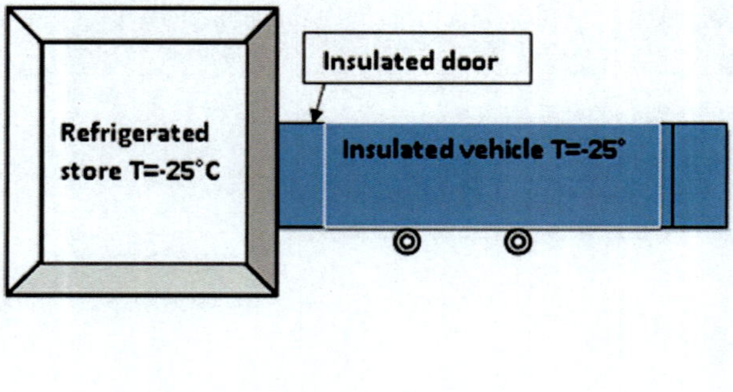

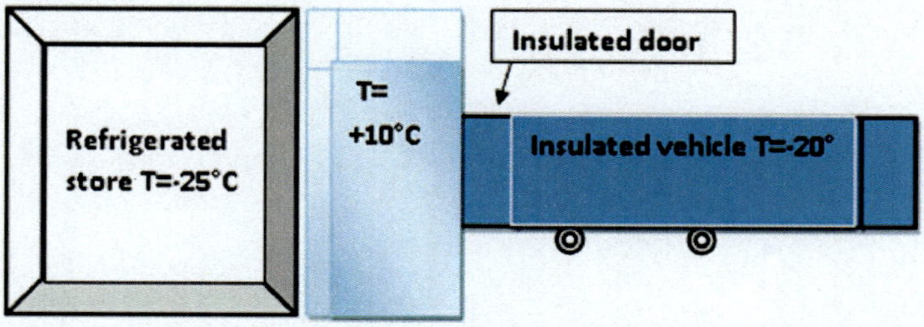

Fig. 3 Case 2

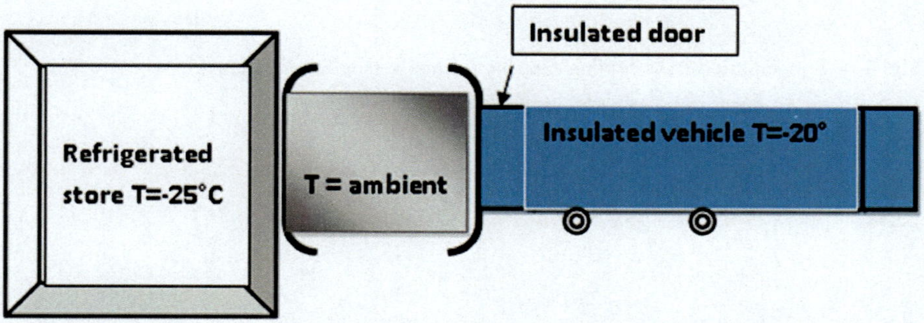

Fig. 4 Case 3

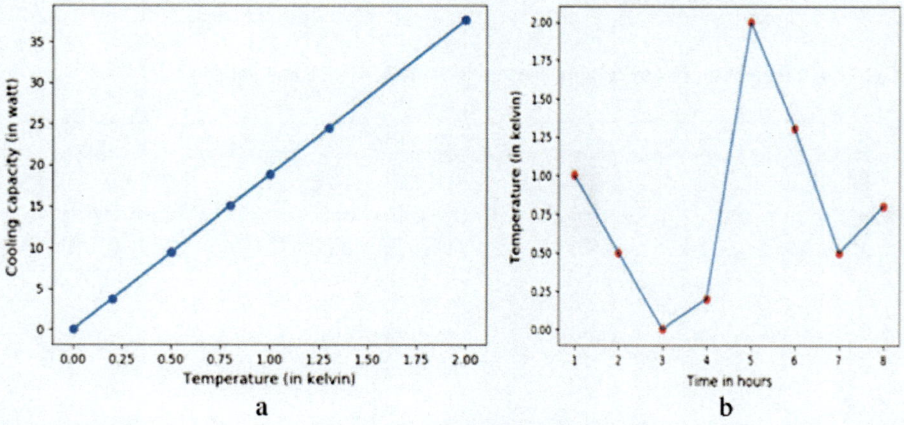

Fig. 5 **a** Temperature versus cooling capacity **b** Time versus Temperature

very difficult to maintain the correct storage temperature. Generally, these factors are responsible for this:

During loading and unloading of the vehicle supplementary heat is introduced due to the time required for the loading/unloading operation. This time must be the lowest possible. The level of protection of the cargo during these operations is equally

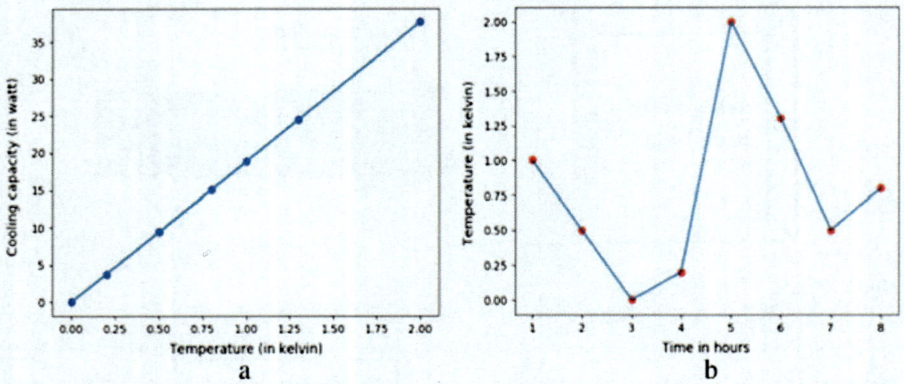

Fig. 6 **a** Temperature versus cooling capacity **b** Time versus Temperature

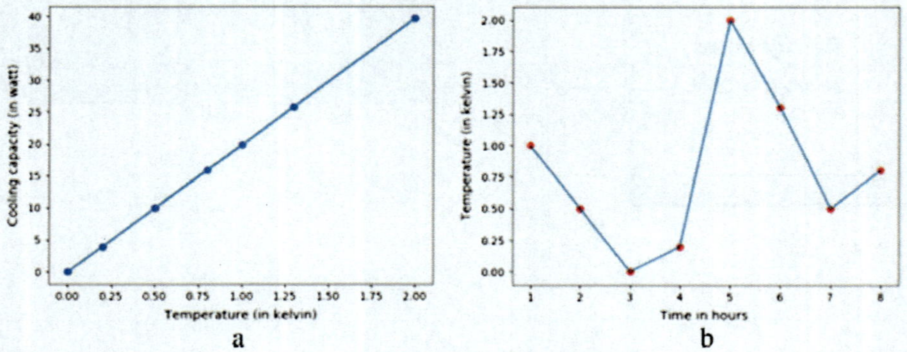

Fig. 7 **a** Temperature versus cooling capacity **b** Time versus Temperature

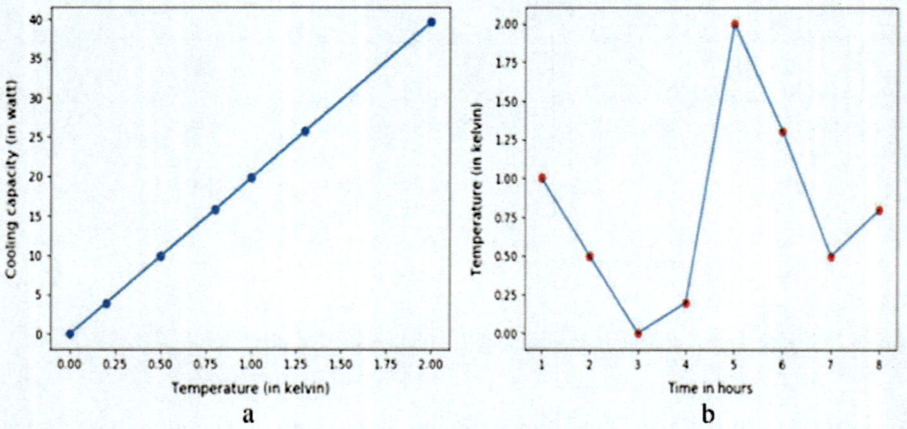

Fig. 8 **a** Temperature versus cooling capacity **b** Time Vs Temperature

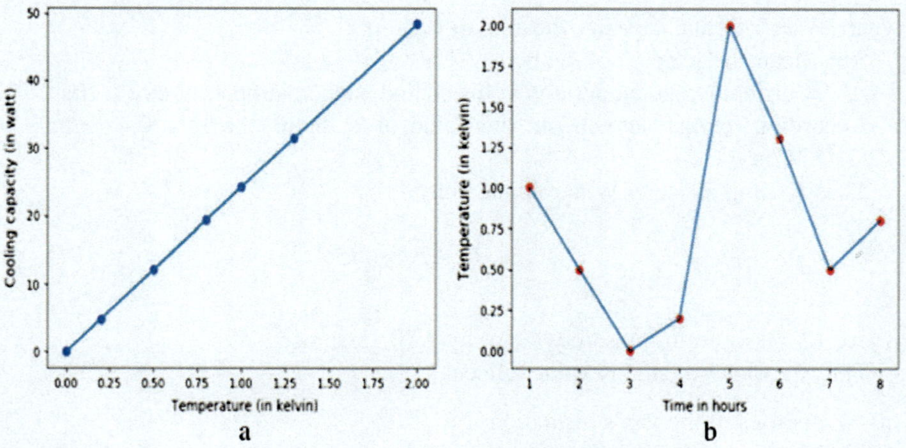

Fig. 9 **a** Temperature versus cooling capacity **b** Time versus Temperature

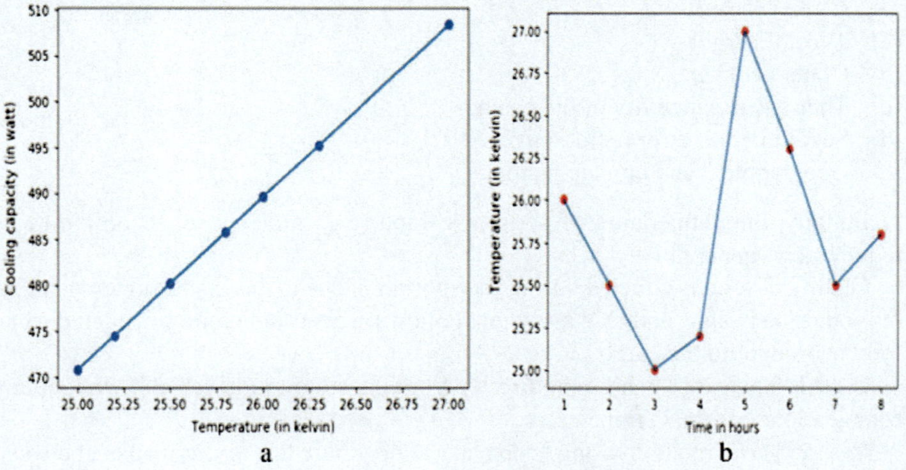

Fig. 10 **a** Temperature versus cooling capacity **b** Time versus Temperature

important. Closed contact between foods and the lateral walls and sun irradiation on the external surface causes an increase in temperature. Defrosting has a more severe effect on foods compared to refrigerated stores (more restricted space for the coils, more humid air inlet, i.e., internal dimensions of cargo) [38].

The essential component of any system to transport frozen food is an insulated container. The overall coefficient of heat transfer, which represents the insulating capacity of the equipment, is defined as

$$K = WS^{-1}\Delta T^{-1} \tag{1}$$

where W = Thermal capacity (heating or cooling).

S = Mean surface area of the body.

ΔT = difference in temperatures (inside and outside temperature of reefer).

According to ATP agreement, the value of K should be K < 0.4 W/m^{2}k or –272.75 W/m^2 °C.

Thus cooling capacity W is calculated by

$$W = K.S.\Delta T.1.75 \quad (2)$$

where 1.75 is a minimum safety factor by ATP agreement.

Thus the experimental K value reflects

(a) Properties of the van's insulation;
(b) Van's design; and
(c) Van's manufacturing process.

And cooling capacity depends on

(a) Type of cargo;
(b) Container size;
(c) Thermal conductivity of the panel;
(d) Set temperature for reefer unit; and
(e) Geographical and climate factors.

Initially, the temperature of perishable food (e.g., milk) is set according to its requirement, generally at 4 °C.

Quality of food is affected during transportation due to the change in temperature of food, its pH value, and CO_2 emission. Following case studies are considered when food is brought to the cargo [36].

Case 1: The cargo is brought directly from the store to the vehicle, without any change in the external temperature.

Case 2: The cargo is crossing a room at a temperature that is intermediate between those of the store and the external ambient.

Case 3: The cargo is brought through the ambient temperature.

Table 3 shows models of cold reefers applied for the above case studies.

Available database of milk cold chain is modified to calculate the cooling capacity of the vehicle. The perishable food is considered as milk, so its initial temperature

Table 3 Cold vehicle models

Van no	Model	Mean surface area (m^2)	K value (W/m^{2}k)
1	Mercedes sprinter 308CDI SWB	26.9	0.4 W/m^{2}k
2	Ford Transit Truck (mid-roof, long wheel-based)	28.3	0.4 W/m^{2}k
3	Ford Transit/Therma Truck (mid-roof, long wheel-based)	34.6	0.4 W/m^{2}k

is assumed as 4 °C when the local travel starts from the depot. The total time of traveling is considered 8 h with 20 times the open frequency and an open duration of 5 min.

6 Results and Discussion

Regression analysis and Random Forest algorithms are implemented.

Inputs for the algorithms:

1. Total time (8 h)
2. The Initial temperature of milk (4 °C)
3. Change in temperature after each visit
4. Difference in temperature
5. Heat transfer coefficient (K) is considered as 0.4 W/m^{2}k.
6. The Surface area of three different vehicles is considered.

The Output of algorithms:

1. Time–Temperature relationship
2. Temperature-Cooling capacity relationship

Algorithm: Regression Analysis: Case 1, Vehicle no 1.
Algorithm: Random Forest: Case 1, Vehicle no 1.
Algorithm: Regression Analysis: Case 2, Vehicle no 1.
Algorithm: Random Forest: Case 2, Vehicle no 1.
Algorithm: Regression Analysis: Case 3, Vehicle no 1.
Algorithm: Random Forest: Case 3, Vehicle no 1.

7 Conclusion

Existing literature is not sufficient to determine perishable food quality. The focus is more on predictive data analytics for evaluating the quality of food after transportation based on data mining techniques using machine learning algorithms. To this end, Random Forest and Regression Analysis algorithms have been implemented in python to relate Time—Temperature data along the chain. It is observed that the Random Forest algorithm is outperformed than Regression Analysis. Vehicle characteristics are equally responsible for temperature abuse in the cold chain.

References

1. NABARD (National Bank for Agriculture and Rural Development) report: Cold Chain Technologies, Transforming Food Supply Chain, May 2017
2. Ndrahaa N, Hsiaoa HI, Vlajicb J, Yang MF. Time-temperature abuse in the food cold chain: review of issues, challenges, and recommendations, Elsevier publication 05/02/2018. Review article
3. Mercier S, Villeneuve S, Mondor M, Uysal I (2017) Time-temperature management along the food cold Chain: a review of recent developments. Comprehensive Rev Food Sci Food Saf, 16
4. Accorsi R, Bortolini M, Baruffaldi G (2017) Internet-of-Things paradigm in food supply chain control and management, Elsevier publication International Conference on Flexible Automation and Intelligent Manufacturing, FAIM2017, pp 27–30 June 2017, Modena, Italy
5. Chudasama R, Dobariya S, Patel K, Lopes H (2017) DAPS: Dairy analysis and prediction system using technical indicators. IEEE publication, conference paper
6. Emenike CC, Van Eyk NP, Hoffman AJ (2016) Improving Cold Chain Logistics through RFID temperature sensing and Predictive Modelling. IEEE International Conference on Intelligent Transportation Systems, Nov
7. Wolfert S, Ge L, Verdouw C, Bogaardt MJ (2017) Big data in smart farming—a review. Elsevier J Agricul Sys 153:69–80
8. Jeble S, Dubey R, Childe SJ, Papadopoulos T, Roubaud D, Prakash A (2018) Impact of big data & predictive analytics capability on supply chain sustainability. Int J Log Manag, March
9. Chandra AA, Lee SR (2014) A method of WSN and sensor cloud system to monitor cold Chain logistics as part of the IoT technology. Int J Multimedia Ubiquit Eng 9(10):145–152
10. Pant RR, Prakash G, Farooqui JA. A framework for traceability and transparency in the dairy supply Chain networks. Elsevier Journal XVIII Annual International Conference of the Society of Operations Management (SOM-14)
11. Stergiou F (2018) Effective management and control of the cold chain by application of Time-Temperature Indicators (TTIs) in food packaging, Review article. J Food Clin Nutr 1(1) February
12. Khanuja, G.S., Sharath, D.H., Nandyala, S., and Palaniyandi, B.: Cold Chain Management Using Model-Based Design, Machine Learning Algorithms, and Data Analytics, SAE Technical Paper 2018–01–1201, 2018.
13. Kawtar H (2017) Machine learning applications in supply chains: an emphasis on neural network applications, IEEE
14. Brown DE, Abbasi A, Lau RYK. Predictive Analytics, 1541–1672/15/ © 2015 IEEE Intelligent Systems
15. Sun H, Jiang G, Kong Q, Chen Z, Li X (2016) Design of real-time monitoring system on raw milk transport process. Int J Multimedia Ubiquit Eng 11(4):335–342
16. Subburaja M, Babub TR, Subramanian RS (2015) A study on strengthening the operational efficiency of dairy supply Chain in Tamilnadu, India,XVIII Annual International Conference of the Society of Operations Management Elsevier publication Procedia—Social and Behavioral Sciences 189:285–29
17. Ilie-Zudor E, Kemeny Z, Buckingham C. Advanced predictive-analysis-based decision support for collaborative logistics networks, Supply Chain Management, an International Journal 20:369–388
18. Verdouw CN, Wolfert J, Beulens AJM, Rialland A (2016) Virtualization of food supply chains with the internet of things, Elsevier publication. J Food Eng 176:128–136
19. Indumathi N, Vijaykumar K (2018) Well-organized milk distribution monitoring system based on Internet of Things (IoT). Int Res J Eng Technol (IRJET) 05(07), July, e-ISSN: 2395–0056 p-ISSN: 2395–0072
20. Wang N, Zhang N, Wang M (2006) Wireless sensors in agriculture and food industry-recent development and future perspective. Computers and Electronics in Agriculture Elsevier Publication, 50

21. Novaes AGN, Lima Jr OF, de Carvalho CC, Bez ET (2015) Thermal performance of refrigerated vehicles in the distribution of perishable food. Pesquisa Operacional 35(2):251–284, ©2015 Brazilian Operations Research Society
22. Kitinoja L (2013) Use of cold chains for reducing food losses in developing countries, PEF White Paper No. 13–03, The Postharvest Education Foundation (PEF) December
23. Cold Chain Development Centre, National Horticulture Board. Technical Standards and Protocol for the Cold Chain in India, 2010
24. Zakeri A, Saberi M, Hussain OK, Chang E (2018) An Early Detection System for Proactive Management of Raw Milk Quality: An Australian Case Study 6:169–3536, 2018 IEEE
25. Lu S, Wang X (2016) Toward an intelligent solution for perishable food cold Chain management, IEEE
26. Hsu C-I, Hung S-F, Li H-C (2007) Vehicle routing problem with time-windows for perishable food delivery, Journal of Food Engineering, Elsevier
27. Maarten L. A. T. M. Hertog, Uysal I (2013) Shelf life modeling for first-expired-first-out warehouse management, Philosophical Transactions of the Royal Society
28. Lange B, Priesemann C, Geiss M, Lambrecht A (2016) Promoting food security and safety via cold Chains. International Zusammenarbeit (GIZ) GmbH, Eschborn, December
29. Shina SJ, Wooa J, Rachuria S (2014) Predictive analytics model for power consumption in manufacturing, selection and peer-review under responsibility of the International Scientific Committee of the 21st CIRP Conference on Life Cycle Engineering, Science Direct
30. Agrawal R (2018) Using Machine learning to transform supply chain management, White paper, Tata Consultancy Services (TCS)
31. Bhardwaj A, Mor RS, Singh S, Dev M (2016) An investigation into the dynamics of supply chain practices in dairy industry: a pilot study. Industrial Engineering and Operations Management Detroit, International Conference
32. Zage D, Glass K, Colbaugh R (2013) Improving supply chain security using big data, 978–1–4673–6213–9/13 ©2013 IEEE 254 ISI 2013
33. Mack M, Dittmer P, Veigt M, Kus M, Nehmiz U, Kreyenschmidt (2014) Quality tracing in meat supply chains. Philosophical Transactions of the Royal Society
34. Smola A, Vishwanathan SVN (2008) Introduction to machine learning, eBook, University Press
35. Shalev-Shwartz S, Ben-David S (2014) Understanding machine learning: from theory to algorithms, eBook, Cambridge University Press
36. Dasgupta A, Nath A (2016) Classification of machine learning algorithms. Int J Innov Res Adv Eng (IJIRAE) 3(03), March, ISSN: 2349–2763
37. Chakurkar P, Shikalgar S, Mukhopadhyay D (2017) An internet of things (IoT) based monitoring system for efficient milk distribution. 2017 International Conference on Advances in Computing, Communication, and Control (ICAC3), Mumbai, pp 1–5
38. Ajay K (2014) Generation of shelf life equations of cauliflower. Int J Agricul Food Sci Technol 5:15–26

FPGA-Based Implementation of Artifact Suppression and Feature Extraction

Shrikant Kumbhar, Anand D. Darji, and Harikrishna M. Singapuri

Abstract Accuracy of biomedical signal diagnostics highly depends on effective noise removal and feature extraction. In this paper, authors have discussed the methods for suppression of such noise/artifacts from biomedical signal. Various feature extraction algorithms for biomedical signals have been simulated and implemented using Field Programmable Gate Array (FPGA). Low-complexity and low-power hardware implementation is the prime constraints for bio-implantable Application-Specific Integrated Circuit (ASIC) development. Focus of this paper is to maintain the trade-off between adequate accuracy and low complexity. FPGA implementation demonstrates feature extraction using Haar wavelet transform gives better trade-off between accuracy and complexity of the hardware.

Keywords Artifact · Baseline wandering · FPGA · Haar wavelet transform · Powerline noise · QRS complex

1 Introduction

Artifact can be defined as any wrong interpretation in the representation and perception and consideration of signal because of involved measuring equipments and device equipments and device (equipment) and methods. Artifacts are any artificial product; a structure or appearance that is not natural, but is because of manipulation. Work is inclined for achieving the low-complexity implementation of biomedical signal denoising and its feature extraction module. Initial or primary diagnosis can be done with lesser accuracy with mobile instrumentation, equipments. Biomedical

S. Kumbhar · A. D. Darji · H. M. Singapuri (✉)
Sardar Vallabhbhai National Institute of Technology, Surat, Gujarat, India
e-mail: hari4singa@gmail.com

S. Kumbhar
e-mail: thewallshrikantkumbhar@gmail.com

A. D. Darji
e-mail: add@eced.svnit.ac.in

Z. Patel et al. (eds.), *Advances in VLSI and Embedded Systems*, Lecture Notes in Electrical Engineering 676, https://doi.org/10.1007/978-981-15-6229-7_10

signal such as Electrocardiogram (ECG) has the morphological features, which can be correctly predicted with some acceptable approximations and inaccuracies.

1.1 Noise Classification

Nature and properties of noise seen in biomedical signals such as ECG, Electroencephalogram (EEG), Electromyography (EMG) signals are having variety across these signals. Noise can be classified based on its occurrence consistency and its contributor.

- Physiologic noise: Noise caused because of human body involvement is called as physiological noise (artifacts). Physiological noise (artifacts) changes in the interested signal because of other physiological processes happening in the surrounding of the body. Most common noises under physiological noise category are eye activity, heart pumping and myo-activity-related noise, muscle tension signals measured using EMG, and heart-related signals.
- Non-physiological noise: Noises such as powerline noise, noise due to electric equipment in vicinity, mechanically produced effects due to ventilators, circulatory pumps, etc come under non-physiological noise. Removing of instrumentation noise is very difficult after certain extent, but it can be reduced through careful circuit design and higher quality measuring equipments.

Powerline interference noise, baseline wander, electromyography noise are persistent in nature. Variations in electrode-skin impedance and activities like patient's movements and breathe cause baseline wandering. Noises such as electrode popup or contact noise, electrode motion artifact, and instrumentation noise occur in a burst. Burst noise is typically classified as a White Gaussian Noise (WGN) which is present on subset of leads for a small time duration. Low-frequency noise such as baseline wandering and other fixed frequency noise such as powerline noise in biomedical signals are addressed in different techniques during filtering them out.

1.2 Need of Feature Extraction

Feature extraction transforms raw signals into more informative signatures or fingerprints of a system. Feature of signal is meaningful content of signal in the context of its application. Large number of data or information is converted to simplified format by maintaining sufficient accuracy. This will suppress the efforts of computation time, task of noise removal, and power efficiency improvement. However, choice of feature extraction and noise removal algorithm has to be made based on type of biomedical signal and has different techniques for each of them.

Artifacts and noises present in biomedical signals degrade performance of systems and hence need to be removed before performing feature extraction. Researchers have proposed many methods to perform preprocessing. But these algorithms are highly complex and some of them do not give reasonable performance. Hence, primary objective is to develop a method which is simple to implement and gives reasonable performance and medical expert can draw correct primary diagnosis.

In Sect. 2, proposed methodology is explained. Section 3 comprised of simulation results and implementation details. Results and comparison gives the comparison between the work presented in paper and previously published efforts based on improvement in Signal-to-Noise Ratio (SNR) and hardware savings.

2 Proposed Methodology

Biomedical signal undergoes denoising step before moving toward feature extraction of the same. Desired level accuracy of biomedical signal is governed by the application or purpose for which that biomedical signal is recorded. Method proposed here is divided into stages largely differentiated into two parts which are "Denoising of biomedical signal" and "Feature extraction of biomedical signal". Each type of biomedical signal has different sets of filters and feature extraction blocks as mentioned in Fig. 1.

2.1 Preprocessing Block for Signal Denoising

Preprocessing block includes the removal of common noise appearing in all types of biomedical signal such as baseline wandering, powerline noise interference. Baseline wandering is the low-frequency noise which can be removed by high-pass filter. Some researchers have used Hamming window to remove noise, we can prefer windowing method for simplicity. Moving average filtering also used to smoothen the signal like ECG since ECG morphology holds the key in determining ECG feature. Finite Impulse Response (FIR) filters can be realized fully parallel or partly serial based on timing goals and hardware resources available.

2.1.1 FIR Filtering

FIR filters are easy to implement on hardware provided that filter coefficients are readily available for computation. FIR filters realized here are with partly serial topologies. The limited resources are available. Fully parallel topology consumes equal number of multipliers as that of order of filter as shown in Fig. 2a. For every tap in FIR filter, a dedicated multiplier is available. Implementation of FIR filter can be done in a variety of serial architectures to obtain the desired speed/area trade-off. Two extreme filter implementation styles are fully parallel and fully serial which is represented in paper [1]. Although fully serial occupies very low area, it requires a

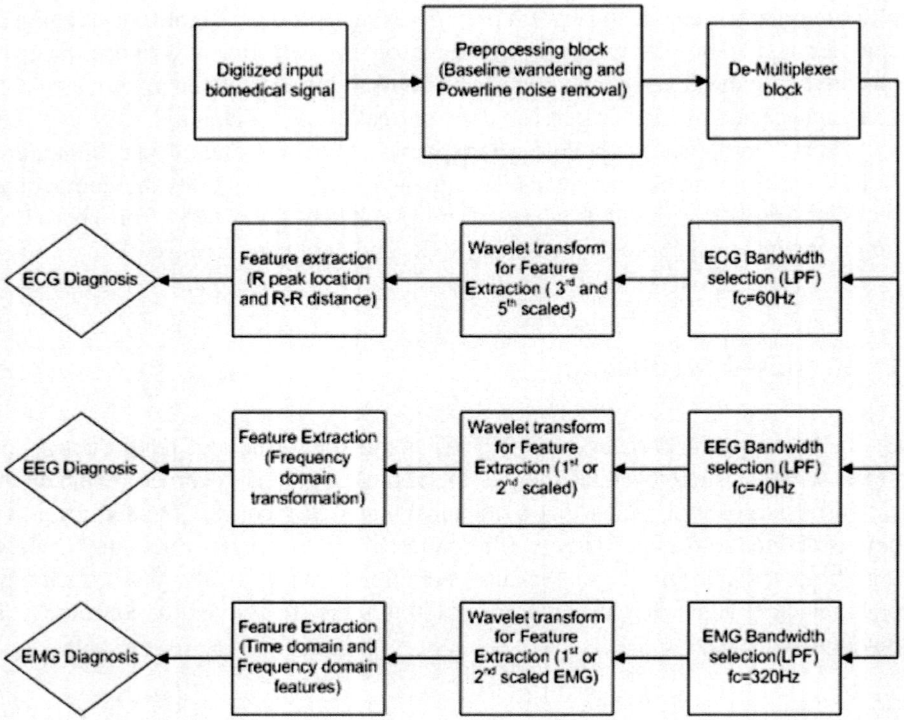

Fig. 1 Block diagram for proposed method

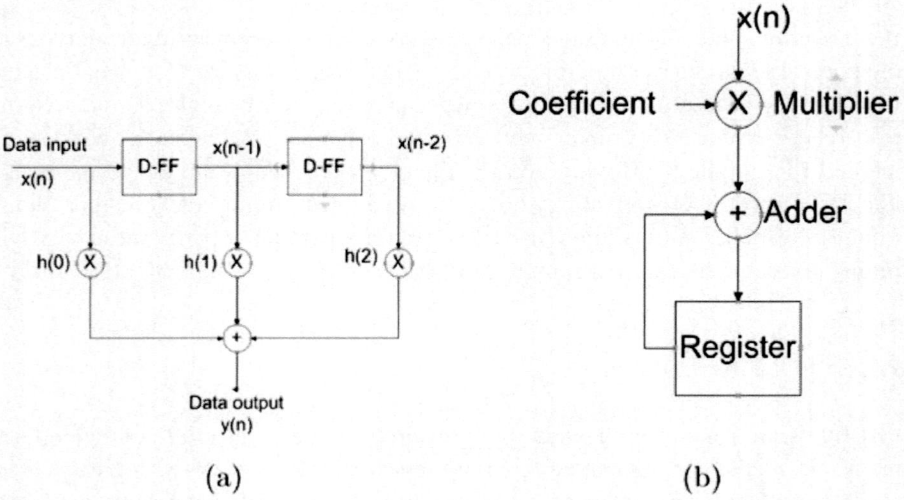

Fig. 2 Structures for FIR filter. **a** Fully parallel. **b** Fully serial

higher clock rate to function. On the other hand, fully parallel structure consists most of chip area with better performance. Comparison between fully parallel and fully serial FIR filter has been performed in the implementation section of this paper.

2.1.2 Moving Average Filtering

Moving average filters are nothing but low-pass filters which are used here in this methodology for smoothening the ECG data. Smoothening of data is directly proportional to window length of moving average filter [2, 3]. For digital filter in DSP, the moving average is generally used mainly due to easiness of structure in terms of understanding and usage. Despite its simplicity, the moving average filter is most selected for reduction of random noise while maintaining a sharp step response.

2.2 Feature Extraction Block

2.2.1 Feature Extraction of ECG Signal

Biomedical signal such as ECG has the QRS complex as a major feature. QRS complex is high-frequency component. P and T waves have low-frequency range. Discrete wavelet transform is used to decompose the signal into low-frequency and high-frequency components. Detail coefficient and approximate coefficients are formed after decomposing the signal through mother wavelet [4]. Haar wavelet has been chosen as it has minimum complexity/hardware requirements along with the fact that the accuracy and reliability of the features extracted from Haar wavelet transform are satisfactorily and almost equivalent to other complex mother wavelets.

2.2.2 Feature Extraction of EMG Signal

EMG signal has time-domain and frequency-domain features. EMG signal is decomposed into first scale to have optimal approximation. MATLAB signal processing toolbox is used to calculate some of the features like total power, peak power frequency, variance of signal, mean of signal, peak amplitude of signal, total power spectral density spread, etc. (Fig. 3).

2.2.3 Feature extraction of EEG signal

Haar wavelet decomposition of EEG signal helps to analyze the EEG signal in frequency domain. Wavelet decomposition has the inherent property of low-pass filtering [5, 6].

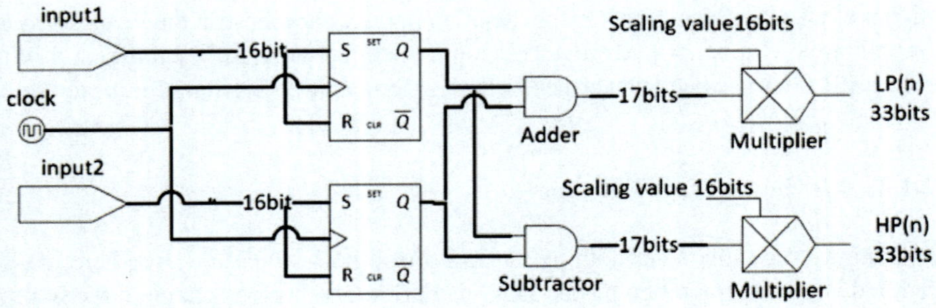

Fig. 3 Haar wavelet transform implementation

3 Simulations and Hardware Implementation Results

For performance evaluation, MATLAB simulations are carried out. Based on MATLAB simulation appropriate wavelet transform is chosen for RTL design. RTL code is written for "Spartan 3E starter board," which comes under Spartan 3E family. Targeted device is "xc3s500e-4-fg320." It has 50 MHz crystal oscillator on board which ultimately provides system clock of 50 MHz. RTL and MATLAB simulation results are mentioned in this section. For simulation and analysis purpose, one and two cycles of ECG signal are used from the database available at "www.physionet.org." Author in [2] has implemented filter algorithm with length of 8 and [3] uses 3, 4, 5 points for averaging filter. Algorithms proposed by [2, 3] have been implemented and compared. In this paper, filter has been implemented using length of 10 which gives improved performance. The order of FIR filter has been decided by MATLAB-based simulation results. Author has simulated and analyzed the results for different order lengths. For optimum length implementation, the order of the filter turns out to be 10 which gives promising results. The values of the FIR coefficients for FPGA implementation have been derived from FDATool by providing filter constraints (Figs. 4, 5, 6, 7, 8, 9, 10, and 11).

3.1 Comparison Between Moving Average Filter and FIR Filter for Biomedical Signals

Denoising of ECG is performed using moving average filter and/or FIR filter. Both the filters are compared based on following terms. According to [7], FIR filter is suggested for ECG noise suppression. Moving average filter is useful for signal with less variations whereas FIR filter is mostly used for biomedical applications for noise suppression or noise cancellation.

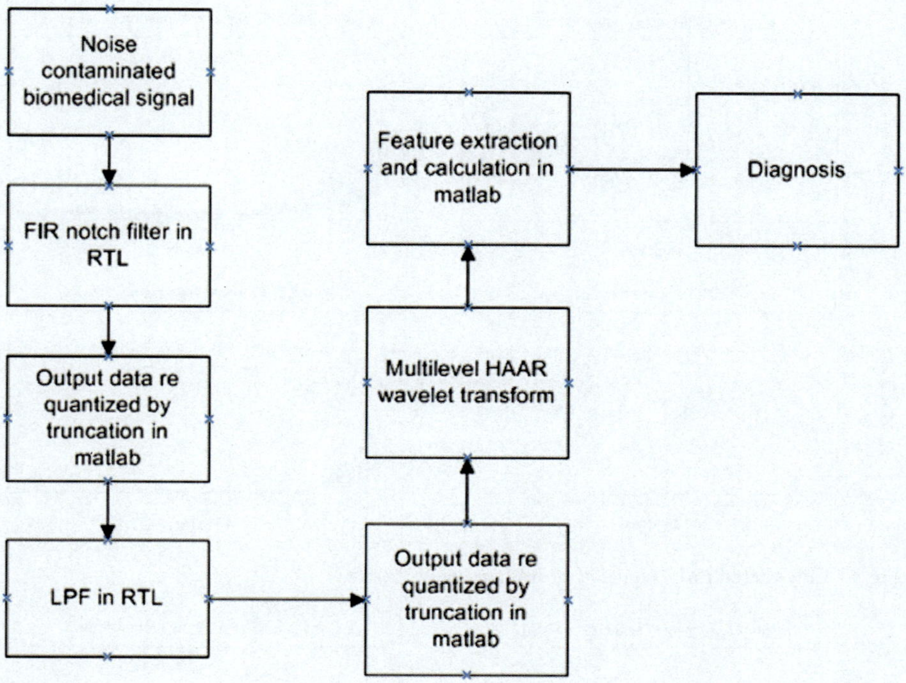

Fig. 4 Proposed system in RTL blocks

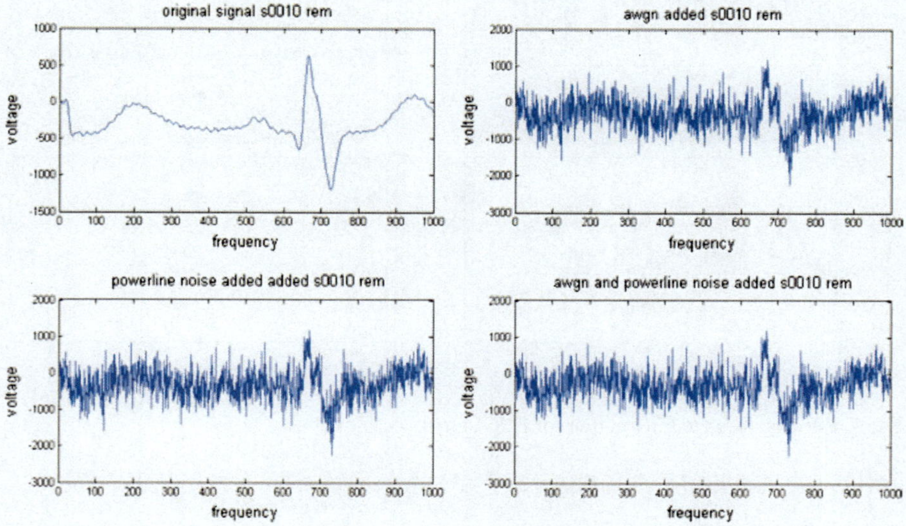

Fig. 5 Signal preparation

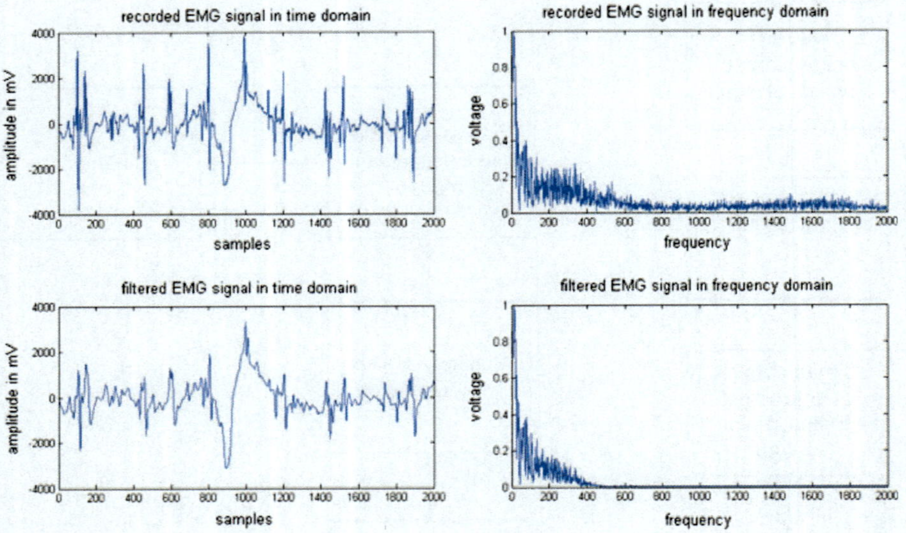

Fig. 6 LPF filtered EMG signal at cutoff frequency 320 Hz

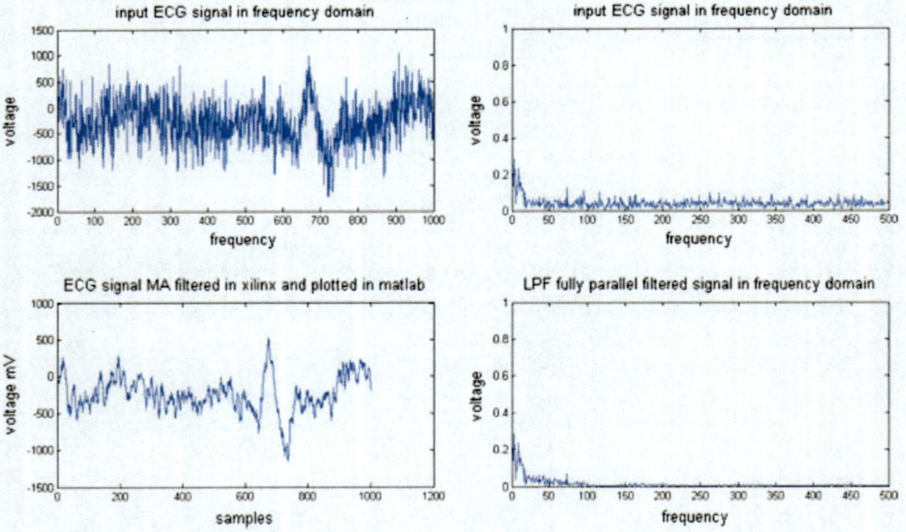

Fig. 7 Moving average filter output for ECG signal

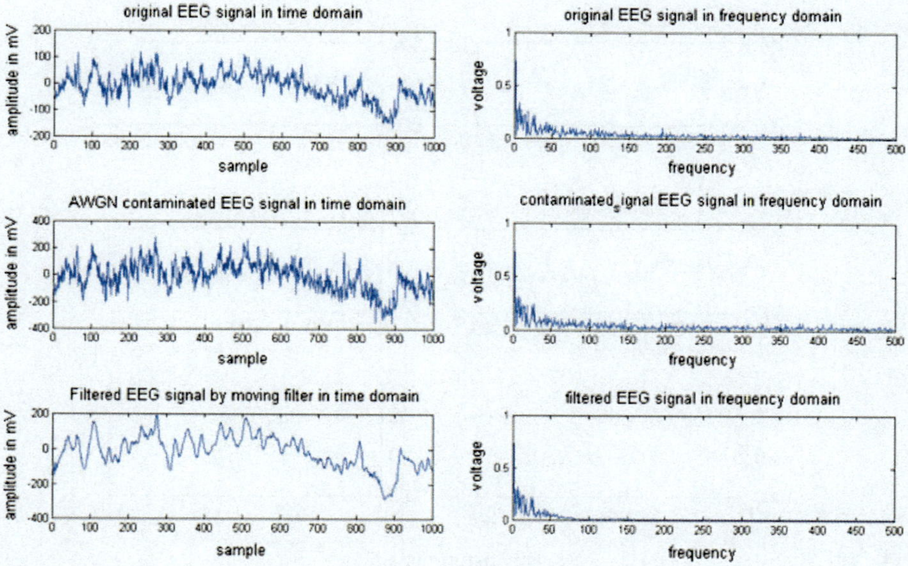

Fig. 8 Moving average filter output for EEG signal

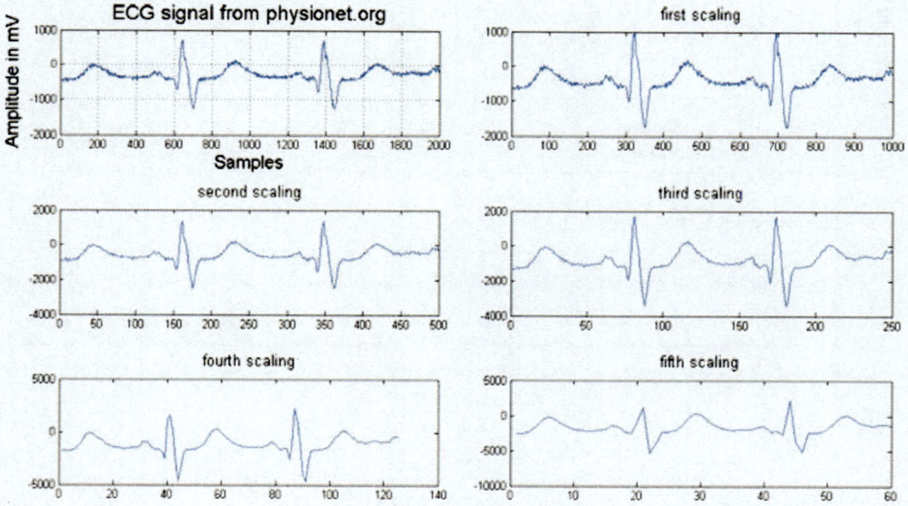

Fig. 9 Multi-level scaled Haar wavelet transform of ECG

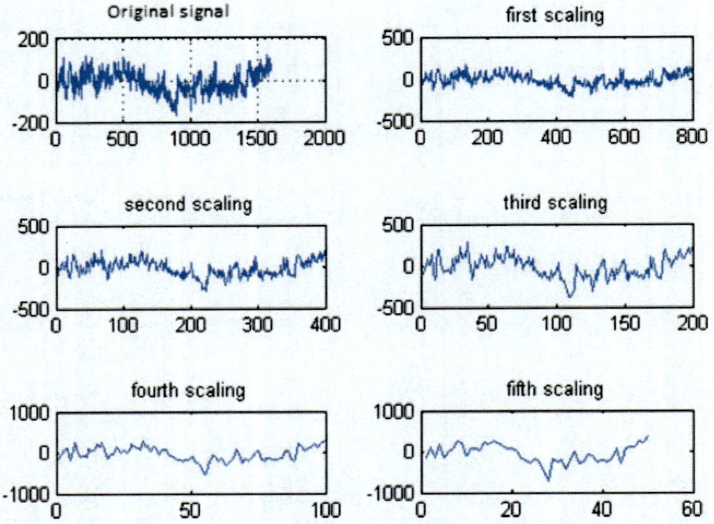

Fig. 10 Multi-level scaled Haar wavelet transform of EEG

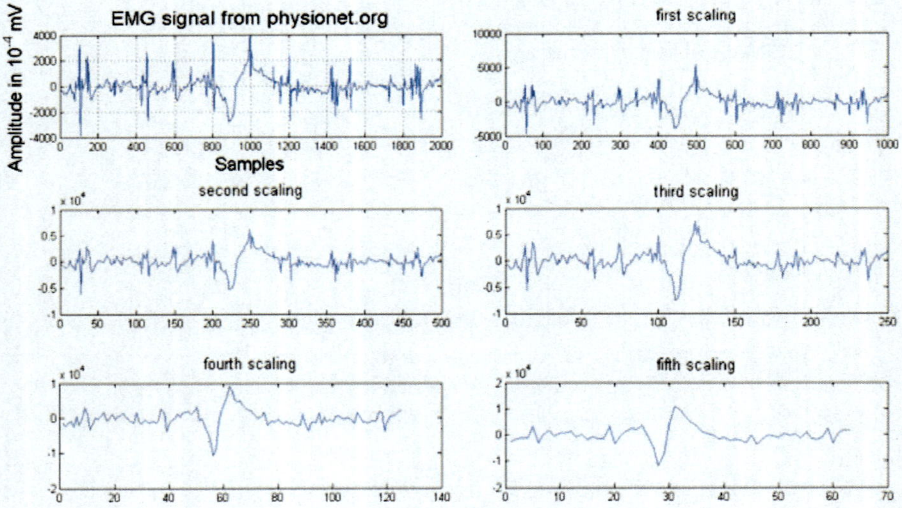

Fig. 11 Multi-level scaled Haar wavelet transform of EMG

Table 1 gives the comparison between FIR filter and moving average filter based on SNR. Table 2 gives comparison for hardware utilization. From Tables 1 and 2, it can be observed that FIR filter provides better performance in terms of SNR and power dissipation but FIR filter is having higher resource utilization as compared to moving average filter.

Table 1 Comparison between moving average filter and FIR filter based on signal-to-noise ratio (SNR)

Filter behavior: LPF	FIR filter (fully parallel) (10 coeff.)	Moving average filter (window size 10)
SNR initial (db)	0.6577	0.6577
SNR final (db)	19.4579	7.3891
Total stop band power before filtering	1 X	1 X
Total stop band power after filtering	0.01822 X	0.02962 X

Table 2 Comparison between moving average filter and FIR filter based on resource utilization

	FIR filter (fully parallel) (10 coeff)	Moving average filter (window size 10)
No. of slices	259	59
Slice flip flops	193	73
4-ip LUT's	454	86
Bonded IOB	52	52
18X multipliers	10	1

Table 3 Comparison between different mother wavelets based on R-peak location accuracy

R-peak locations in	sym4 (sample index)	Haar (after cascade filtering)	db1 (sample index)	db2 (sample index)	db4 (sample index)
Contaminated signal	635,1379	635,1379	635,1379	635,1379	635,1379
Wavelet transformed signal	640,1384	636,1380	636,1383	638,1382	642,1386

3.2 *Comparison Between Different Mother Wavelets Based on Location of R-Peak Accuracy*

Different mother wavelets give different positions of R-peak in contaminated signal as well as in transformed signal which is compared in Table 3. Based on level of accuracy and degree of complexity, the specific wavelet is chosen for specific application. Table 4 shows the data of R-peak locations in signal filtered by moving average and Haar wavelet. It is clearly visible that Haar wavelet transform detects the R-peak in signal with same accuracy compared to other principle wavelet transforms with lesser hardware utilization due to its simplicity in architecture. Table 5 gives the hardware utilization for Haar wavelet transform implemented on FPGA.

Table 4 Comparison between moving average filtered signal and Haar wavelet transformed signal based on R-peak location accuracy

R-peak locations in	MA (sample index)	Haar (sample index)
Contaminated signal	635,1379	635,1379
Wavelet transformed signal	640,1385	636,1380

Table 5 Hardware utilization of Haar wavelet transform

Logic utilization	Haar wavelet transform
No. of slices	36
No. of four input LUTs	34
No. of bonded IOBs	100
No. of multiplier 18 × 18	2

4 Conclusion

Proposed algorithm is primarily focused to process the biomedical signal for suppression of noise and for the feature extraction so that initial diagnosis can be done. Selection of Haar wavelet transform over other complex mother wavelets on the cost of accuracy is to minimize the computational overhead. This can be targeted to real-time first level diagnosis. Moving average filter smoothens the biomedical signal such as ECG. This will effectively help to achieve better accuracy with simple mother wavelet such as Haar wavelet. Cascading of LPF, notch filter, and HPF is used in order to effectively denoise the biomedical signals such as EEG, ECG, and EMG. With similar kind of noises present in almost all biomedical signals, the same circuitry can be used for removing such noises. Significant percent of slices are saved in moving average filter implementation as compared to FIR filter comprising of 10 coefficients for the noise suppression.

References

1. Sundaram K, Marichamy, Pradeepa (2016) FPGA based filters for EEG preprocessing. In: Second international conference on science technology engineering and management (ICONSTEM), pp 572–576
2. Ueno J, Chiu HK, Lin CH, Lin YC, Huang LR, Gong CSA (2015) ECG noise thresholding based on moving average. In: IEEE international conference on consumer electronics, Taiwan, pp 188–189
3. Pandey V, Giri VK (2016) High frequency noise removal from ECG using moving average filters. In: International conference on emerging trends in electrical electronics sustainable energy systems (ICETEESES), pp 191–195

4. Singh Chaudhary M, Kapoor RK, Sharma AK (2014) Comparison between different wavelet transforms and thresholding techniques for ECG denoising. In: International conference on advances in engineering technology research (ICAETR-2014), pp 1–6
5. Vaneghi FM, Oladazimi M, Shiman F, Kordi A, Safari MJ, Ibrahim F (2012) A comparative approach to ECG feature extraction methods. In: Third international conference on intelligent systems modelling and simulation, pp 252–256
6. Saraswat S, Srivastava G, Shukla S (2016) Decomposition of ECG signals using discrete wavelet transform for wolff Parkinson white syndrome patients. In: International conference on micro-electronics and telecommunication engineering (ICMETE), pp 361–365
7. Gonzalez-Fernandez R, Mulet-Cartaya M, Lopez-Cardona JD, Lopez-Rodriguez R (2015) A mobile application for cardiac rhythm study. In: Computing in cardiology conference (CinC), pp 393–396
8. Mironovova M, Bla J (2015) Fast Fourier transform for feature extraction and neural network for classification of electrocardiogram signals. In: Fourth international conference on future generation communication technology (FGCT), pp 1–6
9. Zou Y, Han J, Xuan S, Huang S, Weng X, Fang D, Zeng X (2015) An energy efficient design for ECG recording and r-peak detection based on wavelet transform. IEEE Trans Circuits Syst II Express Briefs 62(2):119–123

Test Time Reduction Using Power-Aware Dynamic Clock Allocation to Scan Vectors

Harikrishna Parmar and Usha Mehta

Abstract As circuit size increases with scale down in technology, the time required to test the circuits also increases. Scheduling of the cores is a very effective technique to reduce the test time of a system-on-chip (SoC) in the given power budget. As the frequency is relative to the power and the test time, by controlling the test clock frequency, the power consumption and the test time per core can be adjusted to yield an optimal solution to the test scheduling problem. In traditional methods, the fixed test clock frequency is applied to all the test vectors of a given core in case of power-aware test scheduling. Whereas in the proposed plan, a power-aware dynamic frequency allocation is done to individual scan vector of the core. Session-based test scheduling scheme is implemented to reduce test time. Results show the improvement up to 58% with session-based test scheduling is achieved over the existing solution for the benchmark SoCs with minor bit and area overhead.

Keywords Dynamic clock allocation · Power-aware testing · SoC testing · Test scheduling

1 Introduction

Deep submicron technology is quickly dominating to extremely complex, million transistor chips. By this technology advancement, a system is consolidated into a chip so-called system on chip (SoC). SoCs are implemented by embedding complex circuits such as DRAMs, ASICs, digital logic and analog and mixed-signal designs. Because of technology scaling, accommodation of the significant number of cores is now feasible on a chip, but that increases SoC complexity, and hence it rises test data volume and test time [1]. Therefore reducing test time is the prime motive in

H. Parmar (✉) · U. Mehta
EC Department, Nirma University, Ahmedabad, India
e-mail: hk_parmar@yahoo.co.in

U. Mehta
e-mail: usha.mehta@nirmauni.ac.in

Z. Patel et al. (eds.), *Advances in VLSI and Embedded Systems*, Lecture Notes in Electrical Engineering 676, https://doi.org/10.1007/978-981-15-6229-7_11

SoC testing research. Many techniques have been developed like test scheduling optimization, test architecture design, and optimization, etc.

SoC testing can be done in three ways: 1. Test Access Mechanism (TAM) design and optimization, 2. Wrapper optimization, and 3. Test scheduling.

TAM carries test patterns from the test pattern source to the device under test and takes a test response from the device under test to the test pattern sink [2]. Core test wrapper makes the interface between design under test (core) and its environment. It interfaces the input–output of a core to the rest of the IC and the test access mechanism [2, 3].

Once an appropriate test transportation system and test translation system have been established, the next big challenge for SoC is the test schedule. It gives the detail about the order in which various cores are tested. It also ensures that there is no resource conflict among cores [3].

The power consumption of the core during the test is usually higher than the functional mode of operation [4]. Parallel testing of a core will reduce test time, but at the same time, it will increase switching activities which lead to higher power consumption. So the power-aware test scheduling technique is required for efficient power management.

Recently, test time and test power are formulated as a function of the test clock frequency and achieve test time minimization for a predefined power budget, where test clock frequency was customized for each test sessions [5]. In this paper, a new idea is presented to customize the test clock frequency of each test pattern in a core in such a way that power dissipation does not go beyond the predefined power budget.

The whole paper is organized as follows. Section 2 discusses the prior work on SoC testing. Section 3 describes the problem statement. Dynamic clock allocation technique and power-aware test scheduling technique is shown in Sects. 4 and 5. Section 6 describes the hardware realization of the proposed approach. Result discussion is done in Sect. 7, and finally, Sect. 8 concludes.

2 Prior Work

Power dissipation during the test is always higher than the normal mode of the operation. So test scheduling is done in such a way that it does not cross the limit of the given power budget. Thus, different power-aware test scheduling techniques were developed to test the SoCs.

One such procedure was proposed in [6], which is based on a polynomial-time algorithm to reduce SoC test time. This method describes minimizing test area overhead by merging the test sessions of two different cores. Scheduling of the core is done with desirability matrix which is calculated from the area penalties. With a similar approach, a new technique is proposed in [7] which states a Polynomial-time algorithm based on precedence constraint. Precedence constraints force a partial order among the cores in SoC. In 2003, a technique proposed in [8] handles precedence constraints along with test conflicts which occurs because of unit testing with multiple test set, cross core testing, sharing of TAMs, and hierarchical SoCs. Handling of conflicts and precedence constraints are based on the order in which the test has to be applied. Another approach to handle this structural conflict is proposed in [9]. The genetic algorithm is used here for test application conflict graph to minimize SoC test time.

Test scheduling technique for minimal energy consumption under power constraints is proposed in [9]. In this approach, switching activities occurring in the overlapping region of the subcircuits is taken into consideration. They have also shown that energy is saved if the test units share common subcircuits or they are executed in the same test sessions. Power-aware test scheduling based on simulated annealing is presented in [9]. In this technique, a testing scheme is partitioned to minimize the number of idle slots. In [9], another approach was presented which stated that conventional power-aware test scheduling technique does not generate the thermal safe solution. Wake up, and leakage power cannot be neglected in deep submicron technology. So he proposed partitioned based thermal and power-aware test scheduling technique.

Genetic algorithm based heuristic technique for power-aware test schedule was proposed in [9]. Here, wrapper configuration, which is represented as a rectangle with height equal to test time and width corresponding to some TAM channels, is used as a core testing solution. Rectangles are placed using best-fit heuristic in such a way that they minimize test time.

Test scheduling with the multi-clock domain is first discussed in [9]. Here, the virtual TAM is used to solve the frequency gap between each core and a given ATE. Virtual TAM is also used here to minimize the power consumption of the core during the test.

Recently, many techniques have been proposed on multi-frequency SoC testing. In [9], the author presents an idea of the multi-frequency test environment; this is the biggest hurdle for SoC manufacturing capability. Many interdependent entities like TAM design; multi-frequency interface configuration, bandwidth matching, and power-aware test scheduling are well incorporated in the test framework. In [9], a new approach is proposed based on deterministic and pseudo-random test pattern to minimize test application time in the multi-clock domain test framework. A heuristic is defined to decide the core to be used for concurrent test and order of test patterns sequence for each core. In another approach, a heuristic on multi-frequency test schedule for the sessionless system is proposed in [9]. Two heuristics were proposed here, one for preemptive test scheduling and another one for non-preemptive test scheduling. In [9], the author has proposed a new multiplexer based architecture to send test data at a different frequency to the core. In a similar approach presented in [5], the author has proposed an idea of minimizing test time by varying test clock frequency. Traditional ILP model was modified by adding the variable frequency parameter. In [9], the author has proposed a method of power-aware test scheduling by scaling supply voltage and test clock frequency. Here, the scaling is done per session if its session-based test scheduling else dynamically if itis sessionless test scheduling. Exact and heuristic algorithm was discussed here for solving optimization. In [9], the author has suggested to increase the test clock frequency of core to P_{max} (maximum power budget of SoC). Test time is much optimized here. In a similar approach [9], the author has suggested to bound test clock frequency to a certain range. In his approach, he has suggested to test core with three clock frequencies $f_T/2$, f_T, and $2f_T$ where f_T is normal test clock frequency. Cores are assigned to any one of three frequencies and then test scheduling is done.

In all the previously published methods, no one has touched the issue of changing the test clock frequency of individual test vectors in a core. In this paper, a new idea is presented based on customizing the test clock frequency of the test vectors in a given core to minimize test time. The details are presented in the following sections.

3 Problem Statement

Letus consider,

N_c—Number of cores in a given SoC.
L—Length of the scan chain.
W—TAM width.
S_c—Number of the scan chain in a given core.
n—Number of test patterns in a given core.
P_{max_avg}—Maximum average power of the core.
P_i—Test power of test vector i.
P_{max}—Maximum power budget of SoC.
f_{func}—Functional frequency of a core.

Determine the efficient algorithm of dynamically allocating the test clock frequency to the different test vectors of the core and schedule the core for a given power budget in such a way that it minimizes the overall test time.

4 Dynamic Clock Allocation

This section describes the methodology used in dynamically allocating test clock frequency to the test vectors of the core during scan in and scan out operation [9].

Example 1 Letus assume that core C1 contains four test patterns which are denoted as T1, T2, T3, and T4 and four output responses O1, O2, O3, and O4. Each test pattern has three scan chains which are 3 bits, 2 bits, and 3 bits long. Also assume that switching between the consecutive test vectors T1-T2-T3-T4 is (1, 7, 5, 7) which is stored in an array ‘Sw’. Here, Sw(2) = is the switching between test vector T1 and T2, Sw(3) = is the switching between test vector T_2 and T_3, Sw(4) = is the switching between test vector T_3 and T_4. Switching Sw(1) is calculated when a very 1st test vector (T1) in the sequence will pass through the initial output response. Here, the initial output response is all 1 s or all 0 s depending on the last bit of the very 1st test vector in the sequence to reduce the switching.

Now, letus find maximum switching among all switching number stored in an array 'Sw' and store that in a variable 'SW_{max}'.

$$SW_{\max} = \max(Sw) \tag{1}$$

Maximum average power which is drawn for a single test vector is defined for each core in SoC. Power consumed by other test vectors is not known. Here, we are considering the dynamic power which is proportional to the switching activity in a test vector.

Now, take SW_{max} as a reference and find out the power of the remaining test vectors of the core.

$$P_i = Sw_i * P_{\max\_avg}/SW_{\max}, \forall i, \leq 1 \leq i \leq n \tag{2}$$

Increase in test clock frequency is decided by introducing a new term called "Power factor—PF," which is defined as

$$PF_i = P_{\max}/P_i, \forall \text{i}, 1 \leq \text{i} \leq \text{n} \tag{3}$$

However, for each core there exists a maximum functional frequency $f_{func,}$ which decides power constraints, e.g., the maximum power limit and structural constraints like critical path, etc.

So, while selecting the test clock frequency, two points need to be kept in mind: (1) Allowed test clock frequency should not go beyond functional frequency $f_{func.}$ Functional frequency is 3 to 4 times higher than the normal test clock frequency f_T [9]. (2) Power dissipation of core should not go beyond given power budget P_{max}.

Based on these criteria, we are modifying Eq. 3 by putting an upper limit of test clock frequency which is shown in Sect. 3.

$$PF_i = \min(f_{func.}/f_T, P_{\max}/P_i), \forall \text{i}, 1 \leq \text{i} \leq \text{n} \tag{4}$$

So, the new test clock frequency is given as

$$f_{NTi} = PF_i * f_T, \forall \text{i}, 1 \leq \text{i} \leq \text{n} \tag{5}$$

Improved test results for Example 1 are shown in Table 1.

Column 1 in Table 1 shows the test vector sequence. Switching among consecutive test vectors are shown in column 2. Test power calculated for all test vectors from Eq. 2 is shown in column 3. Test time of individual test cube is the maximum of all test vectors in a cube which is shown in column 4. Maximum power limit P_{max} is taken as 500 mw. Newly allocated test clock frequency from Eq. 5 is shown in column 5. Updated test power and test time are shown in column 6 and 7.

For Example 1, the test clock frequency is normalized to 1 MHz. Here, with normal test clock, the test time is 12 us, whereas with scaled frequency, the test time is 5.25 us which are shown in column 4 and 7, respectively. Reduction in test time

Table 1 Improved test time for Example 1

Test vector	Switching	Test power (From Eqn 2)(mw)	Test time (us)	Newly allocated freq (from Eqn-5)	Updated test power (mw)	Updated test time (us)
T1	1	35	3	4f	140	0.75
T2	7	250	3	2f	500	1.5
T3	5	178	3	2f	356	1.5
T4	7	250	3	2f	500	1.5
			12			5.25

with scaling the test clock frequency of the test vectors is the main motive of this proposed method.

Here, it is assumed that "set up and hold times" are not violated during this operation. For synchronization, scan in and scan out process is done on the same test clock frequency. For example, in Example 1, scan in of T2 test vector and scan out of O1 is done on the same test clock frequency.

5 Power-Aware Test Scheduling

In the previous section, we have discussed the test time optimization for the single core. In SoC, there are multiple cores which need to be scheduled. ILP-based core scheduling for SoC is shown in this section.

In the traditional method, the test time of the core is calculated from the equation given below.

$$t_i = (1 + \max(S_i, S_o))n + \min(S_i, S_o) \tag{6}$$

where S_i—the length of the scan in test vector, S_o—length of Scan out vector, t_i—test time of core i, n—total test patterns of the core

This equation states that all test patterns will get the same customized test clock frequency. For example, if the power factor is 2, then all the test vectors in the core will have two times higher test clock frequency.

But in our proposed method, instead of giving the same frequency to all test patterns of the core, the variable frequency is given to different test patterns of the core to minimize test time. So, we are modifying test time Eq. 6, which is given as

$$t_i = n + \sum\nolimits_{j=1}^{n} (\max(S_i, S_0))/PF_i + \min(S_i, S_0)/PF_L, \ \forall i, \ 1 \le i \le N_c \tag{7}$$

where PF_L is the Power factor of the last test vector.

5.1 Session-Based Power-Aware Test Scheduling

In session-based test scheduling, each core is assigned to one and only one session. To represent this, a new binary variable X_{ij} is introduced, where i represent the core and j represents the session.

$$\sum_{j=1}^{S} X_{ij} = 1,\ \forall i,\ 1 \le i \le N_c \tag{8}$$

where S—total test session, N_c is the number of cores. $X_{ij} = 1$, if core i is assigned to session j and $X_{ij} = 0$, otherwise.

Since it is the session-based test scheduling, the total power consumption of the cores scheduled in a session must not be greater than P_{max}.

$$\sum_{i=1}^{N_c} P_i * X_{ij} \le P_{\max},\ \forall j,\ 1 \le j \le S \tag{9}$$

After scheduling each core in a different session, the maximum test time of the cores scheduled on each session is stored in an array R.

$$R_j = \max_{(1 \le i \le N_c)} (t_i * X_{ij}),\ \forall j,\ 1 \le j \le S \tag{10}$$

Total test time 'T' of the SoC is the addition of maximum test time of the cores scheduled for each session.

$$T = \sum_{j=1}^{S} R_j,\ \forall j,\ 1 \le j \le S \tag{11}$$

The primary objective of this power-aware test scheduling is to minimize testing time 'T'.

$$Min\, T \tag{12}$$

The pseudo-code for the above ILP problem is shown below:

Pseudo-code for the algorithm is given in Fig. 1. The algorithm starts by defining the number of cores in SoC, the maximum average power of the core, and test clock frequencies. Test power and test time of each core are calculated using Eqs. 2 and 7, respectively. Initially, two lists l1 and l2 are defined which contains unscheduled and scheduled cores, respectively. Once a session is started, the algorithm keeps assigning the cores to the session which is shown inline 9. Cores are placed in a session and then test power of all the cores in a session are added together to check if it remains under the maximum power budget value P_{max} or not. If the session's total

The pseudo-code for the above ILP problem is shown below :

Define the number of cores, test power of the cores, a maximum power budget of the SoC
Calculate test power of each test vectors in the core using equation 2.
Calculate test time of each core using equation 7. Store the test time in an array.
Define array R_i which takes the maximum test time of the core in a particular session i.
Make list l1, which contains all unscheduled cores along with test time.
Make list l2, which contains all scheduled cores
Start a new session for scheduling
While list l1 is not empty **do**
  Place the core from list l1 to list l2
    If $\sum P_i < P_{max}$
      Update maximum test time of that session and store it in array R
      Update session's total power
    Else
      Remove the core from list l2 and place it again in list l1
      If no further cores are possible to accommodate
        Repeat procedure from step 7
      Else
        Repeat procedure from step 8
      End if
    End if
End While
Repeat procedure from step 6
$T = \sum R_i$, add maximum test time of all session together

  If further optimization is possible
    Restart the algorithm from step 3
  Else
    ILP will be stopped and store final test time T

Fig. 1 Pseudo-code for ILP about session-based scheduling

power does not exceed beyond P_{max} then update test time and session's power which is shown inline 10–12. If the session's total power exceeds beyond P_{max} then remove the recently added core and try new core which is shown inline 10 and 14–19. If no further cores are possible to accommodate in a session then start a new session and repeat all procedure from step 8. At last, the maximum test time of all session is added together which gives the final test time of the SoC. If further optimization is possible then repeat scheduling from step 4.

6 Hardware Realization

Test clock generation technique is shown in Fig. 2. Automatic test equipment (ATE) has all the test data to test SoC. Serial binary counter, three D flip flop, AND gates, 2

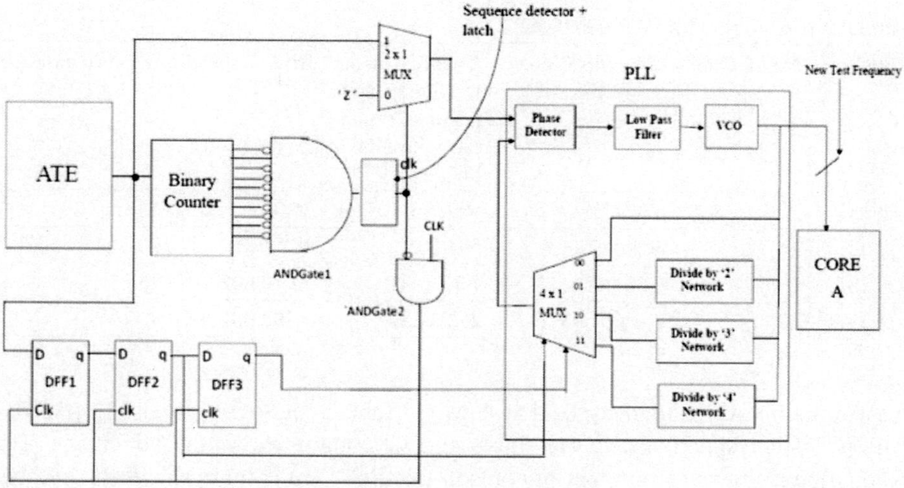

Fig. 2 Hardware for dynamic clock generation

× 1 multiplexer, PLL, sequence detector, and 4 × 1 multiplexer are added as Design for testability (DFT) hardware to test the SoC.

Phase-locked loop is used to make a test clock faster. Here, the slower test clock is multiplied by phase-locked loop and then given to the core which requires faster test clock. As shown in Fig. 2, three divide by N network (divide by 2, divide by 3, and divide by 4) is used in the feedback path of the PLL to multiply the output frequency of PLL. According to frequency selection bits, the multiplexer will select one frequency from four different frequency generated in the feedback path. The output frequency of PLL (new test frequency) is now 'N' times (frequency factor times) faster than the input frequency. One of the test vectors from core A is given this frequency.

Serial data from ATE is given to the serial binary counter, 2 × 1 multiplexer, and D flip flop 1. Serial data given to D flip flop 1 is shifted to another D flip flops on each normal test clock CLK. As soon as data comes to the serial binary counter, it counts up by 1. Sequence detector is connected to the output of ANDGate1. Sequence detector is designed here to detect sequence "010." The output of the sequence detector goes to the input of the 2 × 1 mux selection line and one input of ANDGate2. If the sequence is not matched to "010" then normal test clock is applied to three D flip flop and input to PLL remains high impedance state 'Z'. As sequence "010" is matched, the sequence detector output will go high and lathed. Data on D flip flops are latched and D flip 2 and 3 data is given to 4 × 1 mux to select test frequency (Output data of D flip flop 2 and 3 are used as a selection line for the 4 × 1 multiplexer to select one frequency out of four frequency generated from different "Divide by N" network.). 2 × 1 mux will allow bitstream to pass to PLL and then to core A with the frequency selected by 4 × 1 mux.

Here four frequencies f, 2f, 3f, and 4f are used for the test operation. So, two bits are padded with each test vectors to identify the frequency of operation. So, there will

Table 2 Bits overhead for different SoC

SoC	No. of Core	Total bits needed to test the SoC	Total bits along with frequency factor selection bits to test the SoC	Bits overhead	% Bits overhead
d695	10	574,431	609,731	35,300	5.79
g1023	14	405,679	507,805	102,126	20.11
p22810	29	6,661,179	7,731,841	1,070,662	13.85
p93791	32	27,866,593	29,061,597	1,195,004	4.11

be a minor bit overhead in test data as shown in Table 2 for SoC d695, g1023, p22810, and p93791, respectively. In that Table, the 1st column shows the SoC name. The 2nd column shows the number of cores in that SoC. 3rd column shows the number of bits needed to test the SoC. Total bits along with frequency factor selection bits are put in column 4. Bits overhead and percentage bits overhead is shown in column 5 and 6 respectively.

7 Result Discussion

Here, the proposed algorithm is implemented on windows core i3 processor with 1.00 GHz frequency and 4 GB RAM. The algorithm is applied to the widely used ITC'02 benchmark SoC d695, g1023, p22810, and p93791 with TAM width = 32. Details like number of the scan chain, number of test patterns, etc. per benchmark core are shown in [9]

Leakage current in the circuit causes the static power. To measure the static power we need a physical circuit. Since the internal architecture of the cores used in SoCs g1023, p22810, and p93791 are not publicly available, we cannot calculate static power. So, in all benchmark SoC simulation, we are considering dynamic power only.

Since, test vector and output response of SoC d695, g1023, p22810, and p93791 are not publicly available, we have generated it through MATLAB random function and put 80% don't care bits. Don't care bits in every test vectors are filled with '1' and '0' using MT fill algorithm [9].

The simulation is done on MATLAB 14 and LPSOLVE. Comparison of the proposed algorithm with a reference method [19–26] is shown in Table 3 with SoC d695, g1023, p22810, and p93791. Table 3 shows the simulation results for session-based test scheduling. In Table 3, Column 1 shows the name of SoC. Column 2 shows the number of Cores in SoC. Since the power consumption of each core is not mentioned in the ITC'02 benchmark database, it is taken from [8]. Here, the maximum power limit P_{max} is set as 50% of the gross of total power consumption of all cores which is shown in column 3. The test time of each core in SoC, for

Table 3 Comparison of test time

SoC	No. of cores	Max. power budget (mw)	Test time of Ref method [26] (us)	Test time with proposed method (us)	% Reduction
d695	10	3325	12,456	8049	35.3
g1023	14	1500	24,753	14,226	42.5
p22810	29	9965	105,588	44,347	58.0
p93791	32	39,098	311,716	261,363	16.5

the technique presented in [9], is also not given, so we have implemented the same algorithm again in MATLAB and result is shown in Column 4. Test time with the proposed method is shown in Column 5. Percentage reduction in test time is shown in Column 6. Experimental results show that up to 58% reduction in test time is achieved through session-based test scheduling which shows the effectiveness of this method.

8 Conclusion

Here, it is shown that test time reduction is achieved by assigning appropriate test clock frequency to each pattern of the core with a minor bit overhead. While customizing test clock frequency, care is taken that test clock frequency does not go beyond functional frequency and power dissipation of the core does not go beyond maximum power budget. ILP-based method is used to solve the scheduling of the core in SoC. Experimental results on the ITC'02 benchmark shows that the ILP-based session-based test scheduling method reduces test time up to 58% which shows the effectiveness of this method

References

1. Aikyo T. "Issues on SOC Testing in DSM Era" Design automation conference Japan (2000).
2. Aikyo T (2000) Issues on SOC Testing in DSM Era. Design automation conference Japan
3. Parmar H & Mehta U, "An improved algorithm for TAM optimization to reduce test application time in core based SoC", *IEEE International WIE Conference on Electrical and Computer Engineering (WIECON-ECE)*, 2015
4. Parmar H, Mehta U (2015) An improved algorithm for TAM optimization to reduce test application time in core based SoC. IEEE International WIE Conference on Electrical and Computer Engineering (WIECON-ECE)
5. Chakrabarty K, Iyengar V , & Chandra A "Test resource partitioning for system- on-a-chip" *Frontiers in Electronic Testing,* . (2002).
6. Chakrabarty K, Iyengar V, Chandra A (2002) Test resource partitioning for system- on-a-chip. Frontiers in Electronic Testing

7. Girard P, Nicolici N, & Wen X. "Power-aware testing and test strategies for low power devices." *Springer, 2010.*
8. Girard P, Nicolici N, Wen X (2010) Power-aware testing and test strategies for low power devices. Springer
9. Vijay S , Agrawal V & Agrawal P, "Optimal Power-Constrained SoC Test Schedules With Customizable Clock Rates". *IEEE International SOC Conference,* 2012.
10. Vijay S, Agrawal V, Agrawal P (2012) Optimal power-constrained SoC test schedules with customizable clock rates. IEEE International SOC Conference
11. Ravikumar C, Verma A & Chandra G, "A Polynomial-Time Algorithm for Power Constrained Testing of Core Based Systems". *Eighth Asian Test Symposium,*1999.
12. Ravikumar C, Verma A, Chandra G (1999) A polynomial-time algorithm for power constrained testing of core based systems. Eighth Asian Test Symposium
13. Iyengar V & Chakrabarty K, "System-on-a-Chip Test Scheduling With Precedence Relationships, preemption, and Power". *IEEE Transactions on Computer-Aided Design of Integrated Circuits and Systems,* (pp. 1088–1094), 2002.
14. Iyengar V, Chakrabarty K (2002) System-on-a-chip test scheduling with precedence relationships, preemption, and power. IEEE transactions on computer-aided design of integrated circuits and systems, pp 1088–1094
15. Pouget J, Larsson E , & Peng Z , "SOC test time minimization under multiple constraint". *12th Asian Test Symposium,* (pp. 312–317), 2003.
16. Pouget J, Larsson E, Peng Z (2003) SOC test time minimization under multiple constraint. 12th Asian Test Symposium, pp 312–317
17. Skarvada J (2006a) Test Scheduling for SOC under Power Constraints. Czech Republic, IEEE Design and Diagnostics of Electronic Circuits
18. Skarvada J (2006b) Test scheduling for SOC under power constraints. Czech Republic, IEEE design and diagnostics of electronic circuits
19. Schuele T & Stroele A, "Test Scheduling for Minimal Energy Consumption under Power Constraints". *19th IEEE VLSI Test.* CA, USA, 2001.
20. Schuele T, Stroele A (2001) Test scheduling for minimal energy consumption under power constraints. 19th IEEE VLSI Test. CA, USA
21. Harmanani H, Hassan S (2006a) On Power-Constrained System-On-Chip Test Scheduling Using Precedence Relationships. IEEE North-East Workshop on Circuits and Systems, Gatineau, QC, Canada
22. Harmanani H, Hassan S (2006b) On power-constrained system-on-chip test scheduling using precedence relationships IEEE North-East Workshop on Circuits and Systems Gatineau. QC, Canada
23. Yao C, Kewal S (2009) Partition Based SoC Test Scheduling with Thermal and Power Constraints under Deep. IEEE Asian Test Symposium, Taichung, Taiwan
24. C Yao S Kewal 2009 Partition based SoC test scheduling with thermal and power constraints under deep IEEE Asian Test Symposium Taichung, Taiwan
25. Giri C, Sarkar S & Chattopadhyay S, " A Genetic Algorithm based Heuristic Technique for power constrained test scheduling in core", *IEEE International Conference on Very Large Scale Integration.* Atlanta, GA, USA, 2007.
26. Giri C, Sarkar S, Chattopadhyay S (2007) A genetic algorithm based heuristic technique for power constrained test scheduling in core. IEEE International Conference on Very Large Scale Integration. Atlanta, GA, USA
27. Yoneda T, Masuda K, Fujiwara H (2006) Power-Constrained Test Scheduling for Multi-Clock Domain SoCs. Automation and Test in Europe. Munich, Germany, IEEE Design
28. Yoneda T, Masuda K, Fujiwara H (2006) Power-constrained test scheduling for multi-clock domain SoCs. IEEE Design, Automation and Test in Europe. Munich, Germany
29. Zhao D, Huang R, Yoneda T, Fujiwara H (2007a) Power-Aware Multi-Frequency Heterogeneous SoC Test Framework Design with Floor-Ceiling Packing. IEEE International Symposium on Circuits and Systems, New Orleans, LA, USA

30. Zhao D, Huang R, Yoneda T, Fujiwara H (2007b) Power-aware multi-frequency heterogeneous SoC test framework design with floor-ceiling packing. IEEE International Symposium on Circuits and Systems, New Orleans, LA, USA
31. Haghbayan M , Safari S & Navabi Z, " Power Constraint Testing for Multi-Clock Domain SoCs Using Concurrent Hybrid", *IEEE 15th International Symposium on Design and Diagnostics of Electronic Circuits & Systems (DDECS).* Tallinn, 2012.
32. Haghbayan M, Safari S, Navabi Z (2012) Power constraint testing for multi-clock domain SoCs using concurrent hybrid. IEEE 15th International Symposium on Design and Diagnostics of Electronic Circuits & Systems (DDECS). Tallinn
33. Sheshadri V, Agrawal V & Agrawal P, "Power-Aware SoC Test Optimization through Dynamic Voltage and Frequency Scaling". *IEEE 21st International Conference on Very Large Scale Integration, 2013.*
34. Sheshadri V, Agrawal V, Agrawal P (2013) Power-aware SoC test optimization through dynamic voltage and frequency scaling. IEEE 21st International Conference on Very Large Scale Integration
35. Karmakar R, Agarwal A & Chattopadhyay S, "Particle Swarm Optimization Guided Multi-frequency Power-Aware System-on-Chip Test Scheduling Using Window-Based Peak Power Model". *18th International Symposium on VLSI Design and Test (VDAT).* Coimbatore, 2014.
36. Karmakar R, Agarwal A & Chattopadhyay S (2014) Particle swarm optimization guided multi-frequency power-aware system-on-chip test scheduling using window-based peak power model. 18th International Symposium on VLSI Design and Test (VDAT). Coimbatore
37. Vijay S, Agrawal V & Agrawal P, "Power-Aware Optimization of SoC Test Schedules Using Voltage and Frequency Scaling". *Electron Test, 2016.*
38. Vijay S, Agrawal V, Agrawal P (2016) Power-aware optimization of SoC test schedules using voltage and frequency scaling. Electron Test
39. Parmar H & Mehta U, " Power Aware Network on Chip Test Scheduling with Variable Test Clock Frequency", *Ubiquitous Communications and Network Computing* pp 256–264, 2017
40. Parmar H, Mehta U (2017) Power aware network on chip test scheduling with variable test clock frequency. Ubiquitous Communications and Network Computing, pp 256–264
41. Parmar H & Mehta U, "ILP Based Power-Aware Test Time Reduction using On-chip Clocking in NoC Based SoC", Journal of Low Power Electronics & Applications, pp. 1–12, vol. 9, 2019
42. Parmar H, Mehta U (2019) ILP based power-aware test time reduction using on-chip clocking in NoC Based SoC. J Low Power Elect Appl 9:1–12
43. Wu T, Zhou L & Liu H, "Reducing Scan-shift Power Through Scan Partitioning and Test Vector Reordering", *21st IEEE International Conference on Electronics, Circuits and Systems*, France, 2014
44. Wu T, Zhou L, Liu H (2014) Reducing scan-shift power through scan partitioning and test vector reordering. 21st IEEE International Conference on Electronics, Circuits and Systems, France
45. Shlomi S, Eyal S (2008) Frequency and Power Correlation between At-Speed Scan and Functional Tests. ITC, USA
46. S Shlomi S Eyal 2008 Frequency and power correlation between at-speed scan and functional tests ITC USA
47. Marinissen E, Iyengar V & Chakrabarty K, "A set of benchmarks for modular testing of SOCs". *International Test*, (pp. 519–528), 2002
48. Marinissen E, Iyengar V, Chakrabarty K (2002) A set of benchmarks for modular testing of SOCs. International Test, pp 519–528
49. Parmar H & Mehta U, "WTM Based Reordering of Combine Test Vector & Output Response using Dijkstra Algorithm for Scan Power Reduction". *NUiCONE.* Ahmedabad: Nirma University, 2011.Author, F.: Article title. Journal 2(5), 99–110 (2016).
50. Parmar H, Mehta U (2016) WTM based reordering of combine test vector & output response using Dijkstra algorithm for scan power reduction. NUiCONE. Ahmedabad: Nirma University, 2011. Author, F.: Article title. Journal 2(5):99–110

Impact of Spacers in Raised Source/Drain 14 nm Technology Node $In_{0.53}Ga_{0.47}As$-nFinFET on Short Channel Effects

Jay Pathak and Anand D. Darji

Abstract $In_xGa_{1-x}As$ channel materials with $x > 0.53$ have been extensively researched for beyond silicon CMOS (complementary metal oxide semiconductor) logic applications, and now its promising non-Si n-channel material, enabling a continuous scaling of the supply voltage (V_{dd}) while providing a performance improvement. The numerous kinds of literature suggesting that the impact of spacers and raised source/drain in FinFETs for silicon-based device have good arguments in improving the performances. The FinFETs in sub-14 nm technology nodes face severe short channel effects (SCEs), and the same is true for $In_{0.53}Ga_{0.47}As$-based FinFETs. In this paper, we have investigated the impact of nitride spacer in raised source/drain $In_{0.53}Ga_{0.47}As$-FinFETs for improving the SCEs. The improvement of 10% is observed in subthreshold swing while 32% in drain-induced barrier lowering when nitride spacer is used in raised source/drain $In_{0.53}Ga_{0.47}As$-FinFETs.

Keywords InGaAs · Spacers · Raised source/drain · Subthreshold swing · Drain-induced barrier lowering

1 Introduction

According to ITRS (International Roadmap for Semiconductor) CMOS-based devices for next-generation technology nodes, i.e., sub-16 nm digital technology nodes are now totally dependent on 3D multi-gate device such as FinFETs because of improved short channel effects and better electrostatic control of gate, able to achieve ideal drain-induced barrier lowering (DIBL) and subthreshold swing [1]. However, the III-V materials with high electron mobilities, better transport properties, and high electron injection make them best contender for n-channel field effect transistors (FETs) for high-speed and low-power application devices. The $In_xGa_{1-x}As$ channels in sub-16 nm FinFET device are obtaining a lot of attraction due to the

J. Pathak (✉) · A. D. Darji
Sardar Vallabhbhai National Institute of Technology, Surat, India
e-mail: jaypathak050@gmail.com

Z. Patel et al. (eds.), *Advances in VLSI and Embedded Systems*, Lecture Notes in Electrical Engineering 676, https://doi.org/10.1007/978-981-15-6229-7_12

potential of varying the percentage of Indium composition (x) in $In_xGa_{1-x}As$ and achieving a wide range of bandgap and effective mass [2]. $In_{0.53}Ga_{0.47}As$ FinFET has been explored in several literatures and has been fortuitously fabricated with various techniques. The $In_{0.53}Ga_{0.47}As$-based FinFET for sub-16 nm technology node is critical due to several aspects like lattice mismatch with substrate and oxides (high-K), interface traps, change in transport properties of the material beyond 10 nm thickness, etc. Such problems lead to affect SCEs but by incorporating various techniques in the device, suppression in SCEs can be achieved. The importance of raised S/D technique has been raised in future goal of CMOS scaling as it has to keep I_{on} as high as possible while scaling V_{dd}. The raised S/D helps to reduce S/D resistance which is predominant for sub-22 nm technology nodes [3–7]. In this work, the SCEs in $In_{0.53}Ga_{0.47}As$-FinFET at sub-14 nm technology nodes with three different structures are analyzed. The three structures, namely, traditional, raised source/drain (S/D) and raised S/D with spacers are taken into consideration. The work is divided into three sections, Sect. 2 has the description of device structure and simulation methodology. Section 3 exhibits the physical importance of spacer in raised source/drain, while results and analysis of SCEs are described in Sect. 4.

2 Device Structure and Simulation Methodology

The digital performance of InGaAs FinFET is carried out for sub-16 nm technology node with channel length of 14 nm. The detailed structure parameters are prescribed from 2013 ITRS (International Technology Roadmap for Semiconductors) [8]. Table 1 reflects the dimensions of structure for channel length of 14 nm. The combination of InGaAs/HfO_2 is taken as oxide/semiconductor interface, and several techniques have been successfully implemented to integrate HfO_2 on InGaAs [9]. The spacer is introduced with nitride across the uncovered fins of the raised S/D device. The body is doped with beryllium of concentration 1×10^{17} cm^{-3}, while S/D is doped with silicon of concentration 5×10^{19} cm^{-3}. The doping in body and S/D region is uniformly doped throughout the device. The gate of the device is of nickel as it has low contact resistivity. The structures of $In_{0.53}Ga_{0.47}As$-FinFET are shown in Fig. 1.

The device simulation has been performed using Synopsys SDEVICE 3-D TCAD Tool [10]. The device physics used for the characterizing 14 nm channel length FinFETs have been included by calibrating the fabricated device of 50 nm channel length $In_{0.53}Ga_{0.47}As$ nFinFET [11]. The state of the art is similar to that adopted in [12–14]. Doping dependence Arora model and Lombardi High-K model have been used for mobility of inversion charge carriers. The recombination and generations are apprehend by introducing the models like Hurkx, Auger [15, 16], and ShockleyReadHall. The device operates under high electric field at small bias voltage, so the high electric field phenomenon saturates the carrier's velocity and this event is included using the high-field saturation model. The parameters such as velocity saturation, mobility, density of states, etc. for $In_{0.53}Ga_{0.47}As$ material are extracted from

Table 1 Parameters for design of $In_{0.53}Ga_{0.47}As$-FinFETs[8]

Parameters	Value
L_g (nm)	14
EOT (nm)	0.68
W_{fin} (nm)	8.5
H_{fin} (nm)	21
N_{ch} (cm^{-3})	10^{17}
N_{sd} (cm^{-3})	5×10^{19}
V_{dd}	0.6
L_{ext} (nm)	12
L_{sd} (nm)	50
T_{sd} (nm)	10
ϵ_{ox}	22
$\epsilon_{spacers}$	25

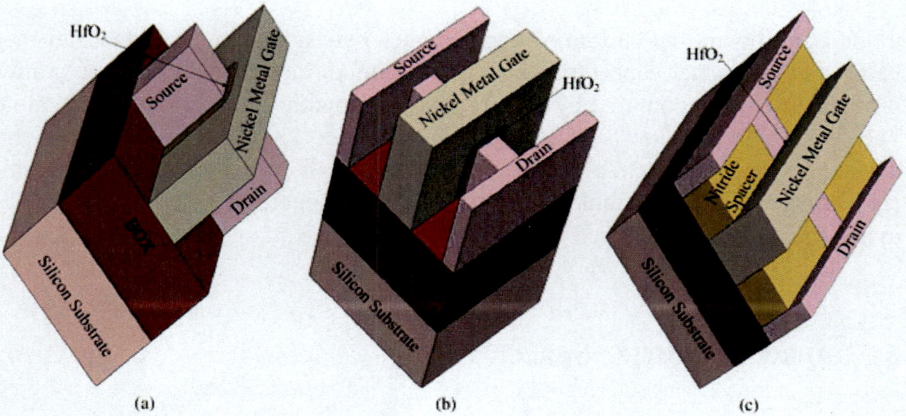

Fig. 1 FinFET structure. **a** Simple $In_{0.53}Ga_{0.47}As$ FinFET. **b** Raised source/drain $In_{0.53}Ga_{0.47}As$ FinFET. **c** Raised source/drain $In_{0.53}Ga_{0.47}As$ FinFET with nitride spacer

the experimental device of 50 nm channel length $In_{0.53}Ga_{0.47}As$ nFinFET [11]. The experimental device of 50 nm channel length $In_{0.53}Ga_{0.47}As$ nFinFET is calibrated with the help of the drift diffusion model and $I_{ds} - V_{gs}$ and $I_{ds} - V_{ds}$ characteristics have been shown in Fig. 2.

Aside from device physics models utilized in adjusting the 50 nm channel length, various models were likewise included, for example, quantization effects and non-parabolicity effects for sub-14 nm $In_{0.53}Ga_{0.47}As$ nFinFETs to deal with quantum confinement effects. Energy bandgap and the band structure of InGaAs that display non-parabolic behavior at various valleys are likewise considered in the device. At 8.5 nm channel thickness affects the quantum mechanical effects, so the fitting parameters in quantum potential parameters (namely, gamma, effective mass) are

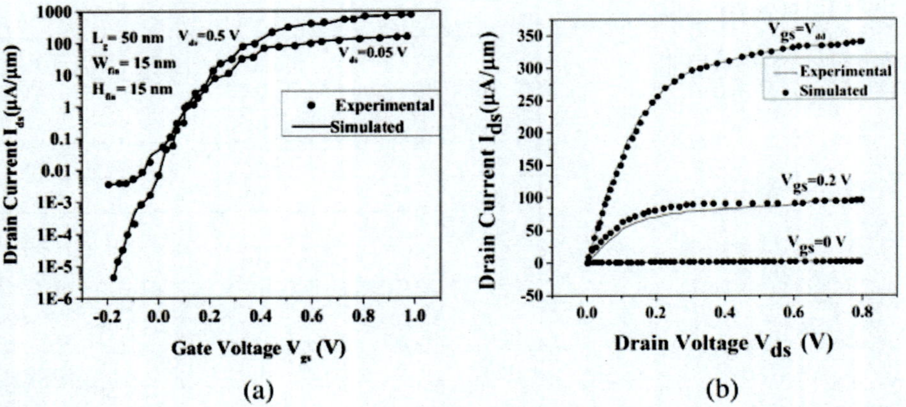

Fig. 2 Calibration of $In_{0.53}Ga_{0.47}As$ nFinFET with $L_g = 50$ nm. **a** $I_{ds} - V_{gs}$ transfer characteristics during different drain bias. **b** $I_{ds} - V_{ds}$ output characteristics during different gate bias [13, 14]

modified to obtain the current–voltage characteristics. The impacts in the energy bands and the carrier concentrations due to quantum mechanical effects at various bias conditions are conjured by density gradient approach and Poisson's equation. The device worked at low (0.05 V) and high drain bias (V_{dd}) using drift diffusion and density gradient approach. The quantum corrected drift diffusion and density gradient models are used for simulation of 14 nm channel length $In_{0.53}Ga_{0.47}As$ FinFETs to obtain various performance parameters of the device [12–14].

3 Impacts of Nitride Spacer

The spacer's engineering in FinFET-based CMOS technology has proved to be the better technique in controlling the short channel effects [17–19]. The effect of spacers in the device has the major impact in charge carrier density due to conduction band energy and electrostatics potential while operating at high biasing. Figure 4a shows the electrostatics potential across the S/D region with and without nitride spacer. The electrostatic potential and conduction band energy at $V_{gs} = V_{dd}$ and $V_{ds} = V_{dd}$ in presence of nitride spacers across S/D region are higher. The field lines at $V_{gs} = V_{ds} = V_{dd}$ are more concentrated in the channel through source region. Introducing spacers leads to additional fringing field which bends the conduction band edge and increases the electron concentration and potential across the channel (Fig. 3). The increment in the electron concentration is due to the extra fringing field effects observed across the source and gate. This fringing electric field is generated from gate through spacer and terminated in the lightly doped channel and changes the potential and electron concentration. The excess of electric field produces severe band bending in conduction band and provides higher electron concentration across channel. The traditional and raised S/D structure does not generate this extra fringing field which

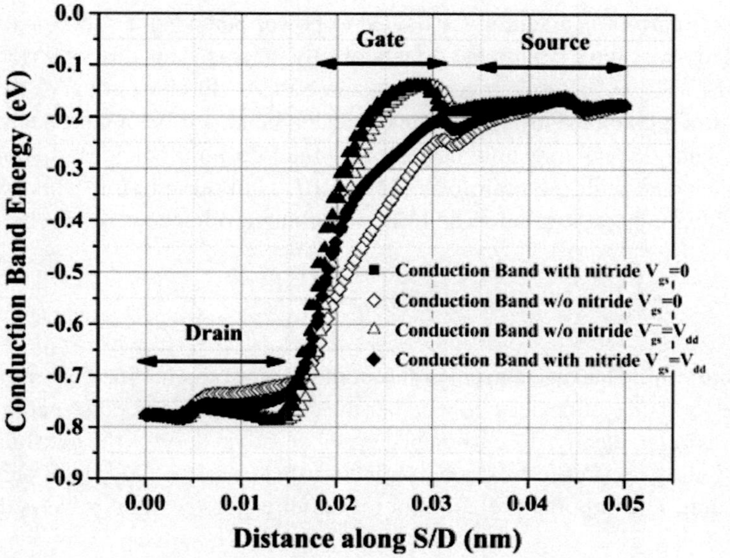

Fig. 3 Conduction band energy along the S/D in on and off state for both devices

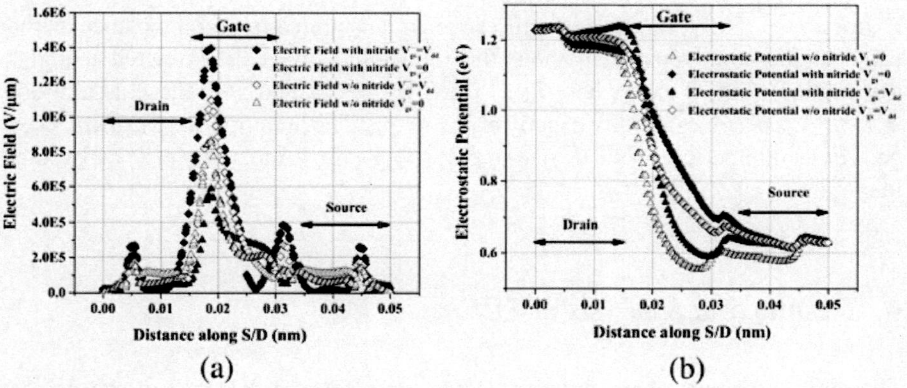

Fig. 4 **a** Electric field profile along the S/D for both devices. **b** Variation of electrostatic potential at the spacer interface and inside the channel (in *on* state)

helps gate to control the channel. The bending helps inversion charge carriers to pass through the channel during *on* regime under gate control. So apart from carrier density, extra bending in CBE across channel region also helps in increasing the charge carrier during *on* regime. The improvement in on-state current is the major achievement when the spacers are introduced in the device. Figure 4b provides the influence of spacer on electrostatic potential in *on* and off state of the device. Thus, by introducing the spacers and its impact on the behavior of conduction band energy, electrostatic potential and electron density concentration at high drain bias and low drain bias improve *on*-state current.

The off current is essential for device in power perspective. The $In_{0.53}Ga_{0.47}As$-FinFET devices have significant effects of off currents. The current observed when $V_{gs} = 0$ for $V_d = V_{dd}$ is known as leakage current. In short channel devices, the source–drain potential strongly affects the band bending over a huge segment of the device. Thus, the subthreshold currents and the threshold voltages of short channel devices change with the drain bias. The DIBL is observed on the behavior of the threshold voltage at saturation and linear region, and hence it is governed through Eq. 1:

$$\text{DIBL} = \frac{\delta V_{th}}{\delta V_{ds}} \tag{1}$$

The subthreshold leakage current is the another key parameter that determines the off-state performance of device and whether the chip's static power dissipation is within tolerable limits. Ideally, the current decreases to zero as soon as the gate bias is below V_{th}. In reality, existence of drain conduction below V_{th} is known as subthreshold conduction. The important relation for the subthreshold swing is governed by Eq. 2:

$$\text{SS} = \frac{dV_{gs}}{d(\log I_{ds})} \tag{2}$$

Extraction of the V_{th} for the device is carried out using the transconductance method [20]. In transconductance method, the threshold voltage is extracted using the assumption of ideal case where $I_{ds} = 0$ for $V_{gs} < V_{th}$, and with these assumptions $d^2 I_{ds}/d^2 V_{gs}$ becomes infinity exactly at $V_{gs} = V_{th}$. The assumption of $d^2 I_{ds}/d^2 V_{gs} = \infty$ is quiet impractical in real device so at $V_{gs} = V_{th}$ the value of $d^2 I_{ds}/d^2 V_{gs}$ becomes maximum.

4 Results and Analysis of SCEs

The current–voltage characteristics decide the electrical behavior of the devices. The $I_{ds} - V_{gs}$ characteristics for all the three devices of $In_{0.53}Ga_{0.47}As$-FinFET are simulated using SDEVICE TCAD tool. Various models for carrier transports, mobility, electrostatic potential, generation and recombination of carriers, etc. have been incorporated to extract the characteristics. The $I_{ds} - V_{gs}$ curves are simulated for both saturation and linear region, saturation curve is extracted at $V_{ds} = 0.6$ V while linear curve is extracted at $V_{ds} = 0.05$ which is reflected in Fig. 5a. The output characteristic, i.e., $I_{ds} - V_{gs}$ is also simulated for all the devices of $In_{0.53}Ga_{0.47}As$-FinFET. The off current characteristic is shown in Fig. 5b for all the devices.

Table 2 has shown the comparison for all the SCEs of devices. Table 2 indicates improvement in SCEs for raised S/D structures with spacers. The improvement of SCEs in presence of spacers can be understood by the phenomenon of the electron concentration and electrostatic potential across channel. The impact of spacers is not limited to on current but also to off currents, DIBL, subthreshold swing, etc. of

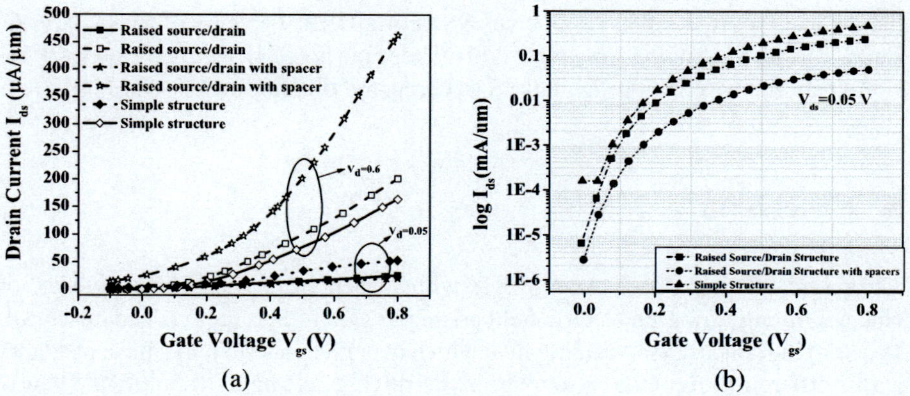

Fig. 5 **a** $I_{ds} - V_{gs}$ characteristics and **b** I_{off} current characteristics of all the $In_{0.53}Ga_{0.47}As$-FinFET devices

Table 2 Performance parameters

	Simple structure	Raised source/drain	Raised source/drain with spacers
I_{on} (μA/μm)	155.3	200.2	450.1
I_{off} (μA/μm)	6	2.5	1.3
DIBL (mV/V)	94	80	66
SS (mV/decade)	125	114.5	103
V_{th} (V)	0.18	0.21	0.216

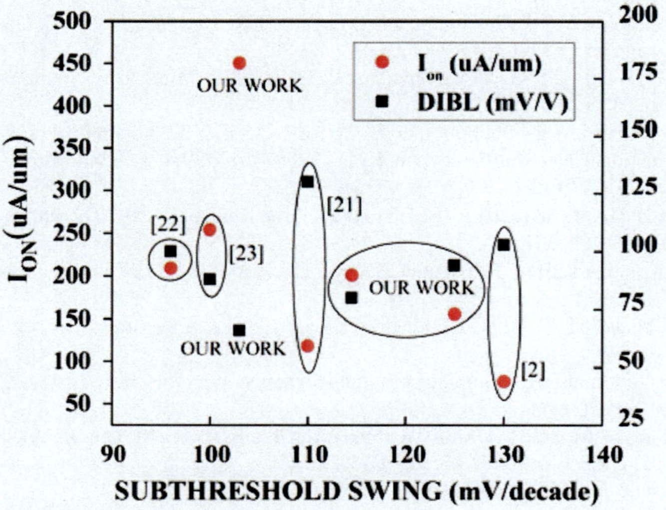

Fig. 6 I_{on} versus SS versus DIBL for [2, 21–23]

the device. Figure 6 reflects I_{on} versus SS versus DIBL benchmark for InGaAs-OI FinFETs with SS ranging from 60 to 200 mV/decade [21–24]. The reported work has significant improvement in I_{on} and SS as compared to other InGaAs-OI FinFETs.

5 Conclusion

The raised S/D $In_{0.53}Ga_{0.47}As$-FinFETs with spacers have good control of gate on channel through fringing electric field arising at gate to S/D region. The raised S/D structure introduces extra fringing field which may give raise to parasitic capacitance across different parts of the region. The extra fringing field helps to control the flow of electron during the off condition. The impact of spacer is observed in subthreshold swing and DIBL of the device. SS and DIBL of raised source/drain with spacer are lowest compared to other devices. SS is as low as 103 mV/decade and DIBL is 66 mV/V. Implementing InGaAs FinFET with spacers is prominent option for high-speed and low-power applications.

Acknowledgments The author would like to thank the Special Manpower Development Program Phase-II (SMDP-II) for VLSI design sponsored by government of India, New Delhi.

References

1. Seoane N, Aldegunde M, Nagy D (2016) Elmessary. Semicond Sci Technol 31(7):75005
2. Thathachary AV, Lavallee G, Cantoro M, Bhuwalka KK, Heo YC, Maeda S, Datta S (2015) IEEE Electron Device Lett 36(2):117
3. Rodwell MJ, Huang CY, Lee S, Chobpattana V, Thibeault B, Mitchell W, Stemmer S, Gossard A (2015) ECS Trans 66(4):135
4. Zota CB, Convertino C, Baumgartner Y, Sousa M, Caimi D, Czornomaz L (2018) 2018 IEEE international electron devices meeting (IEDM), pp 39.4.1–39.4.4. https://doi.org/10.1109/IEDM.2018.8614530
5. Kim J, Huynh HA, Kim S (2017) IEEE Trans Electron Devices 64(5):2072. https://doi.org/10.1109/TED.2017.2685527
6. Gupta M, Kranti A (2018) IEEE Trans Electron Devices 65(6):2406. https://doi.org/10.1109/TED.2018.2823752
7. Rodder M, Yeakley D (1991) IEEE Electron Device Lett 12(3):89. https://doi.org/10.1109/55.75721
8. ITRS (2009) International technology roadmap for semiconductors. http://www.itrs.net/Links/2009ITRS/Home2009.htm/
9. Kim TW, Kwon HM, Shin SH, Shin CS, Park WK, Chiu E, Rivera M, Lew JI, Veksler D, Orzali T et al (2015) IEEE Electron Device Lett 36(7):672
10. Inc Synopsys, View Mountain (2016) CA 2016:
11. Djara V, Deshpande V, Sousa M, Caimi D, Czornomaz L, Fompeyrine J (2016) IEEE Electron Device Lett 37(2):169
12. Pathak J, Darji A (2019) IET Circuits Devices Syst 13(4):428
13. Pathak J, Darji A (2019) Int J Electron 106(10):1514
14. Pathak J, Darji A (2019) J Nanoelectron Optoelectron 14(4):505

15. Hurkx GAM, Klaassen DBM, Knuvers MPG (1992) IEEE Trans Electron Devices 39(2):331
16. Hurkx GAM, de Graaff HC, Kloosterman WJ, Knuvers MPG (1992) IEEE Trans Electron Devices 39(9):2090
17. Tewari S, Biswas A, Mallik A (2016) IEEE Trans Electron Devices 63(6):2313
18. Koley K, Dutta A, Syamal B, Saha SK, Sarkar CK (2013) IEEE Trans Electron Devices 60(1):63
19. Sachid AB, Francis R, Baghini MS, Sharma DK, Bach KH, Mahnkopf R, Rao VR (2008) 2008 IEEE international electron devices meeting (IEEE, 2008), pp 1–4
20. Ortiz-Conde A, García-Sánchez FJ, Muci J, Barrios AT, Liou JJ, Ho CS (2013) Microelectron Reliab 53(1):90
21. Djara V, Deshpande V, Uccelli E, Daix N, Caimi D, Rossel C, Sousa M, Siegwart H, Marchiori C, Hartmann JM, Shiu K, Weng C, Krishnan M, Lofaro M, Steiner R, Sadana D, Lubyshev D, Liu A, Czornomaz L, Fompeyrine J (2015) Proceedings of the symposium on VLSI technology (IEEE, Kyoto, 2015), pp T176–T177
22. Sun X, D'Emic C, Cheng CW, Majumdar A et al (2017) Proceedings of the symposium on VLSI technology (IEEE, Kyoto, 2017), pp T40–T41
23. Luc QH, Yang KS, Lin JW, Chang CC, Do HB, Huynh SH, Ha MTH, Nguyen TA, Lin YC, Hu C et al (2018) IEEE Electron Device Lett 39(3):339
24. Vardi A, Lin J, Lu W, Zhao X et al (2016) Proceedings of the symposium on VLSI technology (IEEE, Honolulu, 2016), pp 1–2

Impact of Multi-Metal Gate Stacks on the Performance of β-Ga_2O_3 MOS Structure

Narendra Yadava and R. K. Chauhan

Abstract Gallium oxide (Ga_2O_3) due to its Ultra-wide bandgap (UWB) becomes the most promising semiconductor material for future-generation electronic applications. The Ga_2O_3 semiconductor possesses excellent material properties such as ultra-wide bandgap of 4.8 eV, ability to withstand breakdown field ranging from 5 to 9 MV/cm along with superior thermal and chemical stability. Also, it has excellent Baliga's figure-of-merit (B-FOM). However, the performance of β-Ga_2O_3 devices is still limited due to the issue related to good ohmic contact materials. In this work, we have studied the Impact of using multi-metal gate stack arrangements on the performance of β-Ga_2O_3 MOS Structure. The performance parameters used in the analysis are I_{Dmax}, I_{Dmin}, I_{ON}/I_{OFF}, g_m, and g_d. It is observed that the Ti/Au metal stack arrangement shows better results among all the metal stack arrangements and hence found to be suitable for high power RF applications with low losses.

Keywords β-ga_2O_3 · Multi-metal stacks · Ultra-wide bandgap semiconductors

1 Introduction

Silicon has reached its theoretical limitations for high power wireless applications [1, 2]. To improve the performance of the high power electronic devices with minimized power losses, it is needed to replace the silicon material by other semiconductor technology. Recently reported wide band gap (E_g) materials like SiC, GaN, and Ga_2O_3 surpass the Si because these materials have the advantage of high breakdown electric field (E_{br}) over the silicon [3]. Ga_2O_3 material with ultra-wide band gap (~5 eV) is found to be the most promising choice for efficient high power electronic applications due to its ability to withstand a breakdown electric field of 8MV/cm [4, 5]. To analyze and suppress the losses in high power devices, Baliga's figure-of-merit

N. Yadava (✉) · R. K. Chauhan
Department of Electronics and Communication Engineering MMMUT, Gorakhpur, India
e-mail: narendrayadava5@gmail.com

R. K. Chauhan
e-mail: rkchauhan27@gmail.com

Z. Patel et al. (eds.), *Advances in VLSI and Embedded Systems*, Lecture Notes in Electrical Engineering 676, https://doi.org/10.1007/978-981-15-6229-7_13

(B-FOM) is used and breakdown field which is one of the key factors in the above FOM plays a crucial role to suppress losses [1].

In high voltage applications, the blocking voltage is required to be very high, i.e., >600 V, and Ga_2O_3 MOSFET already has made improvement in terms of E_c for such applications by overcoming the theoretical limits imposed by other known materials like GaN, SiC, Si, etc. [6]. Ga_2O_3 MOSFET with high mobility of 100 cm^2/Vs [7] provides an opportunity to implement efficient power amplifier combining the advantages of optimized integration of power switches, RF switches, and power amplifiers [8]. However, the performance of β-Ga_2O_3 devices is still limited due to the issue related to good ohmic contact materials. Many researches related to the different gate metal arrangement for Ga_2O_3 MOSFET has already been carried out to overcome the contact metal related issues [9–15]. Higashiwaki et al. investigated the performance of Ga_2O_3 MOSFET having Ti(3 nm)/Pt(12 nm)/Au(280 nm) gate metal stacks of 2 μm in length and obtained low off-state leakage current in the range of few pico-Ampere with good switching performance (Ion-to-Ioff current ratio $>10^{10}$) [10]. Green et al. studied the performance of Ti(20 nm)/Au(480 nm) metal stack gated Ga_2O_3 MOSFET and found that the designed device having MOVPE grown Ga_2O_3 resultant in high field strength of about 3.8MV/cm [11]. Feng et al. studied that the use of Ni(60 nm)/Au(120 nm) metal stack gate electrode in a Ga_2O_3 MOSFET which results in Ion-to-Ioff current ratio $>10^8$ with a high breakdown of 800 V [12].

In this work, we have studied the impact of using different multi-metal gate stacks on the performance of β-Ga_2O_3 MOS Structure. The parameters for the different gate metal stack arrangements are taken from the experimentally reported Ga_2O_3 MOS devices [10–12, 14]. The performance parameters used in the analysis are I_{Dmax}, I_{Dmin}, I_{ON}/I_{OFF}, g_m, and g_d.

2 Description of the Device and Material Parameters

The structure of the Ga_2O_3 MOS devices with different multi-metal gate stacks is shown in Fig. 1.

The channel of the device is a uniformly doped (7e17cm^{-3}) 300 nm thick *n*-type region grown over a semi-insulating single crystal Ga_2O_3 substrate. The multiple Si^+ implantation is used to define source and drain electrode regions of 150 nm shallow box profile. These regions are n-type uniformly doped regions (3e19 cm^{-3}). The source and drain regions are 20 μm apart from each other over which a gate dielectric insulator Al_2O_3 was made to grow followed by deposition of different gate metal arrangement of 2 μm in length on the top. The different gate metal arrangement used in the study includes Al(100 nm), Ti(20 nm)/Au(480 nm), Ni(60 nm)/Au(120 nm), and Ti(3 nm)/Pt-(12 nm)/Au(280 nm). All these data are taken from the experimentally demonstrated Ga_2O_3 MOS devices [10–12, 14].

In Ga_2O_3, the density of states for conduction band (N_c) is evaluated from the electron effective mass ($m_n^* = 0.28m_0$) and is taken ~3.7e18cm^{-3}. The work function for the gate contact of 5.23 eV is taken with electron affinity of 4 eV.

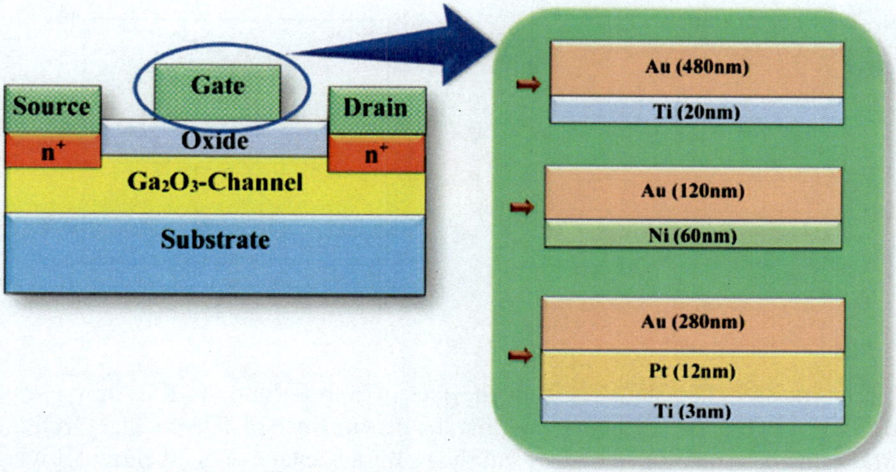

Fig. 1 The structure of the Ga_2O_3 MOS device with different multi-metal gate stacks

3 Results and Discussions

The capacitance versus voltage (C-V) characteristics of the Ga_2O_3 MOS capacitor for different gate electrode metal arrangement is shown in Fig. 2. The value of capacitances for different electrode arrangements is measured at a frequency of 1 MHz. All the relevant plots are extracted from the powerful ATLAS TCAD simulator [16]. In Fig. 2, it is observed that the difference in the value of capacitance for Al, Ti/Au, and Ti/Pt/Au is very small (see the inlet graph). However, the difference in the capacitance value of Ni/Au is large in comparison to all other arrangements. It is because the Ti metal provides good ohmic contacts over the gate oxide (Al_2O_3) in comparison to Ni metal.

Fig. 2 The C-V curve of Ga_2O_3 MOS capacitor for different gate electrode metal arrangement

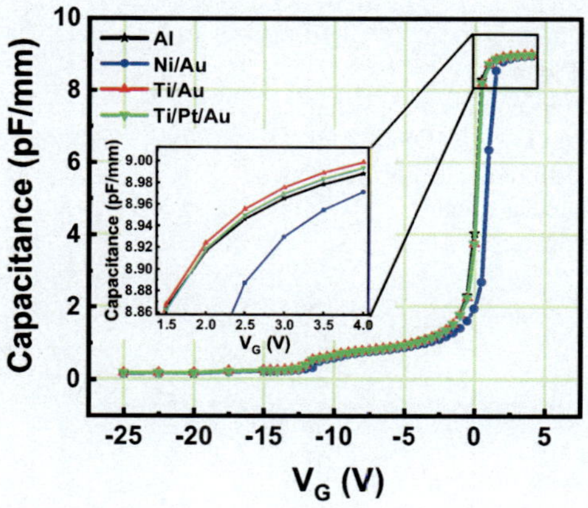

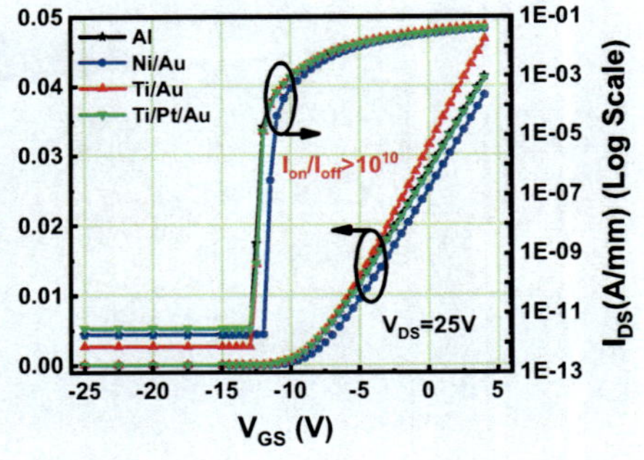

Fig. 3 The transfer characteristic of Ga_2O_3 MOSFET for different gate electrode metal arrangement

The transfer characteristics of the devices in both linear as well as in log scale are shown in Fig. 3. The plot shows that the performance of Ti/Au metal electrode arrangement is superior among all and shows the advantage of good current driving capability. The I_{ON}/I_{OFF} ratio of greater than 10^{10} with OFF-state leakage current of few pico-Amperes is obtained for the all the devices.

Figure 4 represents the plot of transconductance (g_m). The g_m which is an essential parameter to determine the gain of the amplifier is found to be superior in case Ti/Au metal electrode arrangement in comparison to other metal stack arrangements and it is due to the improved value of ON-state current. The transconductance (g_m) can be formulated by using Eq. (1). High g_m is required to obtain improved carrier transport efficiency, and to implement efficient analog/RF amplifiers.

In analog circuits, to achieve high gain, low g_d is required. Figure 5 represents the characteristic of output-conductance (g_d) versus drain voltage (V_{DS}) at V_{GS} equal to −4 V. Lower value of g_d is observed in Ni/Au metal stack arrangement in comparison to any other gate stack arrangements and provides an advantage to implement high gain amplifier. Table 1 summarizes the performance comparison of the obtained results for different metal electrode arrangements in Ga_2O_3 MOSFET. It is observed that the Ti/Au metal stack arrangement shows better results among all the metal stack arrangements. It has the highest value of ON-state current, i.e., 13.1%, 20.7%,

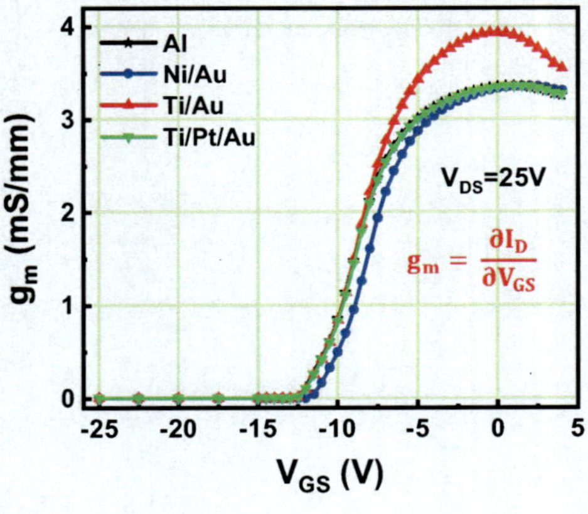

Fig. 4 The transconductance behavior of Ga_2O_3 MOSFET for different gate electrode metal arrangement

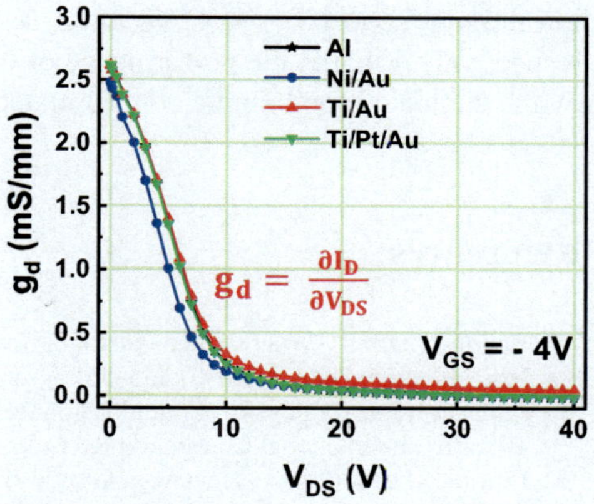

Fig. 5 The output-conductance behavior of Ga_2O_3 MOSFET for different gate electrode metal arrangement

Table 1 Performance comparison of various gate stack arrangements

Gate metal type	I_{Dmax} (mA)	I_{Dmin} (pA)	I_{ON}/I_{OFF}	g_m (mS)	g_d (mS)
Al (100 nm)	41.3	1.8	2.3×10^{10}	3.35	2.62
Ni(60 nm)/Au(120 nm)	38.7	1.8	2.1×10^{10}	3.36	2.48
Ti(20 nm)/Au(480 nm)	46.7	0.7	6.6×10^{10}	3.93	2.61
Ti(3 nm)/Pt(12 nm)/Au(280 nm)	41.2	2.8	1.4×10^{10}	3.36	2.62

and 13.3% superior to that of Al, Ni/Au, and Ti/Pt/Au, respectively. The value of OFF-state leakage current is also found to be lower and it is 96.12%, 96.12%, and 27% lower to that of Al, Ni/Au, and Ti/Pt/Au, respectively.

$$g_m = \frac{\partial I_D}{\partial V_{GS}} \tag{1}$$

$$g_d = \frac{\partial I_D}{\partial V_{DS}} \tag{2}$$

4 Conclusions

The Impact of multi-metal gate stacks on the performance of β-Ga_2O_3 MOS Structure is studied. The Ion/Ioff of greater than 10^{10} with OFF-state leakage of a few pico-Amperes is obtained for the all the devices. It is observed that the Ti/Au metal stack arrangement shows better results among all the metal stack arrangements and hence found to be suitable for high power RF applications with low losses. It has highest value of ON-state current, i.e., 13.1%, 20.7%, and 13.3% superior to that of Al, Ni/Au, and Ti/Pt/Au, respectively. The value of OFF-state leakage current is also found to be lower and it is 96.12%, 96.12%, and 27% lower to that of Al, Ni/Au, and Ti/Pt/Au, respectively. Finally, the performance of β-Ga_2O_3 devices can be further improved by the choice of good ohmic contact materials with good adhesive properties.

References

1. Baliga BJ (1989) Power semiconductor device figure of merit for high-frequency applications. IEEE Electron Device Lett 10:455–457
2. Johnson E (1965) Physical limitations on frequency and power parameters of transistors. In: 1958 IRE International Convention Record, vol 13, pp 27–34
3. Pearton SJ et al (2018) A review of Ga_2O_3 materials, processing, and devices. Appl Phys Rev 5:11301
4. Higashiwaki M, Sasaki K, Kuramata A, Masui T, Yamakoshi S (2012) Gallium oxide (Ga_2O_3) metal-semiconductor field-effect transistors on single-crystal β-Ga2O3 (010) substrates. Appl Phys Lett 100:13504
5. Irmscher K, Galazka Z, Pietsch M, Uecker R, Fornari R (2011) Electrical properties of β-Ga2O3 single crystals grown by the Czochralski method. J Appl Phys 110:63720
6. Green AJ et al (2016) 3.8-MV/cm breakdown strength of MOVPE-grown Sn-doped β-Ga_2O_3MOSFETs. IEEE Electron Device Lett 37:902–905
7. Sasaki K et al (2012) Device-quality β-Ga_2O_3 epitaxial films fabricated by ozone molecular beam epitaxy. Appl Phys Express 5:35502
8. Green AJ et al (2017) β-Ga_2O_3 MOSFETs for radio frequency operation. IEEE Electron Device Lett 38:790–793
9. Yadava N, Chauhan RK (2019) RF performance investigation of β-Ga 2 O 3 /Graphene and β-Ga 2 O 3 /black phosphorus heterostructure MOSFETs. ECS J Solid State Sci Technol
10. Higashiwaki M et al (2013) Depletion-mode Ga_2O_3 metal-oxide-semiconductor field-effect transistors on β-Ga_2O_3 (010) substrates and temperature dependence of their device characteristics. Appl Phys Lett 103:123511
11. Green AJ et al (2016) 3.8-MV/cm breakdown strength of MOVPE-grown Sn-doped β-Ga2O3 MOSFETs. IEEE Electron Device Lett. https://doi.org/10.1109/LED.2016.2568139
12. Feng Z et al (2019) A 800 V β-Ga 2 O 3 Metal–Oxide–semiconductor field-effect transistor with high-power figure of merit of Over 86.3 MW cm −2. Physica Status Solidi. https://doi.org/10.1002/pssa.201900421.
13. Sasaki K, Higashiwaki M, Kuramata A, Masui T, Yamakoshi S (2013) Si-Ion implantation doping in β-Ga2O3and its application to fabrication of low-resistance ohmic contacts. Appl Phys Express 6:86502
14. Oslo UO. FYS2210. University of Oslo, Viktor Bobal
15. Mastro MA et al (2017) Perspective—opportunities and future directions for Ga_2O_3. ECS J Solid State Sci Technol 6:P356–P359
16. ATLAS user's manual (2014) SILVACO Int, Santa Clara, CA

Improved VLSI Architecture of Dual-CLCG for Pseudo-Random Bit Generator

Mangal Deep Gupta and R. K. Chauhan

Abstract In this paper, improved VLSI architecture of dual coupled linear congruential generator (CLCG) for pseudorandom bit generation is proposed to enhance the timing performance with minimum area overhead. To improve its performance, a splitting structure of the adder circuit rather than a single adder is used in linear congruential generator (LCG). Its effect on the design of a dual-CLCG is observed to compute some of the essential performance parameters such as maximum frequency of bit generation, minimum period of clock signal, initial clock latency, output to output latency and area complexity. For Spartan-3E XC500E FPGA based synthesis of proposed architecture with 8, 16 and 32-bit word length, the ISE design suite provided by Xilinx is used.

Keywords Pseudo-random bit generator · Dual-CLCG architecture · Delay · Operating frequency · FPGA · Verilog HDL

1 Introduction

Randomness is a common word which is used broadly in cryptography application to privacy of information associated with various encryption and de-encryption applications in the area of defense, finance, health-care, biometric information, transaction and telecardiology to information are processed, stored, and further protected from interruption [1, 2]. these random numbers can be used to producing large prime numbers for cipher key construction [3] (It is used in, AES algorithm, signatures [4], RSA cryptography [5] and memory encryption [6]). Some other applications of random bit sequence in Monto-Carlo simulations [7], numerical analysis [8], computer programming, decision making, sampling, security services (i.e. confidentiality, authentication, non-repudiation, and data integrity), etc.

M. D. Gupta (✉) · R. K. Chauhan
Department of Electronics and Communication Engineering MMMUT, Gorakhpur, India
e-mail: mangaldeepgcet@gmail.com

R. K. Chauhan
e-mail: rkchauhan27@gmail.com

Z. Patel et al. (eds.), *Advances in VLSI and Embedded Systems*, Lecture Notes in Electrical Engineering 676, https://doi.org/10.1007/978-981-15-6229-7_14

Linear congruential generator (LCG) and linear feedback shifts register (LFSR) are the most common pseudorandom bit generators with low complexity. Problems associated with these random bit sequences, it is insecure due to its linear structure and fail in NIST or randomness test [8, 9]. Blum Blum Shub (BBS) generator is one of the secure and unpredictable key generators is present in [10, 11]. It is highly secure due to a large prime number factorization problem but its design of VLSI architecture is challenging for handling and computing the large prime integer. So, designing an efficient architecture for pseudorandom bit generation is a major challenge to generate random bit in uniform clock time and minimum output to output latency with minimum area overhead. Two optimized architecture of LCGs i.e. the Ranq1 and Ran are proposed in [12]. It is designed by XOR shifting and multiplication with carry generator. Most of them are required high clock latency, the non-uniform clock rate of bit generator and large hardware complexity [13, 14]. To overcome these drawbacks, the coupling of LCG i.e. CLCG was proposed in [15] and [16]. In [17], a coupling of two CLCG architecture is proposed which is denoted by CLCG-1 and CLCG-2. Each architecture generates separates output with different seed values. The coupling in this architecture makes a more secure than single architecture. Although improvement in the secure generation of pseudorandom bit sequence, the CLCG algorithm fails in five major NIST statistical tests [18]. It is failing in desecrate Fourier transform (DFT) test i.e. used to investigate the periodic bit pattern and shows the weak generator. It is generating a one-bit random output only when it meets the condition of inequality equations. So. It cannot be generating pseudorandom bit at every iteration. In [19], proposed a new hardware architecture of the existing dual-CLCG algorithm is developed that generates the pseudorandom bit at a uniform clock rate.

In this paper, we design an improved VLSI architecture of dual-CLCG for pseudorandom bit generator to enhance the timing performance with minimum area overhead. To improve its performance, we design a splitting structure of the adder circuit rather than a single adder i.e. used in the linear congruential generator (LCG). Thereafter, it's FPGA synthesis has been done using Spartan-3E XC500E FPGA. The remaining brief of work is organized as follows: Section-2 presents the proposed architecture of dual-CLCG for random bit generation. Section-3 presents the performance comparison between proposed architecture and previous reported work [19] in terms of maximum frequency of bit generation, minimum period of clock signal, initial clock latency, output to output latency and area complexity (in terms of total flip-flops, number of slices and LUTs). Finally, Section-4 concludes the work.

2 Proposed Architecture of Dual-CLCG

The dual-CLCG method techniques is proposed by Katti et al. [18]. It is dual coupling of four linear congruential generator (LCG) and is define by following recurrence equations:

$$x_{i+1} = [(2^{r1} \times x_i) + x_i + b_1] \bmod 2^n \tag{1}$$

$$y_{i+1} = [(2^{r2} \times y_i) + y_i + b_2] \bmod 2^n \tag{2}$$

$$p_{i+1} = [(2^{r3} \times p_i) + p_i + b_3] \bmod 2^n \tag{3}$$

$$q_{i+1} = [(2^{r4} \times q_i) + q_i + b_4] \bmod 2^n \tag{4}$$

$$z_i = \left\{ \begin{array}{l} 1\ if\ x_{i+1} > y_{i+1}\ and\ p_{i+1} > q_{i+1} \\ 0\ if\ x_{i+1} < y_{i+1}\ and\ p_{i+1} > q_{i+1} \end{array} \right\} \tag{5}$$

Here b_1, b_2, b_3, b_4 are the constant parameter and x_0, y_0, p_0 and q_0 are the initial seeds for above recurrence equation for pseudorandom bit generation. Here r is the positive integer i.e.$1 < 2^r < 2^n$. Final output variable z_i is the random output sequence. It is evaluated based on Eq. (5) in every iteration.

Algorithm 1 To generates pseudorandom bit sequence Z_i using dual-CLCG algorithm
Input: positive integer n and $m = 2^n$

Initialization:
Prime number: $b_1, b_2, b_3, b_4 < m$
Initial value: x_0, y_0, p_0 and $q_0 < m$
Output: Z_i

1. For $i = 0\ to\ k$
2. Evaluate the value of $x_{i+1}, y_{i+1}, p_{i+1}\ and\ q_{i+1}$ using equation (1), (2), (3) and (4).
3. $z_i = \left\{ \begin{array}{l} 1\ if\ x_{i+1} > y_{i+1}\ and\ p_{i+1} > q_{i+1} \\ 0\ if\ x_{i+1} < y_{i+1}\ and\ p_{i+1} < q_{i+1} \end{array} \right\}$
4. Return Z_i;

Algorithm 1 presents the steps or procedure to generates pseudorandom bit sequence using above recurrence Eqs. (1)–(5). It is based on iteration for $i = 0tok$. At $i = 0$, initialize the value of x, y, p and z by x_0, y_0, p_0 and q_0. Evaluation of the value of x_{i+1}, y_{i+1}, p_{i+1} and q_{i+1} in each iteration is done according to Eqs. (1)–(5) and find the randomized bit sequence Z_i.

The design of efficient hardware architecture of dual-CLCG in terms of area, latency, frequency of bit generator and power consumption. In this section we proposed an efficient VLSI architecture of dual-CLCG for random bit generator based on mathematical recurrence relation. In this architecture we focus to minimize the complexity of adder circuit, which is used in each linear congruential generator (LCG) to compute the mathematical equation. In each LCG at every iterations, value of x_i, y_i, p_i and q_i are left shifted by r_1, r_2, r_3 and r_4 bit respectively. These shifted value is added with (x_i, y_i, p_i and q_i) and (b_1, b_2, b_3 and b_4) correspondingly each LCG architecture according to mathematical equation. In this architecture of dual-CLCG, we used the improved structure of adder structure to enhance the performance of overall architecture. We split the adder architecture into two n-bit adder circuit to improve the timing performance and area complexity. It is done according to Eqs. (6)–(9).

$$\begin{aligned} s_{x1}[n-1:0] &= b_1[n-1:0] + x[n-1:0], \\ s_{x2}[n-r_1-1:0] &= s_{x1}[n-1:r_1] + x[n-r_1-1:0] \end{aligned} \tag{6}$$

$$\begin{aligned} s_{y1}[n-1:0] &= b_2[n-1:0] + y[n-1:0], \\ s_{y2}[n-r_2-1:0] &= s_{y1}[n-1:r_2] + y[n-r_2-1:0] \end{aligned} \tag{7}$$

$$\begin{aligned} s_{p1}[n-1:0] &= b_3[n-1:0] + p[n-1:0], \\ s_{p2}[n-r_3-1:0] &= s_{p1}[n-1:r_3] + p[n-r_3-1:0] \end{aligned} \tag{8}$$

$$\begin{aligned} s_{q1}[n-1:0] &= b_4[n-1:0] + q[n-1:0], \\ s_{q2}[n-r_4-1:0] &= s_{q1}[n-1:r_4] + q[n-r_4-1:0] \end{aligned} \tag{9}$$

Figure 1 shows the proposed architecture of dual-CLCG with modified arrangements of adder circuits. Arrangements of adder structure according to Eqs. (6)–(9) is clearly shown in this figure. A 2:1 multiplex is used to initial value and iteration value x, y, p and q. It is control by a signal $START$. This architecture is required two n-bit comparators. It is used to compare the value of (x_{i+1}, y_{i+1}) and (p_{i+1}, q_{i+1}) according to Eq. (5). XOR value of these comparators become a pseudorandom random sequence (z_i).

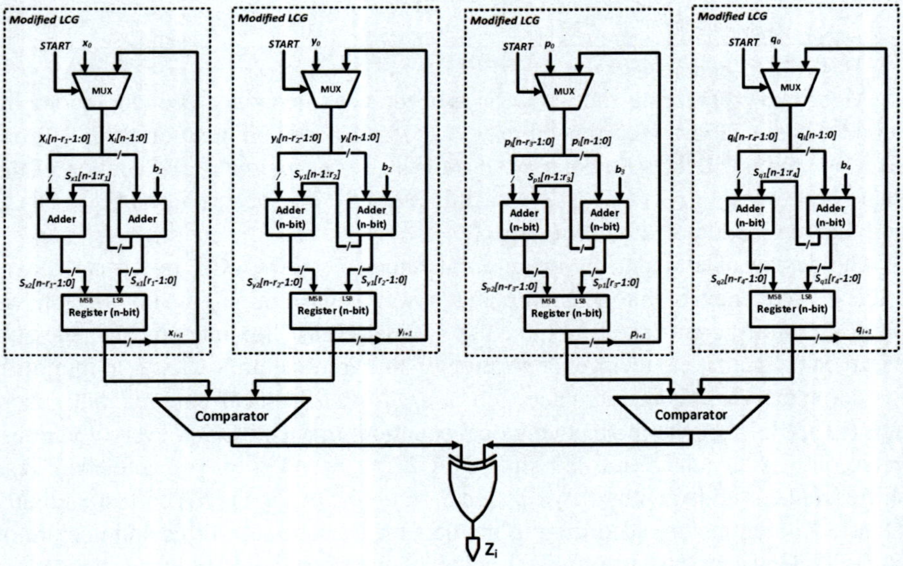

Fig. 1 Proposed architecture of improved dual –CLCG architecture for pseudo random bit generator

The complexity of proposed dual-CLCG architecture required maximum critical path delay which is contributed by two- adder circuits, multiplexer, one comparator and one XOR gates while in previous literature [19], critical path delay is contributed by three-operands modulo adder, multiplexer, shifting operation, one comparator and one XOR gates. The proposed architecture of dual-CLCG utilize the area of four n-bit adder circuits, four (n-r)-bit adder circuits, four 2 × 1 mux, one 1-bit XOR gates, four n-bit register and two n-bit binary comparator. Here A_{adr1} and T_{adr1} represent the area and critical path delay of n-bit adder circuit with two operands. While A_{adr2} and T_{adr2} represent the area and critical path delay of (n–r)-bit adder circuit with two operands. A_{mux} and T_{mux} represent the area and critical path delay of 2 × 1 mux. A_x and T_x represent the area and critical path delay of 1-bit XOR circuit. A_{cmp} and A_{reg} represent the area of n-bit comparator and register.

Area, $A_{\text{proposed dual-CLCG}} = 4(A_{adr1} + A_{adr2} + A_{mux} + A_{reg}) + 2A_{cmp} + A_x$
Critical path delay, $T_{\text{proposed dual-CLCG}} = T_{adr1} + T_{adr2} + T_{mux} + T_{cmp} + T_x$.

3 Result and Discussion

The improve structure of dual-CLCG method for different bit size of 8, 16 and 32-bit are design by Verilog HDL code. Its simulation and synthesis has been done using ISE design suite by Xilinx. Synthesis of this architecture has been done using commercially available Spartan-3E XC500E FPGA. Performance parameter of improved architecture of dual-CLCG for pseudorandom bit generator such as maximum frequency of bit generation, minimum time period of clock signal, initial clock latency, output to output latency and area complexity (in terms of total flip-flops, number of slices and LUTs) are evaluated and compare with previously reported dual-CLCG [19]. It is shown in Table 1. Figure 2 shows the simulation waveform using Verilog test bench for 8-bit word length of proposed architecture. For this simulation we take the prime number ($b_1 = 1, b_2 = 3, b_3 = 141$ and $b_4 = 79$) and initialization of x_0, y_0, p_0, q_0 equal to 1, 2, 14, 3 respectively. left shifting of

Table 1 Synthesis report of proposed architecture of dual-CLCG based on Spartan-3E FPGA

Methods	Size (bit)	Total FFs	Number of slices	Number of LUTs	Maximum frequency (MHz)	Time (T_{CLK}) (ns)	Initial clock latency (T_{CLK})	O/P to O/P latency (T_{CLK})
Dual- CLCG [19]	8	32	71	135	194.74	5.135	1	1
	16	64	167	320	130.78	7.646	1	1
	32	128	385	737	106.09	9.426	1	1
Proposed Dual-CLCG	8	37	61	113	236.577	4.227	1	1
	16	69	117	225	215.566	4.639	1	1
	32	133	236	453	183.051	5.463	1	1

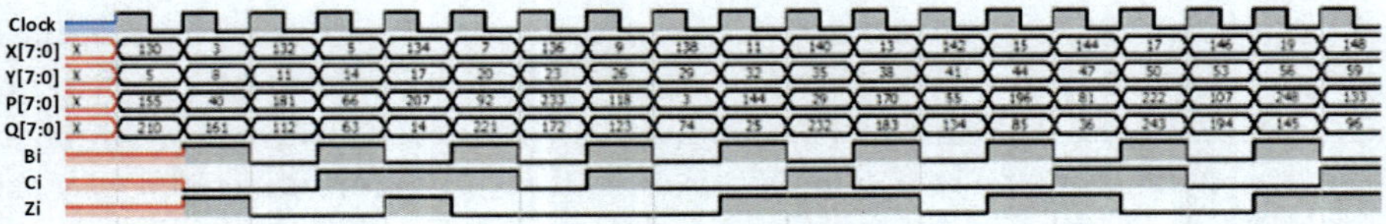

Fig. 2 Simulation waveform of proposed 8-bit dual-CLCG

x_i, y_i, p_i and q_i is done by 7-bit for this simulation. We get the pseudorandom sequence(Z_i = 100100011101100 ....). From Table 1, it is show that improvement in maximum frequency of bit generation using proposed methodology is 21.4%, 64.8%, 72.5% for 8-bit, 16-bit and 32-bit word length respectively, when compare to conventional architecture. Optimum time period of clock signal to meet the all designing constraints is also important for fast generation of pseudorandom bit sequence. We get the improvement in time period of clock signal by 17.6%, 39.3% and 42.0% for 8-bit, 16-bit and 32-bit word length respectively as compare with dual-CLCG [19]. Initial clock latency and output to output latency are one clock cycle as same as previous reported work [19]. We find the area complexity (in terms of total flip-flops, number of slices and LUTs) of proposed architecture is better than previous work as shown in Table 1 and Fig. 3.

4 Conclusion

In this work, an improved VLSI architecture of dual-CLCG for pseudorandom bit generation is studied to enhance the timing performance with minimum area overhead. Modification in the adder structure has been done to improve the speed of pseudorandom bit generator of linear congruential generator (LCG) in the dual-CLCG design with minimum hardware overhead. It uses a split adder block to suppress the unnecessary addition. The results of the proposed architecture are compared with previous dual-CLCG architecture [19]. It is clearly demonstrated that the generation of pseudorandom bit sequences using proposed architecture is faster by 42.0% than architecture available in literature [19]. This work demonstrates that the proposed architecture of dual-CLCG can significantly work at the lower clock time period and it can successfully generate bit sequence at 72.5% higher frequency compared to the earlier reported designs while maintaining the same latency.

Fig. 3 Graphical representation of performance parameter for proposed dual-CLCG architecture and comparison with previous literature [19]. **a** Area complexity. **b** Maximum frequency of bit generation. **c** Minimum possible time-period of clock signal

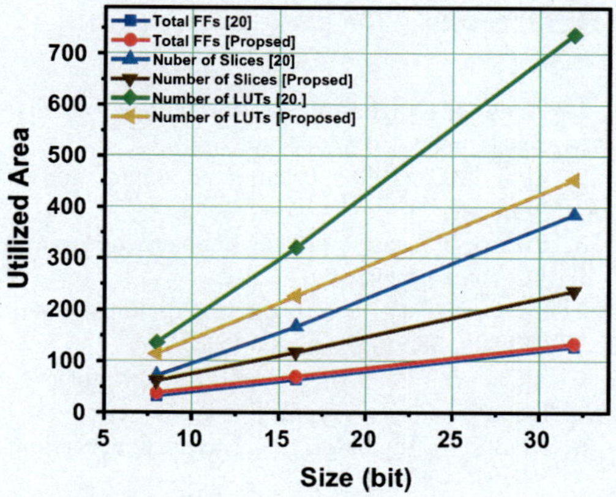

(a)

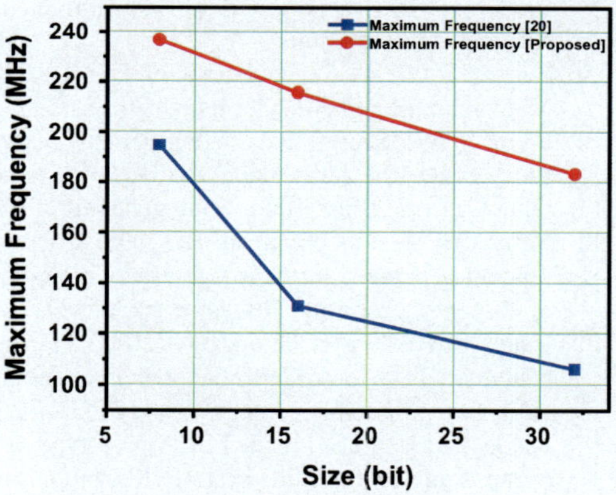

(b)

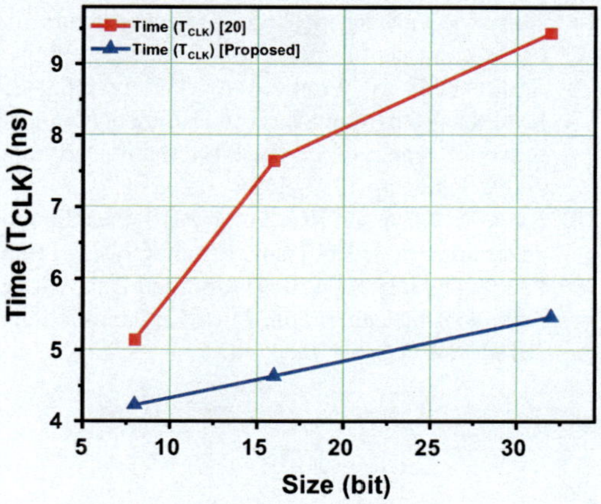

(c)

References

1. Wang X, Canghai Y (2009) Cryptanalysis and Improvement on a cryptosystem based on a chaotic map. Comput Math Appl
2. Luby MG (1996) Pseudorandomness and cryptographic applications. University Press, Princeton
3. Even S, Mansour Y (1997) A construction of a cipher from a single pseudorandom permutation. J Cryptol 10(3):151–161
4. Lamport L (1979) Constructing digital signatures from a one-way function. Technical Report CSL-98, SRI International Palo Alto
5. ElGamal T (1985) A public key cryptosystem and a signature scheme based on discrete logarithms. Adv Cryptol Spring, 10–18
6. Henson M, Taylor S (2014) Memory encryption: a survey of existing techniques. ACM Comput Surv 46(4):53
7. Gentle JE (2003) Random number generation and monte carlo methods. Science & Business Media, Springer
8. Froberg C-E, Frhoberg CE (1969) Introduction to numerical analysis. Addison-Wesley Reading, Massachusetts
9. Zenner E (2004) Cryptanalysis of LFSR-based pseudorandom generators a survey. Univ Mannheim, Mannheim, Germany
10. SternJ(1987) Secret linear congruential generators are not cryptographically secure. In: Proceedingsofhe 28th Annual Symposium Foundations of Computer Science, pp 421–426
11. Ding C (1997) Blum-Blum-Shub generator. IEEE Electron Lett 33(8):667
12. Sidorenko A, Schoenmakers B (2005) Concrete security of the BlumBlum-Shub pseudo-random generator. In: Cryptography and Coding (Lecture Notes in Computer Science), vol 3796. Springer, Berlin, Germany, pp 355–375
13. Banks S,Beadling P, Ferencz A (2008) FPGA implementation of Pseudo random number generators for Monte Carlo methods in quantitative finance. In: International Conference on, IEEE Reconfigurable computing and FPGAs, 2008. ReConFig'08, pp 271–276
14. Panda AK, Ray CK (2015) FPGA prototype of low latency BBS PRNG. In: Proc. IEEE Int. Symp. Nanoelectron. Inf. Syst. (INIS), Indore, India, pp 118–123
15. Lopez pp, Millan ES (2010) Cryptographically secure pseudorandom bit generator for RFID tags. In: Proc. Int. Conf. Internet Technol. Secured Trans, vol. 11, London, UK, pp 1–6
16. Katti RS, Kavasseri RG (2008) Secure pseudo-random bit sequence generation using coupled linear congruential generators. In: Proceedings IEEE International Symposium of Circuits System (ISCAS), Seattle, WA, USA, pp 2929–2932
17. Katti RS, Srinivasan SK (2009) Efficient hardware implementation of a new pseudo-random bit sequence generator. In: IEEE International Symposium on Circuits and Systems, 2009 (ISCAS 2009), IEEE, pp 1393–1396
18. Katti RS, Kavasseri RG, Sai V (2010) Pseudorandom bit generation using coupled congruential generators. In: IEEE Trans. Circuits Syst. II, Exp. Briefs, vol 57, no 3, pp 203–207
19. Panda AK, Ray KC (2019) Modified Dual-CLCG method and Its VLSI architecture for pseudorandom bit generation. In: IEEE Transactions on Circuits and Systems I: Regular Papers, 2019

Low Power Radix-8 Modulo $2^n + 1$ Multiplier Using Modified Weighted Method

Naveen Kumar Kabra and Zuber M. Patel

Abstract To design a residue number system using moduli set $2^n - 1, 2^n, 2^n + 1$. The modulo $2^n + 1$ is the biggest hindrance as it requires $(n + 1)$ number of bits for their representation, while the other two sets use only n bits. In this paper, we present a new algorithm to design radix-8 modulo $2^n + 1$ multiplier using a modified weighted method with light bias. The multiplier uses only $\lfloor n/3 \rfloor + 2$ modulo reduced partial product along with bias term. The new multiplier is also capable to handle zero input. The proposed design is compared with recently proposed technique, and the synthesis result shows 27.14–38.56% less power consumption with almost same area.

Keywords Low power · Modulo multiplier · Parallel algorithm · RNS processor · Modulo arithmetic

1 Introduction

Residue Number System (RNS) is an unconventional and non-weighted number system. It is a carry free number system which has numerous theoretic properties and unique features that can be used to boost up the speed of domain- specific computing [1]. The two unique properties of RNS makes it more attractive and viable solution to implement modern digital Very Large Scale Integration (VLSI).

1. The modular and distributive properties of RNS allows to achieve high performance in the emerging distributed and ubiquitous computing-based application such as cloud, wireless ad hoc networks and application require tolerance against soft error [2].

N. K. Kabra (✉) · Z. M. Patel
Electronics Engineering Department, Sardar Vallabhbhai National Institute of Technology, Surat, India
e-mail: nkkabrasvnit@gmail.com

Z. Patel et al. (eds.), *Advances in VLSI and Embedded Systems*, Lecture Notes in Electrical Engineering 676, https://doi.org/10.1007/978-981-15-6229-7_15

2. Parallelism property of RNS allows to optimise energy performance, especially in those applications that involve repetitive computations and very long word length arithmetic like repeated modulo multiplications in cryptographic algorithm and multiply-and-accumulate operations in signal processing algorithm [3], digital image processing [4], Rivest-Shamir-Adleman (RSA) algorithms [5].

A RNS is defined as a set of relatively prime moduli. A relatively prime moduli is a set of integers such that the Greatest Common Divisor (GCD) of any two numbers in the moduli set is 1. If P denotes the module set, then

$$\begin{gathered} P = \{P_1, P_2, P_3, \ldots, P_n\} \\ GCD(P_i, P_j) = 1 \;\forall\; i \text{ and } j \text{ expect } i \neq j \end{gathered} \tag{1}$$

Any integer X in the residue class Z_M; $X \in Z_M$, (modulo M ring), where $M = \{P_1, P_2, P_3, \ldots, P_n\}$, has a unique n-tuple representation given by

$$X \xrightarrow{RNS} (X_1, X_2, X_3 \ldots, X_n) \tag{2}$$

where $X_i = X \bmod P_i$ is called the ith residue of X.

Similar to unsigned number, signed number system also has a RNS n-tuple representation of any integer in $(-\frac{M}{2} to \frac{M}{2} - 1)$ for M is even and $(-\frac{M-1}{2}$ to $\frac{M-1}{2})$ for M is odd; where X_i is defined as

$$X_i = \begin{cases} X \bmod P_i, & \text{if } X > 0 \\ (M - |X|) \bmod P_i, & \text{otherwise.} \end{cases}$$

To design a RNS system based on balance word length most often used moduli set $2^n - 1, 2^n, 2^n + 1$. Out of these, modulo $2^n + 1$ is the biggest hindrance as it requires $(n + 1)$ number of bits for their representation while other two sets uses only n bits. Thus, the design of modulo $2^n + 1$ multiplier requires more logical efforts comparison to other modulo multiplier because it uses $(n + 1)$ bit operand. To address this, a variety of techniques are proposed in different literatures [6–16]. Depending on the representation of Multiplier (B) and Multiplicand (A), most of the techniques come under the following three category

1. Both in weighted representation [6–9]
2. Both in diminished-1 representation [10–15]
3. One in diminished-1 and another in weighted representation [16].

Multiplier proposed in [6], was based on category 1 form except 2^n, which is represented as an all zeros vector. Modulo $2^n + 1$ multiplier based on a binary multiplier was described in [7]. Modulo $2^n + 1$ multiplier using weighted-binary representation suggested by [8], exploits the redundancy in the representation. The partial product matrix was divided into four groups and simplified considering the fact that the bits from only one group can be different from 0s. Thus, the partial product matrix was reduced to an $n \times n$ bit array. In [9], the correction terms of [8], were further simplified and represented as a single additional partial product. However, in comparison to [10–13, 15], the designed proposed for diminished-1 in [17], is the most prevalent for modulo $2^n + 1$ multiplier designs. Based on encoding scheme used, the diminished-1 multiplier can be categorised into: (a) Non-encoded multipliers [10, 11] (b) Radix-4 Booth encoded multipliers [12–14]. In [12], the number of Modulo Reduced Partial Products (MPPA's) were reduced to $\lceil n/2 \rceil + 1$ and length of each partial product was $n + 1$ bits. However, the summation of partial products was performed in an integer Carry Save Adder (CSA) array that outputs S and C vector of the length $n + \log_2 n$ bits that needed to be reduced further. The modulo reduction entailed two additional Carry Save Adder's (CSAs) in series with the CSA array. This technique for Modulo Reduced Partial Product Accumulator (MPPA) has been superseded by the use of the regular Complimentary End Around Carry-Carry Save Adder (CEAC-CSA) tree in [13, 14]. In [13], a unified architecture for modulo $2^n + 1$ multiplier using diminished-1 and weighted-binary number representations was proposed.

Radix-8 Booth encoded modulo $2^n + 1$ multipliers with area -power efficiency were proposed in [18, 19]. In [18], the weighted method was used for the representation of both operands. The design has problem related to bias constant. The aggregate bias due to hard multiple and partial product is composed of multiplier-dependent dynamic bias and multiplier-independent static bias. The bias was represented by hardwired the output of booth encoder and constant 1s at appropriate bit position with at most two level of AND gates. The total number of modulo reduced partial product, including bias imposed, was equal to $\lfloor \frac{n}{3} \rfloor + 6$. The proposed architecture was implemented using $\lfloor \frac{n}{3} \rfloor + 1$ Booth Encoder (BE), $n(\lfloor \frac{n}{3} \rfloor + 1)$ Booth Selector (BS), $n(\lfloor \frac{n}{3} \rfloor + 4)$ Full Adder (FA) and a parallel prefix $2^8 + 1$ adder. While in [19], the diminished-1 method was used to represent both operands. The method to generate partial product similar properties was used as in [18]. Thus, the design of the booth selector was similar to [18]. A Modified Booth Encoder (MBE) was proposed to perform booth encoding. But the algorithm is not capable to handle zero input. The architecture was implemented using $\lfloor \frac{n}{3} \rfloor + 1$ MBE, $n(\lfloor \frac{n}{3} \rfloor + 1)$ BS, $[(n + 1)(\lfloor \frac{n}{3} \rfloor) - (n - 1)]$ FA, $n - \frac{n}{4}$ Half Adder (HA), three 2-input logic gates and a diminished-1 $2^8 + 1$ adder. The main contribution of the proposed algorithm was as follows:

1. The number of $(2^n + 1)$ modulo reduced partial product including bias was reduced to $\lfloor \frac{n}{3} \rfloor + 2$. In addition to that it has a '1' constant that was handled with help of 2-input three gates.
2. A new Hard Multiple Generator (HMG) was proposed using a modified structure of parallel-prefix operator. But the proposed design has a limitation related to the generation of carry signal. The odd carry signals were used to generate the sum bits at both even and odd positions.

The rest of the paper is organised as follows: Sect. 2 gives the mathematical expression of the proposed algorithm of modulo $2^n + 1$. The proposed architecture of each section such as Modified Booth Encoder* (MBE*), Modified Booth Selector* (MBS*), partial product accumulation stage and finally modulo $2^n + 1$ adder is presented in Sect. 3. The implementation analysis and result verifying the effectiveness of proposed architecture is presented in Sect. 4. Finally, the paper is concluded in Sect. 5.

2 Modified Weighted Modulo $2^n + 1$ Algorithm

In this section, we describe a new algorithm to design a radix-8 modulo $2^n + 1$ multiplier using Modified Weighted Method (MWM) with light bias approach. The major contribution in algorithm is as follows:

1. Reduction in number of modulo reduced partial product.
2. New approach to generate the bias term without effecting number of modulo reduced partial product terms.
3. Architecture is capable of handling zero input without much effecting area requirement.

The proposed algorithm uses only $\lfloor n/3 \rfloor + 2$ number of modulo reduced partial product. To implement the architecture $\lfloor \frac{n}{3} \rfloor + 1$ MBE*, $n(\lfloor \frac{n}{3} \rfloor + 1)$ MBS*, $[(n + 1)(\lfloor \frac{n}{3} \rfloor) + 2]$ FA, $n - (\lfloor \frac{n}{3} \rfloor + 1)$ HA and diminished-1 $2^8 + 1$ adder is used.

Assume that 'A' and 'B' are multiplicand and multiplier of $(n + 1)$ bit wide (Aϵ[0, 2^n] and Bϵ[0, 2^n]) and their product 'PD' is also of $(n + 1)$ bit wide. It can be given as

$$\begin{aligned} PD &= AB + B - B \\ &= (A + 1)B - B \end{aligned}$$

After taking modulo $2^n + 1$ of above equation, it can be describe as:

$$\left| PD \right|_{2^n+1} = \left| |(A + 1)B|_{2^n+1} - B \right|_{2^n+1} \tag{3}$$

Here, we identified two different cases according to number of bits of A and B

1. For n is odd
2. For n is even

2.1 When n Is Odd

Let multiplier $B = 7$ bits ($B = b_6b_5b_4b_3b_2b_1b_0$) then in weighted representation B can be written as

$$\begin{aligned} B &= 2^6b_6 + 2^5b_5 + 2^4b_4 + 2^3b_3 + 2^2b_2 + 2^1b_1 + 2^0b_0 \\ &= -2^8b_8 + 2^8b_8 - 2^7b_7 + 2^7b_7 + 2^6b_6 + 2^5b_5 + 2^4b_4 + 2^3b_3 + 2^2b_2 + 2^1b_1 + 2^0b_0 \\ &= 2^8b_8 - 2^7b_7 + 2^6(-4b_8 + 2b_7 + b_6 + b_5) + 2^3(-4b_5 + 2b_4 + b_3 + b_2) \\ &\quad + 2^0(-4b_2 + 2b_1 + b_0) \end{aligned} \tag{4}$$

According to lemma of modulo $2^n + 1$ [20]

$$|2^n|_{2^n+1} = |-1|_{2^n+1}$$

$$|2^{nk+i}|_{2^n+1} = \begin{cases} +2^i, \text{ if } k \text{ is even;} \\ -2^i, \text{ if } k \text{ is ood.} \end{cases}$$

After comparison of above equation with standard lemma of modulo $2^n + 1$ along with number of bits n the Eq. 4 can be written as

$$|B|_{2^n+1} = \left|(b_n - 2b_{n+1} + b_0 + 2b_1 - 4b_2) + \sum_{i=1}^{2}(b_{3i-1} + b_{3i} + 2b_{3i+1} - 4b_{3i+2})2^{3i}\right|_{2^n+1}$$

$$|B|_{2^n+1} = \left|[(b_0 + b_n) + 2(b_1 + b_{n+1}) - 4(b_2 + b_{n+1})] + \sum_{i=1}^{\lfloor \frac{n}{3} \rfloor}(b_{3i-1} + b_{3i} + 2b_{3i+1} - 4b_{3i+2})\right|_{2^n+1} \tag{5}$$

Now after applying boolean identities of OR gate with respect to binary addition, Eq. 5 can be written as

$$\begin{aligned} |B|_{2^n+1} &= \left|[(b_0 \vee b_n) + 2(b_1 \vee b_n) - 4(b_2 \vee b_n)] + \sum_{i=1}^{\lfloor \frac{n}{3} \rfloor}(b_{3i-1} + b_{3i} + 2b_{3i+1} - 4b_{3i+2})2^{3i}\right|_{2^n+1} \\ &= \left|[b^*_{-1} + b^*_0 + 2b^*_1 - 4b^*_2] + \sum_{i=1}^{\lfloor \frac{n}{3} \rfloor}(b_{3i-1} + b_{3i} + 2b_{3i+1} - 4b_{3i+2})2^{3i}\right|_{2^n+1} \end{aligned}$$

Thus, the final equation in case of odd number of bits become

$$\left|B\right|_{2^n+1} = \left|\sum_{i=0}^{\lfloor\frac{n}{3}\rfloor}(b^*_{3i-1} + b^*_{3i} + 2b^*_{3i+1} - 4b^*_{3i+2})2^{3i}\right|_{2^n+1} \tag{6}$$

$$\left|B\right|_{2^n+1} = \left|\sum_{i=0}^{\lfloor\frac{n}{3}\rfloor}(b_i^{*_{MB}})2^{3i}\right|_{2^n+1} \tag{7}$$

where for $i = 0$

$$b_0^{*_{MB}} = b^*_{-1} + b^*_0 + 2b^*_1 - 4b^*_2;$$

for $1 \leq i \leq \lfloor\frac{n}{3}\rfloor$

$$b_i^{*_{MB}} = b_{3i-1} + b_{3i} + 2b_{3i+1} - 4b_{3i+2};$$

along with $b^*_{-1} = 0$, $b^*_0 = b_0 \vee b_n$, $b^*_1 = b_1 \vee b_n$ and $b^*_2 = b_2 \vee b_n$.

2.2 When n Is Even

Similarly, assume multiplier $B = 8$ bits ($B = b_7b_6b_5b_4b_3b_2b_1b_0$) then in weighted representation B can be written as

$$\begin{aligned} B &= 2^7b_7 + 2^6b_6 + 2^5b_5 + 2^4b_4 + 2^3b_3 + 2^2b_2 + 2^1b_1 + 2^0b_0 \\ &= -2^8b_8 + 2^8b_8 + 2^7b_7 + 2^6b_6 + 2^5b_5 + 2^4b_4 + 2^3b_3 + 2^2b_2 + 2^1b_1 + 2^0b_0 \\ &= 2^8b_8 + 2^6(-4b_8 + 2b_7 + b_6 + b_5) + 2^3(-4b_5 + 2b_4 + b_3 + b_2) \\ &\quad + 2^0(-4b_2 + 2b_1 + b_0) \end{aligned} \tag{8}$$

After comparison of the above equation with standard lemma of modulo $2^n + 1$ as described in case-1 along with number of bits n, the Eq. 8 can be written as

$$\begin{aligned} \left|B\right|_{2^n+1} &= \left|(-b_n + b_0 + 2b_1 - 4b_2) + \sum_{i=1}^{2}(b_{3i-1} + b_{3i} + 2b_{3i+1} - 4b_{3i+2})2^{3i}\right|_{2^n+1} \\ &= \left|(b_n - 2b_n + b_0 + 2b_1 - 4b_2) + \sum_{i=1}^{2}(b_{3i-1} + b_{3i} + 2b_{3i+1} - 4b_{3i+2})2^{3i}\right|_{2^n+1} \\ &= \Bigg|[(b_0 + b_n) + 2(b_1 - 2b_n) - 4b_2] \\ &\quad + \sum_{i=1}^{2}(b_{3i-1} + b_{3i} + 2b_{3i+1} - 4b_{3i+2})2^{3i}\Bigg|_{2^n+1} \end{aligned}$$

Now using identity of boolean algebra of binary subtraction $x - y = x \oplus y - 2\overline{x}y$ and $b_0 + b_n = b_0 \vee b_n$ for $i \doteq n$ above equation can be rewrite as

$$\begin{aligned}|B|_{2^n+1} &= \Big|[(b_0 \vee b_n) + 2(-2\overline{b_1} \cdot b_n + (b_1 \oplus b_n) - 4b_2] \\ &\quad + \sum_{i=1}^{2}(b_{3i-1} + b_{3i} + 2b_{3i+1} - 4b_{3i+2})2^{3i}\Big|_{2^n+1} \\ &= \Big|[(b_0 \vee b_n) - 4\overline{b_1} \cdot b_n + 2(b_1 \oplus b_n) - 4b_2] \\ &\quad + \sum_{i=1}^{2}(b_{3i-1} + b_{3i} + 2b_{3i+1} - 4b_{3i+2})2^{3i}\Big|_{2^n+1} \\ &= \Big|[(b_0 \vee b_n) - 4(\overline{b_1} \cdot b_n + b_2) + 2(b_1 \oplus b_n)] \\ &\quad + \sum_{i=1}^{2}(b_{3i-1} + b_{3i} + 2b_{3i+1} - 4b_{3i+2})2^{3i}\Big|_{2^n+1}\end{aligned}$$

similarly, replacing arithmetic operator by OR operator in the above equation

$$\begin{aligned}|B|_{2^n+1} &= \Big|[(b_0 \vee b_n) - 4(\overline{b_1} \cdot b_n \vee b_2) + 2(b_1 \oplus b_n)] \\ &\quad + \sum_{i=1}^{2}(b_{3i-1} + b_{3i} + 2b_{3i+1} - 4b_{3i+2})2^{3i}\Big|_{2^n+1}\end{aligned}$$

$$|B|_{2^n+1} = \Big|(b^*_{-1} + b^*_0 + 2b^*_1 - 4b^*_2)2^0 + \sum_{i=1}^{2}(b_{3i-1} + b_{3i} + 2b_{3i+1} - 4b_{3i+2})2^{3i}\Big|_{2^n+1}$$

Thus, the final equation in case of even number of bits become

$$|B|_{2^n+1} = \Big|\sum_{i=0}^{\lfloor \frac{n}{3} \rfloor}(b^*_{3i-1} + b^*_{3i} + 2b^*_{3i+1} - 4b^*_{3i+2})2^{3i}\Big|_{2^n+1} \tag{9}$$

$$|B|_{2^n+1} = \Big|\sum_{i=0}^{\lfloor \frac{n}{3} \rfloor}(b_i^{*_{MB}})2^{3i}\Big|_{2^n+1} \tag{10}$$

where for $i = 0$

$$b_0^{*_{MB}} = b^*_{-1} + b^*_0 + 2b^*_1 - 4b^*_2;$$

for $1 \leq i \leq \lfloor \frac{n}{3} \rfloor$

$$b_i^{*MB} = b_{3i-1} + b_{3i} + 2b_{3i+1} - 4b_{3i+2};$$

along with $b^*_{-1} = 0$, $b_0^* = b_0 \vee b_n$, $b_1^* = b_1 \oplus b_n$ and $b_2^* = b_2 \vee \overline{b_1} \cdot b_n$.

Now from Eqs. 6, 7, 9 and 10 multiplier bit B can be Booth encoded as

$$|B|_{2^n+1} = |B|_{2^n+1} = \left| \sum_{i=0}^{\lfloor \frac{n}{3} \rfloor} (b_i^{*MB}) 2^{3i} \right|_{2^n+1} \tag{11}$$

where for $i = 0$

$$b_0^{*MB} = b^*_{-1} + b_0^* + 2b_1^* - 4b_2^*;$$

for $1 \leq i \leq \lfloor \frac{n}{3} \rfloor$

$$b_i^{*MB} = b_{3i-1} + b_{3i} + 2b_{3i+1} - 4b_{3i+2};$$

for n equal odd number of bits

$b^*_{-1} = 0$, $b_0^* = b_0 \vee b_n$, $b_1^* = b_1 \vee b_n$ and $b_2^* = b_2 \vee b_n$.

for n equal even number of bits

$b^*_{-1} = 0$, $b_0^* = b_0 \vee b_n$, $b_1^* = b_1 \oplus b_n$ and $b_2^* = b_2 \vee \overline{b_1} \cdot b_n$.

Booth encoder is designed in such a way that it is valid for both cases. Now put value of Eq. 11 into Eq. 3:

$$\begin{aligned}
\left| PD \right|_{2^n+1} &= \left| (A+1) \left| \sum_{i=0}^{\lfloor \frac{n}{3} \rfloor} (b_i^{*MB}) 2^{3i} \right|_{2^n+1} - B \right|_{2^n+1} \\
&= \left| \left| \sum_{i=0}^{\lfloor \frac{n}{3} \rfloor} (b_i^{*MB}) 2^{3i} (A+1) \right|_{2^n+1} - B \right|_{2^n+1} \\
\left| PD \right|_{2^n+1} &= \left| \left| \sum_{i=0}^{\lfloor \frac{n}{3} \rfloor} PP_i \right|_{2^n+1} - B \right|_{2^n+1}
\end{aligned} \tag{12}$$

where

$$\begin{aligned}
PP_i &= (b_i^{*MB}) 2^{3i} (A+1) \\
&= b_i^{*MB} 2^{3i} A + b_i^{*MB} 2^{3i}
\end{aligned}$$

The partial product (PP_i) is generated according to lemma presented in [20]. Table 1 shows partial product to be generated with respect to Booth encoded bits. It is clear from table that for each Booth encoded bit a partial product is generated along with

Table 1 Partial product generation of proposed modulo $2^n + 1$ multiplier using modified weighted method

a_n	b_{31+2}	b_{3i+1}	b_{3i}	b_{3i-1}	PP_i	CR_i
0	0	0	0	0	$\underbrace{111\ldots}_{(n-3i)}\underbrace{000}_{(3i)}$	$2^{3i}+1$
0	0	0	0	1	$a_{n-1-3i}\ldots a_0\overline{a_{n-1}}\ldots\overline{a_1 a_{n-3i}}$	+1
0	0	0	1	0	$a_{n-1-3i}\ldots a_0\overline{a_{n-1}}\ldots\overline{a_1 a_{n-3i}}$	+1
0	0	0	1	1	$a_{n-1+(3i+1)}\ldots a_0\overline{a_{n-1}}\ldots\overline{a_1 a_{n-(3i+1)}}$	+1
0	0	1	0	0	$a_{n-1-(3i+1)}\ldots a_0\overline{a_{n-1}}\ldots\overline{a_1 a_{n-(3i+1)}}$	+1
0	0	1	0	1	$h_{n-1-(3i)}\ldots h_0\overline{h_{n-1}}\ldots\overline{h_1 h_{n-3i}}$	+1
0	0	1	1	0	$h_{n-1-(3i)}\ldots h_0\overline{h_{n-1}}\ldots\overline{h_1 h_{n-3i}}$	+1
0	0	1	1	1	$a_{n-1-(3i+2)}\ldots a_0\overline{a_{n-1}}\ldots\overline{a_1 a_{n-(3i+2)}}$	+1
0	1	0	0	0	$\overline{a_{n-1-(3i+2)}}\ldots a_0 a_{n-1}\ldots a_1 a_{n-(3i+2)}$	+1
0	1	0	0	1	$\overline{h_{n-1-(3i)}}\ldots h_0 h_{n-1}\ldots h_1 h_{n-(3i)}$	+1
0	1	0	1	0	$\overline{h_{n-1-(3i)}}\ldots h_0 h_{n-1}\ldots h_1 h_{n-(3i)}$	+1
0	1	0	1	1	$\overline{a_{n-1-(3i+1)}}\ldots a_0 a_{n-1}\ldots a_1 a_{n-(3i+1)}$	+1
0	1	1	0	0	$\overline{a_{n-1-(3i+1)}}\ldots a_0 a_{n-1}\ldots a_1 a_{n-(3i+1)}$	+1
0	1	1	0	1	$\overline{a_{n-1-(3i)}}\ldots a_0 a_{n-1}\ldots a_1 a_{n-(3i)}$	+1
0	1	1	1	0	$\overline{a_{n-1-(3i)}}\ldots a_0 a_{n-1}\ldots a_1 a_{n-(3i)}$	+1
0	1	1	1	1	$\underbrace{111\ldots}_{(n-3i)}\underbrace{000}_{(3i)}$	$2^{3i}+1$
1	X	X	X	X	$\underbrace{111\ldots}_{(n-3i)}\underbrace{000}_{(3i)}$	$2^{3i}+1$

constant term equal to 1, expect for '0000' and '1111', i.e. for scaling factor '0' that introduces constant term equal to $2^{3i} + 1$. Thus, sum of all correction term derived using partial product can be given by

$$CR = \lfloor \frac{n}{3} \rfloor + C_n \tag{13}$$

where $C_n = \sum_{i=0}^{\lfloor \frac{n}{3} \rfloor} C_{3i} 2^{3i}$.

After putting value of Eq. 13 into Eq. 12

$$\left|PD\right|_{2^n+1} = \left|\sum_{i=0}^{\lceil \frac{n}{3} \rceil -1} PP_i + \lceil \frac{n}{3} \rceil + C_n - B\right|_{2^n+1}$$

$$= \left|\sum_{i=0}^{\lceil \frac{n}{3} \rceil -1} PP_i + \lceil \frac{n}{3} \rceil + C_n - b_n 2^n - B_{n-1:0}\right|_{2^n+1}$$

$$\left|PD\right|_{2^n+1} = \left|\sum_{i=0}^{\lceil \frac{n}{3} \rceil -1} PP_i + \lceil \frac{n}{3} \rceil + C_n + b_n - B_{n-1:0}\right|_{2^n+1}$$

Now from lemma $|-X|_{2^n+1} = |\overline{X}+2|_{2^n+1}$ above equation can be written as

$$\left|PD\right|_{2^n+1} = \left|\sum_{i=0}^{\lceil\frac{n}{3}\rceil-1} PP_i + \lceil\frac{n}{3}\rceil + C_n + b_n + \overline{B_{n-1:0}} + 2\right|_{2^n+1}$$

$$= \left|\sum_{i=0}^{\lceil\frac{n}{3}\rceil-1} PP_i + \lceil\frac{n}{3}\rceil + C_n + b_n + 1 + 1 + \overline{B_{n-1:0}}\right|_{2^n+1}$$

$$= \left|\sum_{i=0}^{\lceil\frac{n}{3}\rceil-1} PP_i + \lceil\frac{n}{3}\rceil + C_n^* + 1 + \overline{B_{n-1:0}}\right|_{2^n+1}$$

$$\left|PD\right|_{2^n+1} = \left|S + C + 1\right|_{2^n+1} \tag{14}$$

The Eq. 14 represents final equation of proposed algorithm where $\left|S+C\right|_{2^n+1} = \left|\sum_{i=0}^{\lceil\frac{n}{3}\rceil-1} PP_i + \lceil\frac{n}{3}\rceil + C_n^* + \overline{B_{n-1:0}}\right|_{2^n+1}$ and $C_n^* = 1 + b_n + C_n$.

3 Architecture for Modulo $2^n + 1$ Multiplier Using Modified Weighted Method

This section describes, the proposed architecture of radix-8 modulo $2^n + 1$ multiplier based on the modified weighted method as described in Sect. 2. The proposed architecture is shown in Fig. 1. The architecture mainly consist of Modified Booth encoder* (MBE*), Modified Booth Selector* (MBS*), Modified Booth Selector Compliment* (MBSC*) (star (*) is used to differentiate proposed architecture from standard nomenclature MBE, MBS and MBSC), Compliment End Around Carry-Carry Save Adder (CEAC-CSA) tree and a diminished-1 modulo $2^n + 1$ adder. MBE* examines successive overlapping of 4 bits of multiplier B as $\{b_{3i+2}, b_{3i+1}, b_{3i}, b_{3i-1}\}$ and encodes each as an element of the set $\{\pm4, \pm3, \pm2, \pm1, \pm0\}$. Each MBE* block produced 5 bits named (LSB to MSB) as ON (First position bit), TW (Second position bit), TH (Third position bit), SG (Signed position bit) and HG (Hard Multiple position bit). The 5 bits along with the multiplicand A are used to form PP (Partial Product). The partial product is produced by MBS* block. To generate a partial product with the help of MBE* and MBS*, straight binary method is used which is explained in Sects. 3.1 and 3.2. The correction term (C_{3i}) is required to represent only when $a_n = 0$ or $b_n = 0$ or $b_i^{*MB} = 0$ which is generated by Correction Term Generator (CTG). The design of CTG depends on the output of MBE* block. The generated partial products are accumulated with the help of CEAC-CSA tree, and finally the modulo product is generated with help of diminished-1 modulo adder.

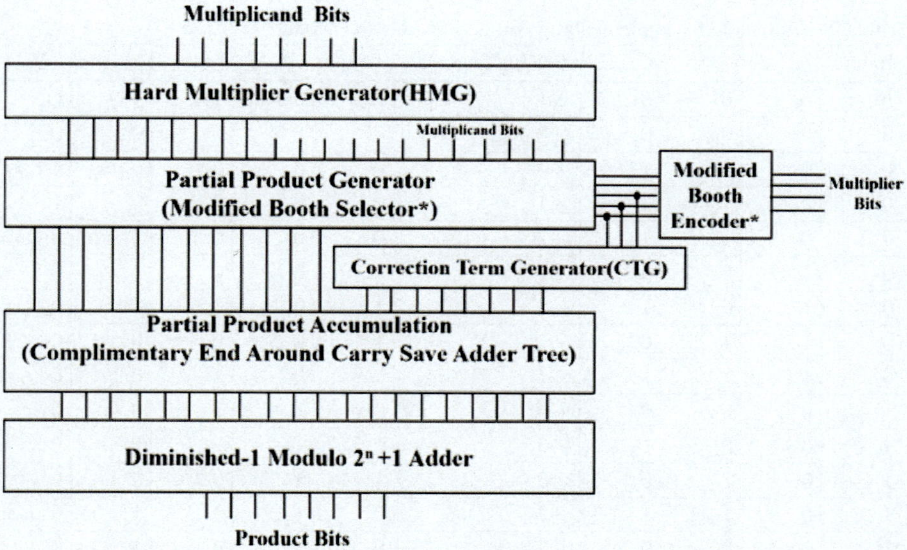

Fig. 1 Basic block diagram of proposed architecture of modulo $2^n + 1$ multiplier

3.1 *Implementation of Modified Booth Encoder**

Truth Table 2 shows a possible combination of successive overlapping of multiplier bits for booth encoding in signed representation except input combination generates the scaling factor ±3. The block diagram and logical representation of MBE* under the above approach is shown in Fig. 2a, b. It uses a 5-bit bus approach. Equations 15 to 19 show boolean equation of logical diagram. The 1-bit input $\overline{a_n}$ is introduced into MBE* to handle the case when nth bit of A input equals to '1' which shows that the 8-bit input is zero.

$$HG = (b_{(3i+2)} \oplus b_{(3i+1)}) \cdot (b_{(3i)} \oplus b_{(3i-1)}) \tag{15}$$

$$SG = b_{(3i+2)} \tag{16}$$

$$ON = (b_{(3i+2)} \odot b_{(3i+1)}) \cdot (b_{(3i)} \oplus b_{(3i-1)}) \tag{17}$$

$$TH = (b_{(3i+2)} \oplus b_{(3i+1)}) \cdot (b_{(3i+1)} \odot b_{(3i)}) \cdot (b_{(3i)} \odot b_{(3i-1)}) \tag{18}$$

$$TW = (b_{(3i+1)} \oplus b_{(3i)}) \cdot (b_{(3i)} \odot b_{(3i-1)}) \tag{19}$$

3.2 *Implementation of Modified Booth Selector**

According to Table 1 explained in Sect. 2, to generate modulo reduced partial product a complimented left circular shift of A and -A are different. Thus, two kinds of Booth selector blocks are proposed named as MBS* and MBSC*. For ith partial product,

Table 2 Truth table for implementation of modified booth encoder*

Input				Scaling factor	Output				
b_{3i+2}	b_{3i+1}	b_{3i}	b_{3i+1}	SF	HG	SG	TH	TW	ON
0	0	0	0	+0	0	0	0	0	0
0	0	0	1	+1	0	0	0	0	1
0	0	1	0	+1	0	0	0	0	1
0	0	1	1	+2	0	0	0	1	0
0	1	0	0	+2	0	0	0	1	0
0	1	0	1	+3	1	0	0	0	0
0	1	1	0	+3	1	0	0	0	0
0	1	1	1	+4	0	0	1	0	0
1	0	0	0	−4	0	1	1	0	0
1	0	0	1	−3	1	1	0	0	0
1	0	1	0	−3	1	1	0	0	0
1	0	1	1	−2	0	1	0	1	0
1	1	0	0	−2	0	1	0	1	0
1	1	0	1	−1	0	1	0	0	1
1	1	1	0	−1	0	1	0	0	1
1	1	1	1	−0	1	0	0	0	0

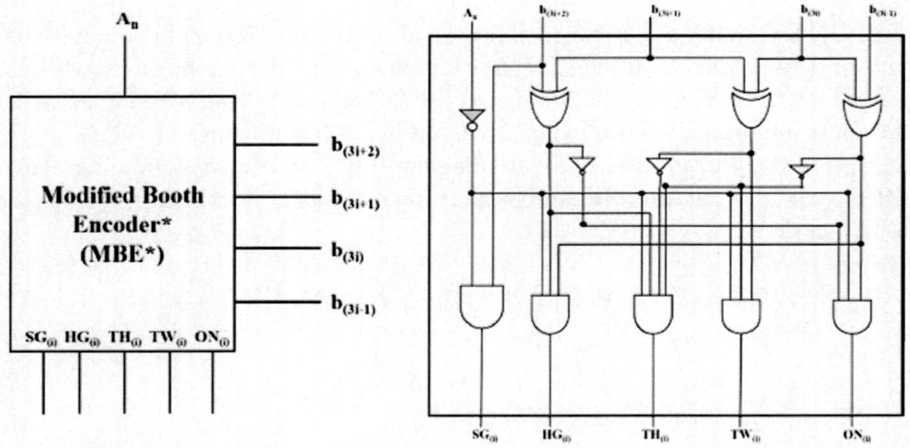

(a) Block diagram

(b) Logical diagram

Fig. 2 Implementation of proposed modulo $2^n + 1$ modified booth encoder* using signed representation

Table 3 Truth table for implementation of modified booth selector*

HG	SG	TH	TW	ON	PP_i	$\overline{PP_i}$
0	0	0	0	0	1	0
0	0	0	0	1	a_i	$\overline{a_i}$
0	0	0	1	0	a_{i-1}	$\overline{a_{i-1}}$
1	0	0	0	0	h_i	$\overline{h_i}$
0	0	1	0	0	a_{i-2}	$\overline{a_{i-2}}$
0	1	1	0	0	$\overline{a_{i-2}}$	a_{i-2}
1	1	0	0	0	$\overline{h_i}$	h_i
0	1	0	1	0	$\overline{a_{i-1}}$	a_{i-1}
0	1	0	0	1	$\overline{a_i}$	a_i
0	0	0	0	0	1	0

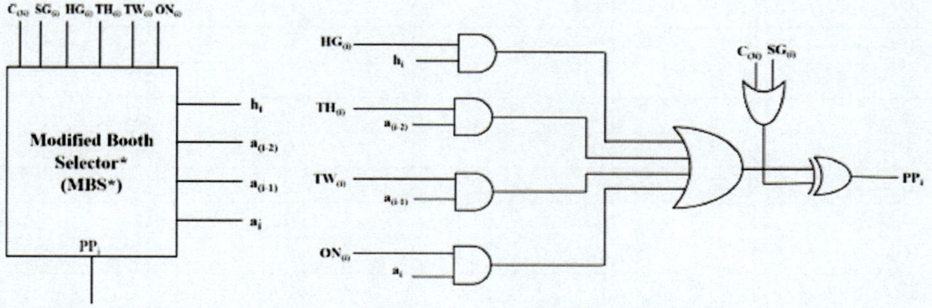

(a) Block diagram (b) Logical diagram

Fig. 3 Implementation of proposed modulo $2^n + 1$ modified booth selector*

there are 3i MBSC* blocks and (n-3i) MBS* blocks will be used. Truth Table 3 shows the bit wise partial product generation for different Booth encoded bits. Figure 3a, b shows a block diagram and logical representation of MBS* block while Fig. 4a, b shows a block diagram and logical representation of MBSC* block, respectively.

3.3 *Implementation of Correction Term Generator (C_{3i})*

This section describes generation of correction term (C_{3i}) required when any one of condition $a_n = 0$ or $b_n = 0$ or $b_i^{*MB} = 0$ will happen. From Table 1 correction term for proposed algorithm will be in form of $0c_{i+3}00c_i00c_600c_300c_0$ where $c_i \epsilon$ [0, 1]. Truth Table 4 shows correction term of 0th partial product and it can be seen that C_i is equal to 1 only when four output of Booth encoder HG,TH,TW and ON is equal to 0. Thus, CTG is implemented by 4 input NOR GATE as shown in Fig. 5a. Figure 5b shows correction term generator for 8-bit modulo $2^n + 1$ multiplier.

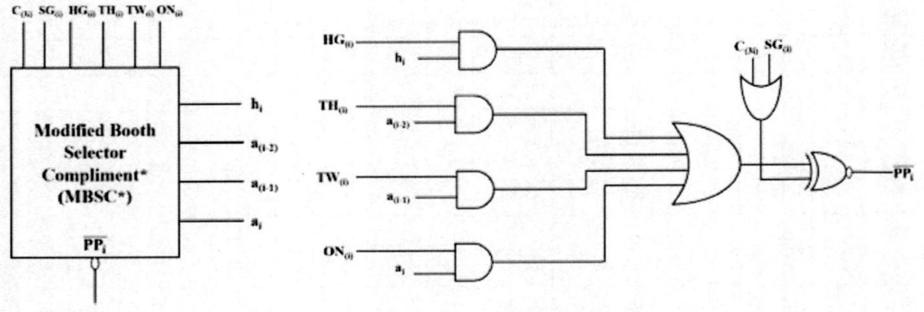

(a) Block diagram (b) Logical diagram

Fig. 4 Implementation of proposed modulo $2^n + 1$ modified booth selector compliment*

Table 4 Truth table for implementation of correction term generation

HG	SG	TH	TW	ON	C_i
0	0	0	0	0	1
0	0	0	0	1	0
0	0	0	1	0	0
1	0	0	0	0	0
0	0	1	0	0	0
0	1	1	0	0	0
1	1	0	0	0	0
0	1	0	1	0	0
0	1	0	0	1	0
0	0	0	0	0	1

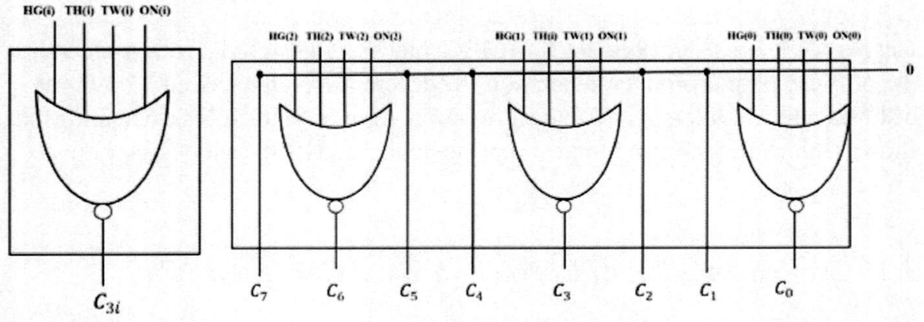

(a) Logical diagram of n-bit CTG (b) CTG for 8-bit modulo $2^n + 1$ multiplier

Fig. 5 Implementation of proposed modulo $2^n + 1$ correction term generator for modified weighted method

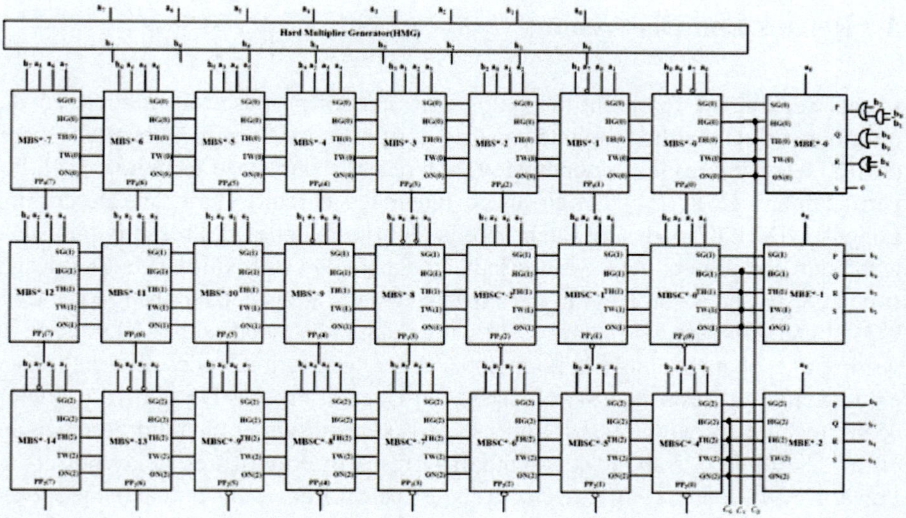

Fig. 6 Proposed architecture of partial product generator for modified weighted method

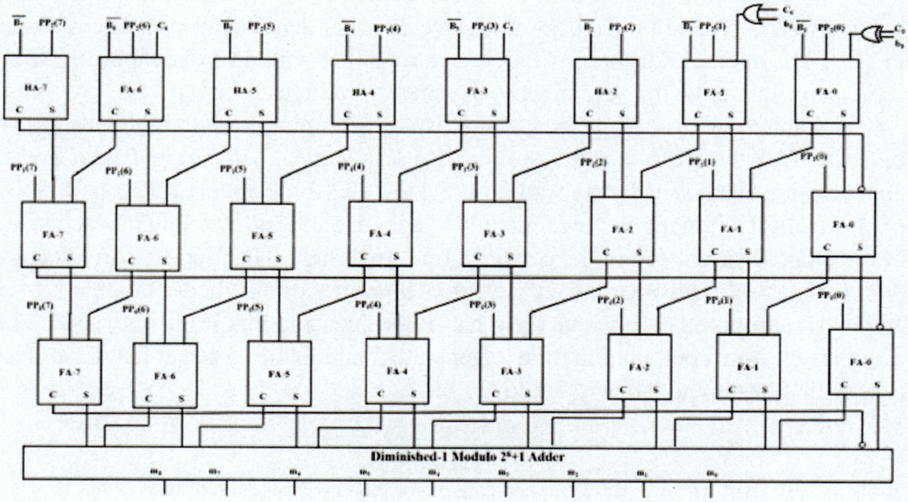

Fig. 7 Proposed architecture of partial product accumulator for modified weighted method

All partial products are accumulated with help of CEAC-CSA tree while constant values will naturally be added and the accumulated partial product reduced to two n-bit operands. The CSA tree is usually constructed with full adder's (FAs). A stage that takes input correction term can be constructed using half adder's (HAs) and FA as shown in Fig. 7. Finally, diminished-1 modulo $2^n + 1$ adder accepts n-bit inputs and generates the final product in weighted representation. The 8-bit modulo $2^8 + 1$ multiplier is shown in Figs. 6 and 7.

4 Results and Discussions

In this Section, we represent the synthesis results of proposed for modulo $2^n + 1$ multiplier using modified weighted method at Application Specific Integrated Circuit (ASIC) level and results are compared with recently published technique [19]. For verification, VHSIC (Very High Speed Integrated Circuit) Hardware Description Language (VHDL) codes for the proposed algorithm along with recently published technique are written, then synthesised and simulated using xilinx ISE 14.7 design suite at 8, 16 and 32 bit length. The expected results are generated using MATLAB CODE and compared simultaneously with results generated with VHDL code by writing a test bench using file i/o method. After verification, codes are synthesised using genus synthesis solution of cadence at General Process Design Kit (GPDK)-45 nm technology with 0.9 volt under the worst-condition at medium optimization effort. The Tables 5, 6 and 7 show synthesised result in terms of area, delay, power and Power Delay Product (PDP), respectively. It is clear from Table 5, that the proposed architecture consumes 4.84% more area at 8-bit operand length. However, at larger operand size, area penalty is negligible. In fact, at 32-bit operand length, it shows 0.88% of area saving. It is because of generation of correction term (bias) in a simple way, as well as its addition at CSA tree stage does not demands any extra circuits as in [19]. The another advantage of architecture is that with an extra input in MBE* stage helps to handle the zero input with same overall multiplier area.

The Table 6, shows the power comparison of the proposed architecture. It can be seen from the table that the design uses less static, dynamic and total power consumption that lies in the range of 2.11–5.15%, 27.15–38.58% and 27.14–38.56%, respectively. The saving in power mainly occurs due to less switching at each stage of the proposed architecture, especially at CSA tree stage. The proposed architecture does not use any extra circuit to generate final inputs for diminished-1 modulo $2^n + 1$ adder. The proposed architecture also has a disadvantage that it fails at speed. The design uses more computation time to generate final modulo product but it uses less energy that lies in the range of 20.26–33.82%.

Table 5 ASIC performance parameter comparison for modulo $2^n + 1$ multiplier

No of bits	Cell area (μm^2)			Delay (ps)		
	Proposed	[19]	Area saved (%)	Proposed	[19]	Delay increase (%)
8	542	517	−4.84	4527	4264	+6.17
16	1646	1639	−0.43	11,052	10,261	+7.70
32	5622	5672	+0.88	20,253	18,506	+9.44

Table 6 Power performance parameter comparison for modulo $2^n + 1$ multiplier

No of bits	Static power (nw)			Dynamic power (nw)			Total power (nw)		
	Proposed	[19]	Power saved (%)	Proposed	[19]	Power saved (%)	Proposed	[19]	Total power saved (%)
8	12.08	12.34	2.11	10077	14017.28	28.11	10089.08	14029.62	28.09
16	36.36	37.65	3.43	37499.65	61057.12	38.58	37536.01	61094.77	38.56
32	120.52	127.07	5.15	166941.52	229165.14	27.15	167062.04	229292.21	27.14

Table 7 Power delay product (PDP) comparison for modulo $2^n + 1$ multiplier

No of bits	PDP (10^{-15}j)		
	Proposed	[19]	Energy saved (%)
8	45.67	59.82	23.65
16	414.85	626.89	33.82
32	3383.50	4243.28	20.26

5 Conclusion

A new algorithm to design modulo $2^n + 1$ multiplier using a modified weighted method has been presented. Algorithm significantly reduces the number of modulo reduced partial product with light bias approach. The proposed architecture is able to handle zero input by introducing an extra input to modified booth encoder*. The synthesis result shows improvement in power and PDP at all bit length. The improvement in design is archived at the expense of an increase in delay. Hence, the proposed architecture can be used in low power digital signal processors.

References

1. Taylor FJ (1984) Residue arithmetic a tutorial with examples. IEEE Comput 17(5):50–62. https://doi.org/10.1109/MC.1984.1659138
2. Watson RW, Hastings CW (1966) Self-checked computation using residue arithmetic. Proc IEEE 54(12):1920–1931. https://doi.org/10.1109/PROC.1966.5275
3. Conway R, Nelson J (2004) Improved RNS FIR filter architectures. IEEE Trans Circuits Syst II Express Briefs 51(1):26–28. https://doi.org/10.1109/TCSII.2003.821524
4. Wang W, Swamy MNS, Ahmad MO (2004) RNS application for digital image processing. In: 4th IEEE international workshop on system-on-chip for real-time applications, pp 77–80. https://doi.org/10.1109/IWSOC.2004.1319854
5. Yen SM, Kim S, Lim S, Moon SJ (2003) Rsa speedup with chinese remainder theorem immune against hardware fault cryptanalysis. IEEE Trans Comput 52(4):461–472. https://doi.org/10.1109/TC.2003.1190587

6. Zimmermann R (1999) Efficient VLSI implementation of modulo ($2^n \pm 1$) addition and multiplication. In: Proceedings 14th IEEE symposium on computer arithmetic (Cat. No. 99CB36336), pp 158–167. https://doi.org/10.1109/ARITH.1999.762841
7. Hiasat A (1992) New memoryless, mod ($2^n \pm 1$) residue multiplier. Electron Lett 28(3):314–315. https://doi.org/10.1049/el:19920194
8. Wrzyszcz A, Milford D (1993) A new modulo $2^n + 1$ multiplier. In: Proceedings of the 1993 IEEE international conference on computer design: VLSI in computers and processors, ICCD'93, pp 614–617. https://doi.org/10.1109/ICCD.1993.393303
9. Vergos HT, Efstathiou C (2007) Design of efficient modulo $2^n + 1$ multipliers. IET Comput Digital Tech 1(1):49–57
10. Wang Z, Jullien G, Miller CW (1996) An efficient tree architecture for modulo $2^n + 1$ multiplication. J VLSI Signal Process 14:241–248
11. Efstathiou C, Vergos HT, Dimitrakopoulos G, Nikolos D (2005) Efficient diminished-1 modulo $2^n + 1$ multipliers. IEEE Trans Comput 54(4):491–496. https://doi.org/10.1109/TC.2005.63
12. Ma Y (1998) A simplified architecture for modulo ($2^n + 1$) multiplication. IEEE Trans Comput 47(3):333–337. https://doi.org/10.1109/12.660169
13. Sousa L, Chaves R (2005) A universal architecture for designing efficient modulo $2^n + 1$ multipliers. IEEE Trans Circuits Syst I Regul Pap 52(6):1166–1178. https://doi.org/10.1109/TCSI.2005.849143
14. Chen JW, Yao RH (2010) Efficient modulo $2^n + 1$ multipliers for diminished-1 representation. IET Circuits Devices Syst 4(4):291–300. https://doi.org/10.1049/iet-cds.2009.0284
15. Chang CH, Molahosseini AS, Zarandi AAE, Tay TF (2015) Residue number systems: a new paradigm to datapath optimization for low-power and high-performance digital signal processing applications. IEEE Circuits Syst Mag 15(4):26–44. https://doi.org/10.1109/MCAS.2015.2484118
16. Curiger AV, Bonnenberg H, Kaeslin H (1991) Regular VLSI architectures for multiplication modulo ($2n + 1$). IEEE J Solid-State Circuits 26(7):990–994. https://doi.org/10.1109/4.92018
17. Leibowitz L (1976) A simplified binary arithmetic for the fermat number transform. IEEE Trans Acoust Speech Signal Process 24(5):356–359. https://doi.org/10.1109/TASSP.1976.1162834
18. Muralidharan R, Chang CH (2012) Area-power efficient modulo $2^n - 1$ and modulo $2^n + 1$ multipliers for $\{2^n - 1, 2^n, 2^n + 1\}$ based RNS. IEEE Trans Circuits Syst I Regul Pap 59(10):2263–2274. https://doi.org/10.1109/TCSI.2012.2185334
19. Mirhosseini SM, Molahosseini AS, Hosseinzadeh M, Sousa L, Martins P (2017) A reduced-bias approach with a lightweight hard-multiple generator to design a radix-8 modulo $2^n + 1$ multiplier. IEEE Trans Circuits Syst II Express Briefs 64(7):817–821. https://doi.org/10.1109/TCSII.2016.2601285
20. Efstathiou C, Pekmestzi K, Axelos N (2011) On the design of modulo $2^n + 1$ multipliers. In: 2011 14th Euromicro conference on digital system design, pp 453–459. https://doi.org/10.1109/DSD.2011.64

Design of Prominent Single-Precision 32-Bit Floating-Point Adder Using Single-Electron Transistor Operating at Room Temperature

Ankur Sharma, Rasika Dhavse, Yash Agrawal, and Rutu Parekh

Abstract The floating-point (FP) addition is the most frequently used FP operation. Here we are using single-electron transistor (SET) for floating-point addition. This research aims to implement a 32-bit binary floating-point adder with IEEE 754 standard using SET. In floating-point arithmetic, FP addition is the most difficult activity and it offers more delay and more power consumption. Here we are comparing SET and CMOS-based (16 nm) floating-point adder. SET-based FP adder consumes very less power and also very less delay. For simulation and verification, CADENCE virtuoso is used. According to our results, SET-based FP addition has 79.70% improvement in power and 97.67% faster than CMOS-based FP.

Keywords CMOS · CADENCE virtuoso · Coulomb blockade · Floating-point addition · Single-electron transistor (SET)

1 Introduction

Due to advances in microelectronics technology, downscaling of the MOSFET devices and many integrated devices achieved the edge of micrometer and nanometer dimensions. The performance deterioration was observed due to the constant decrease in the feature size of the current CMOS technology [1]. The least transistor measurements will reach beneath 10 nm by 2020 according to the new reports of international technology roadmap for semiconductor (ITRS) because of nonstop scaling down of

A. Sharma (✉) · Y. Agrawal · R. Parekh
Dhirubhai Ambani Institute of Information and Communication Technology, Gandhinagar 382 007, Gujarat, India
e-mail: ankursharma1605@gmail.com

Y. Agrawal
e-mail: mr.yashagrawal@gmail.com

R. Parekh
e-mail: rutu_parekh@daiict.ac.in

R. Dhavse
The National Institute of Technology, Surat (NIT Surat) 395007, Gujarat, India

Z. Patel et al. (eds.), *Advances in VLSI and Embedded Systems*, Lecture Notes in Electrical Engineering 676, https://doi.org/10.1007/978-981-15-6229-7_16

the measurements on integrated circuits and semiconductor chips [2]. This is the reason many problems, like short channel effects, costly lithography, and ultrathin gate leakage, will take place. As stated by Moore's Law and the driving of technology in nano-scale routine, new nano-electronic solutions will be desired, overcoming the physical and economic barriers of recent technologies. Among the present devices like TFET, NEMS, SET, CNTFET, and graphene FET, SET is the most promising device that can be hybridized with CMOS technology. SET has many features like small size, low power consumption, and high-speed capabilities [3]. SET is a potential nano-device that can be designed and fabricated to build high-performance logic circuit. Floating-point (FP) adder is a complex operation that needs high memory and processing power. So in this paper we shall exploit SET to perform FP addition and compare the result with nano CMOS circuit.

SET consists of source, drain, gate, back gate, island, and two tunneling junctions. Fig. 1a shows the SET symbol. Here V_{DS} = bias voltage, C_D and C_S = tunneling junction capacitance, C_{G1} and C_{G2} = gate capacitance, R_D and R_S = tunneling

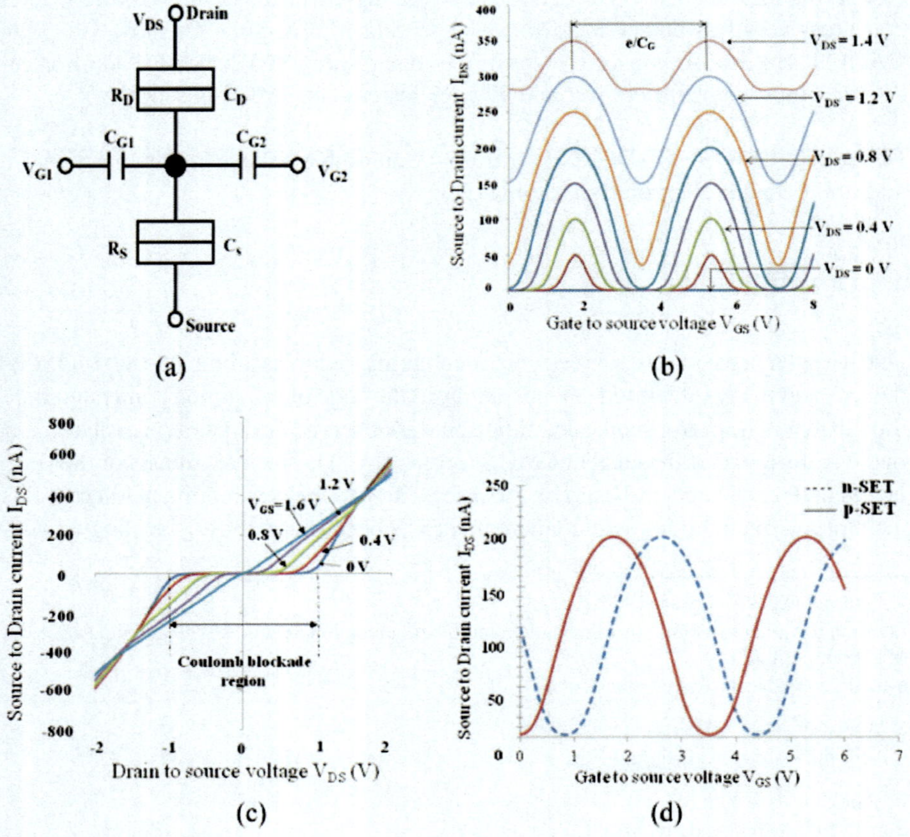

Fig. 1 **a** Single-electron transistor. **b** The I_{DS} versus V_{GS} characteristic of the SET. **c** The I_{DS} versus V_{DS} characteristic of the SET. **d** The I_{DS} versus V_{GS} characteristic of the n-SET and p-SET

junction resistance, V_{G1} and V_{G2} = gate voltage. A detailed information about SET working and modeling is available in the literatures [4, 5]. This work is based on metallic SET which is fabricated using back-end-of-line (BEOL) fabrication process, showing a low thermal budget process [6]. The SET basically works on two principles, Coulomb blockade and quantum mechanical tunneling. The responsibility for SET working as a switching device lies with these two phenomena. It will work as a p-switch and n-switch by adjusting a SET's V_{G2} back gate potential. This combination is known as n-SET and p-SET, replicating n-MOS and p-MOS. It is observed that in the region of the Coulomb blockade, the current is nearly zero. It can be viewed as a simple gate-controlled switch. Switch is 'OFF' when Coulomb blockade region exists, otherwise switch is in 'ON' state. SET and CMOS have complementary properties [6].

Figure 1b is the drain current characteristics at constant V_{DS}, drain to source voltage. With small values of V_{DS}, electrons cannot rise above the single-electron charging energy and cannot tunnel onto the island, so current cannot flow in the device. This is called the Coulomb blockade effect, and leads to the I–V characteristics shown in Fig. 1c. By the SET characteristic, we can conclude that SET will be used as a complementary state, n-SET, and p-SET. Fig. 1d shows I_{DS} versus V_{GS} curve of n-SET and p-SET.

SET parameters are defined in the work presented by Parekh et al. [6]. The SET characteristics are based on the MIB model (Mahapatra–Ionescu–Banerjee) [7]. To compare with bulk CMOS 16-nm, we are using 16-nm BSIM predictive model [8] and implementing it in analog design environment of CADENCE virtuoso [9]. So here we have capacitance $C_D = C_S = C_J$; the control gate capacitance, $C_{G1} = C_{G2} = C_G$; the tuning gate capacitance, $C_{B1} = C_{B2} = C_B$; so the total capacitance is C_Σ ($C_\Sigma = 2C_J + C_G + C_B$) and the tunnel junction resistances, $R_{D1} = R_{S1} = R_{D2} = R_{S2} = R_t$. The parameters are: $C_{G1} = 0.045$aF, $C_{G2} = 0.050$aF, $C_D = 0.030$aF, $C_S = 0.030$aF, $R_t = 1\ M\Omega$, $T = 300$ K, and a CMOS 16 nm parameter is defined by 16-nm BSIM predictive model [8, 10].

2 IEEE 754 Single-Precision Floating-Point Format

A FP number is shown by real numbers in binary format. According to IEEE 754 binary interchange format for single precision, Mantissa (M) of the floating-point binary number is represented by 23 bits, whereas sign (S) has 1 bit and exponent (E) has 8 bits, which is shown in Fig. 2. In this format mantissa's MSB must be '1' and exponent E > 0 and less than 127 in normalized binary FP interchange format [11]. The representation of the IEEE standard 754 32-bit single-precision FP number is shown in Eq. 1.

$$\text{Binary Value} = (-1)\,(\text{Sign bit}) \times 2(\text{Exponent} - 127) \times (1.\ \text{Mantissa}) \quad (1)$$

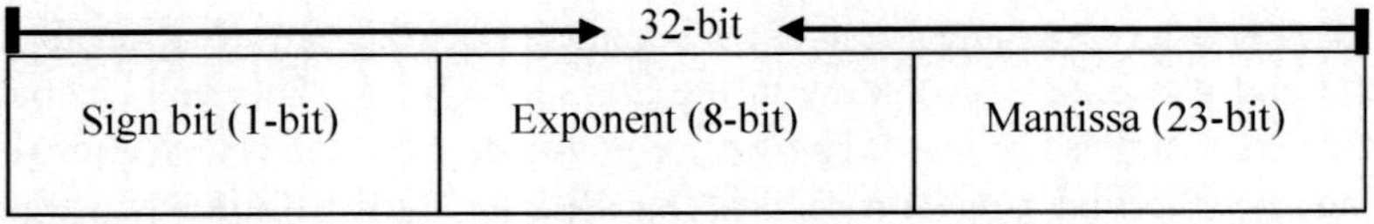

Fig. 2 IEEE 754 single-precision format

An example of single-precision FP binary format:
Suppose a decimal number is 37.5.
In binary, it can be written as:
$1.00101100000000000000000 \times 2^5$.
The IEEE 754 single-precision 32-bit format can be shown as:

0……Sign
10000100……Exponent
00101100000000000000000……Mantissa.

Floating-Point Addition

The most frequently used FP operation is the addition. It accounts for nearly 50% of logical scientific operation in digital signal processing processors, math co-processors, arithmetic operations in embedded processors, and many data processing units. In these processors, every part requires very high numerical stability and precision and is subsequently FP based. In most hardware designs FP unit is one of the most significant convention applications desirable because of accuracy, ease of use, and robustness to quantization errors added in the design. When compared to FP multiplication, FP addition is the more complicated and most troublesome task. FP adder comprises many sub-operation and variable latency. For improving the overall latency of FP adders a lot of works have been reported. In 1985, IEEE issued 754 standards for binary floating-point arithmetic [12]. In this work we have designed and simulated 32-bit FP adder and a detailed explanation on algorithms, architecture-operation, sub-operations, and simulation results is followed in the subsequent section (Fig. 3).

3 Single-Precision 32-Bit FP Adder Algorithm, Architecture, and Operation

3.1 Proposed Floating-Point Addition Algorithm

This paragraph will investigate the FP algorithm. As a part of this proposed algorithm, FP architecture modules are designed and simulated in cadence virtuoso environment including their function, capacity, and use. The proposed algorithm for floating-point addition is shown below [13] and following are the steps.

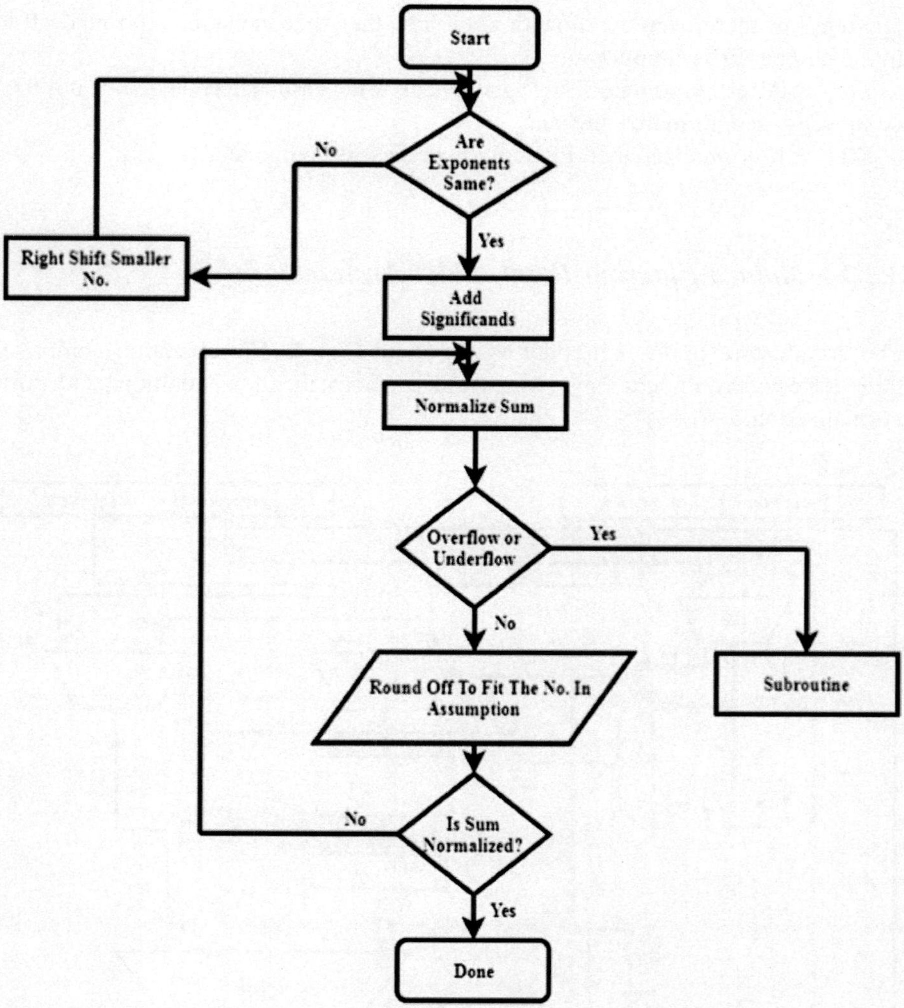

Fig. 3 Floating-point addition algorithm flow

Step 1: First check if the exponents are the same or not.

Step 2: If exponent numbers are not the same, then we right shift the smaller number by exponent difference.

Step 3: If exponents are the same, we add the significands.

Step 4: After addition, we check that sum is normalized or not. If not, then we normalize the sum.

Step 5: After normalization of sum we check that if there is an overflow or underflow in this.

Step 6: If there is an overflow or underflow, it has to go from the subroutine process.

Step 7: If there is no overflow or underflow then the next step is to round off to fit the number in assumptions.

Step 8: After rounding off, we again check if the sum is normalized. If not then go to step 4 and normalize the sum.

Step 9: If normalized then FP addition process is completed.

3.2 Standard Floating-Point Adder Architecture

The architecture of the FP adder is shown in Fig. 4. The operation comprises three noteworthy undertakings which are pre-normalization, addition, and post-normalization.

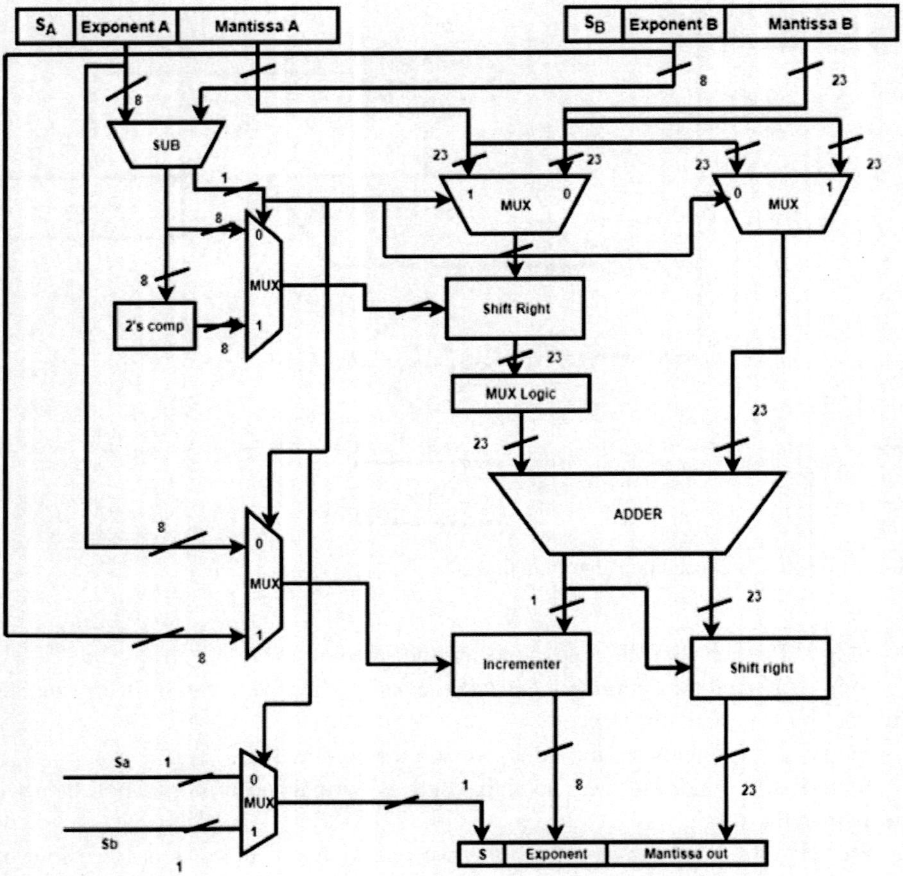

Fig. 4 Floating-point addition architecture

(1) Pre-normalization is composed of exponent difference and right shift the smaller number.
(2) The second task is to add these two numbers.
(3) The third task is post-normalization. In post-normalization, we check if the number is underflow or overflow and then round off the number.

4 Design Consideration of SET and CMOS (16 nm) FP and Simulation Results

Here in this work SET-based FP adder has been compared with CMOS 16 nm FP adder. To simulate, 16 nm CMOS PTM model has been integrated and implemented in the spectre circuit simulator and virtuoso analog design environment of CADENCE [8, 10]. SET parameters are defined in the work by Parekh et al. [6]. The SET is a metallic SET that uses Ti/TiO_X interface at the junctions fabricated on a TEOS layer by the nanodamascene process. This SET can operate at room temperature (300°K). The parameters are: C_B = back gate capacitance = 50zF, junction capacitance (C_D and C_S) = 30zF, C_G = gate capacitance = 45zF, tunnel junction resistance R_T = 1 M, high voltage = 800 mV, low voltage = 0 V, gain = 1, T = 300 K[6]. With this data we have simulated the SET-based FP adder with the following example.

Example: For an example we have taken two numbers 0.6 and 0.1 and their FP representation is shown below.

Addends A = 0.6 = 00111111000110011001100110011010
Addends B = 0 0.1 = 00111101110011001100110011001101
To add these two floating-point representations,
Step 1: If exponent numbers are not the same, then we right shift the smaller number by their difference.
Exponent A = 01111110
Exponent B = − 01111011
00000011(3) places will be shift of Addends B.
Addends B is,
0 01111011 10011001100110011001101
One place shifted and in MSB of mantissa hidden bit is shifted,
0 01111100 11001100110011001100110
Two places shifted,
0 01111101 01100110011001100110011
Three places shifted,
0 01111110 00110011001100110011001 = Addends B (new)
Step 2: Addition
0 01111110 1.00110011001100110011010 (0.6) = Addends A
+ 0 01111110 0.00110011001100110011001 (0.1) = Addends B
0 01111110 1.01100110011001100110011 (Sum)

Step 3: Normalize the output and the result is (get the "hidden bit" to be a 1).
0 01111110 01100110011001100110100

5 Performance Evaluation

Table 1 shows the performance evaluation of single-precision FP adder (32-bit). The proposed 32-bit CMOS at 16 nm technology FP adder is compared with a single-electron transistor-based FP. Here logic '0' is shown as 800 mV, while logic '1' is shown as 0 V. When compared with CMOS 16 nm, SET-based FP adder consumed less power and also very less delay, which is also presented and proved in Table 1. SET current is in nano-amperes (nA) range. In addition, the SET referred in this work is metallic SET. So the parasitics at the interconnection are reduced. Due to this, it offers low power and high speed compared to CMOS (Figs. 5 and 6).

Table 1 Performance analysis of FP

Parameter	CMOS (16 nm)	SET	Improvement (%)
Delay (ps)	904.40	21	97.67
Power (μW)	16.81	3.41	79.70

Fig. 5 Result for 16 nm CMOS floating-point (FP) adder

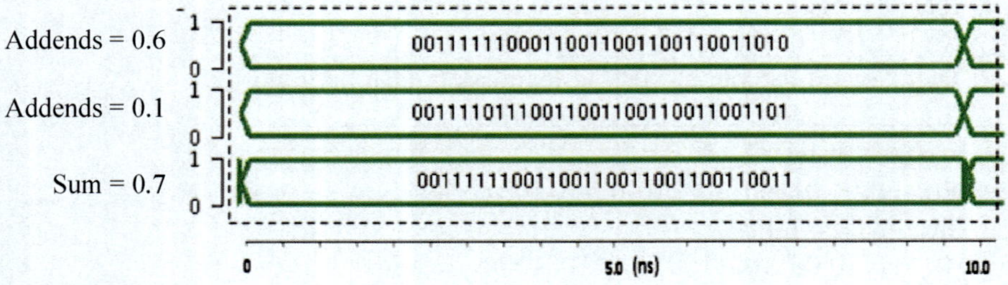

Fig. 6 Result for SET floating point Adder (FP) adder

6 Conclusion

We proposed an efficient single-precision floating-point adder using a single-electron transistor in this paper. The result confirmed that SET-based FP adder is energy efficient than its 16 nm CMOS FP adder. SET-based single-precision 32-bit FP adder design is 79.70% power efficient and 97.67% faster as compared to 16 nm CMOS-based design. Delay and power have been reduced to a significant amount in the case of SET-based FP design. SET is a low current device in nA compared to MOSFET that is in microamperes. Since power is proportional to square of current, it is naturally low compared to CMOS. Then about delay, when a SET is driving the delay is less due to tunneling of the current and the dimensions of quantum dot in few nanometers. So it is faster compared to CMOS where the current is because of drift. But when a SET drives CMOS the delay is more as it has to drive a heavy load with little current. Here we are using the metallic SET circuits, so it can be easily stacked above the CMOS platform presenting a low cost and low thermal budget of highly integrated systems. From these results, we can say that the emerging technology SET can overcome the power consumption and delay issues with present ICs.

References

1. Zardalidis G, Karafyllidis IG (2008) SECS: a new single-electron-circuit simulator. Proc IEEE Trans Circ Syst I Regular Pap 55(9):2774–2784
2. International Technology Roadmap for Semiconductors (ITRS) (2012) Available at https://www.itrs.net/
3. Zhang F, Tang R, Kim YB (2005) SET-based nano-circuit simulation and design method using HSPICE. Proc Microelectron J 36(8):741–748
4. Mahapatra S, Ionescu AM (2006) Hybrid CMOS single-electron-transistor device and circuit design. Artech House Inc.
5. Durrani ZAK (2010) Single-electron devices and circuits in silicon. World Scientific
6. Parekh R, Beaumont A, Beauvais J, Drouin D (2012) Simulation and design methodology for hybrid SET-CMOS integrated logic at 22-nm room-temperature operation. IEEE Trans Electron Devices 59(4):918–923
7. Mahapatra S, Vaish V, Wasshuber C, Banerjee K, Ionescu AM (2004) Analytical modeling of single electron transistor for hybrid CMOS-SET analog IC design. IEEE Trans Electron Devices 51(11):1772–1782
8. Zhao W, Cao Y (2007) Predictive technology model for nano-CMOS design exploration. ACM J Emerg Technol Comput Syst (JETC) 3(1):1
9. Cadence Design Systems. Available https://www.cadence.com
10. Predictive Technology Model (PTM). https://ptm.asu.edu/
11. Rahman A, Khalid M, Islam AS, Rahman M (2014) Optimized hardware architecture for implementing IEEE 754 standard double precision floating point adder/subtractor. In: Proceedings of the IEEE 17th international conference on computer and information technology (ICCIT), pp 147–152
12. IEEE Computer Society. Standards Committee, & American National Standards Institute (1985) IEEE standard for binary floating-point arithmetic, vol 754
13. Hennessy JL, Patterson DA (2011) Computer architecture: a quantitative approach. Elsevier

Image Communication Using Quasi-Cyclic Low-Density Parity-Check (QC-LDPC) Code

Dharmesh Jayantibhai Patel, Pinalkumar Engineer, and Ninad Sunilkumar Bhatt

Abstract Proposed quasi-cyclic low-density parity-check code (QC-LDPC) efficiently communicates image at comparable peak signal-to-noise ratio (PSNR) with less signal-to-noise ratio (SNR). Encoding of the image is done using the Gauss–Jordan elimination method while decoding is done using the min-sum iterative message-passing algorithm using the code length 1152. Hardware design of decoder is based on fully parallel architecture for achieving more throughput.

Keywords Quasi-cyclic low-density parity-check code · Channel coding · Encoding · Min-sum decoding · Bit error rate (BER) · Peak signal-to-noise ratio · Signal-to-noise ratio

1 Introduction

Low-density parity-check code is popular due to its performance near channel capacity. It was invented by R. Gallager in his Ph.D. thesis at MIT in 1963 [8]. It was forgotten due to the unavailability of software and implementation complexity of code. It was rediscovered by D. J. C. Mackay and Neal in the mid-'90s at Cambridge University [12]. It was graphically represented by R. M. Tanner in 1981 using a bipartite graph, also known as tanner graph [14]. It is classified into two categories: (1) random constructed LDPC code and (2) structured constructed LDPC code. Random constructed code perform well than the structured one but implementation complexity is more than structured constructed code [10]. Structured constructed code is widely

D. J. Patel (✉)
Dr. S. & S. S. Ghandhy Government Engineering College, Surat, India
e-mail: djpgecs@gmail.com

P. Engineer (✉)
S. V. National Institute of Technology, Surat, India
e-mail: pje@eced.svnit.ac.in

N. S. Bhatt (✉)
C. K. Pithawala College of Engineering and Technology, Surat, India
e-mail: ninad.bhatt@ckpcet.ac.in

Z. Patel et al. (eds.), *Advances in VLSI and Embedded Systems*, Lecture Notes in Electrical Engineering 676, https://doi.org/10.1007/978-981-15-6229-7_17

used in short and moderate length of code, while the random constructed code is used in long length code such as satellite communication standard, i.e., DVB-S2. The LDPC code was adopted in various wireless standards, such as IEEE 802.11n (Wi-Fi) and IEEE 802.16e (Wi-Max). It is also used in different storage media applications.

It is constructed by sparse parity-check matrix H which contains less number of the nonzero entries in it. Due to that, the computational complexity of an LDPC decoder is less. Different methods of code constructions [11, 13, 16, 17] are used to construct parity-check matrix H of QC-LDPC code. These methods give a different performance of bit error rate (BER) for a different number of iterations. All these methods are limited in girth due to its structured constructions. Performance of BER in the waterfall region can be improved by density evaluation while the error floor can be improved by avoiding short cycle (cycle of four). Good performing code can efficiently perform image communication with less signal-to-noise Ratio (SNR) and less bit error rate (BER) with high peak signal-to-noise Ratio (PSNR) because an improvement in BER improves PSNR [5]. The code construction suggested using Sridara–Fuja–Tanner (SFT) structure and quadratic congruence gives BER of 10^{-5} approximately at 4.5 dB SNR for 20 iterations [16]. The construction suggested in [11] gives BER of 10^{-5} at SNR of 3 dB while the code construction suggested in [13] gives BER of 10^{-5} for 35 iterations at 4.5 dB SNR.

The parity-check matrix H obtained using this kind of constructions is used to make the generator matrix either by linear time encoding method or by Gauss–Jordan elimination method [10]. The generator matrix is used to encode the data. The linear time encoding keep parity-check matrix sparse [10]. The computational complexity using Gauss–Jordan elimination method to reduce parity-check matrix H into systematic form is $O(N^3)$ and complexity of encoder is $O(N^2)$, where N is code length [15]. In [15], other methods were suggested such as encoding based on the structure of the parity-check matrix, encoding by erasure decoding, which are the methods used for block codes. These methods encode the message with linear complexity. Decoding can be done using either by hard-decision decoding or by soft-decision decoding. The accuracy of soft-decision decoding is more compared to hard-decision decoding. The Sum–Product, min-sum and off-set min-sum decoding are the examples of soft-decision decoding. To reduce the computational complexity proposed decoder used min-sum iterative message-passing decoding algorithm. There is a coding gain difference of 0.27–1.03 dB between Sum-Product decoding and min-sum decoding algorithm [2]. According to author[15], the decoding complexity is $O(N)$, which is less compared to the encoder. According to author [4], the BER performance of SPA is improved by 1.8 dB compared to Bit Flipping Algorithm (BFA) of hard-decision decoding. Decoder design might be of bit-serial, partially parallel or fully parallel type [3]. The bit-serial design gives fewer throughput with the use of less resources. The partially parallel design gives throughput, which is more than bit-serial but less than fully parallel design. The fully parallel design gives more throughput with increasing hardware resources [3]. The bit-serial and partially parallel design require intermediate memory blocks. Proposed decoder design is based on fully parallel approach. In [1], partial parallel design was implemented on Virtex-7

FPGA for code length (1152, 2304) and obtained throughput of 300 Mbps for 20 iterations and 400 Mbps for 15 iterations. The design was implemented with the use of BRAM due to which the throughput was reduced and slice LUTs were consumed less. In [1] quantization of (6, 4) was used which is more than our proposed design for the code length 1152. In [6], partially parallel design was implemented for three different rates which gives throughput of 100 Mbps with 12 iterations. This design is based on the offset belief propagation algorithm. A fully parallel LDPC decoder was designed using (4, 1) quantization achieving a throughput of 16.9 Gbps at SNR value of 3.5 dB [3].

1.1 Code Construction

Proposed work construct the structured LDPC code using isolated shifted identity matrices known as quasi-cyclic LDPC code. The LDPC code is constructed for half rate, and it is of (3, 6) regular type LDPC code. Here 3 indicate column weight and 6 indicate row weight of a parity-check matrix H. The minimum distance linearly increased with code length due to a column weight of 3. For the code of column weight less than 3, the minimum distance is logarithmically increased [8]. The more minimum distance will increase the error-correcting capability of the decoder, according to Hamming. Proposed work is done for QC-LDPC code of length 1152. The parity-check matrix H constructed using the proposed method is as follows

$$H = \begin{bmatrix} I_1 & I_2 & I_3 & I_4 & I_5 & I_6 \\ I_6 & I_5 & I_4 & I_3 & I_2 & I_1 \\ I_2 & I_3 & I_4 & I_5 & I_6 & I_7 \end{bmatrix} \quad (1)$$

where I_1 to I_7 are the circularly shifted identity matrices of size p their rows are right shifted rows. It means first row is right shifted by ‘1’ with respect to last row and second row is right shifted with respect to first. The minimum distance of the above constructed code is $(J+1)!$ for (J, L) type regular code [7]. The rate of the code is $1 - \frac{w_c}{w_r}$ where w_c = weight of column and w_r = weight of row. According to that rate of our code is $\frac{1}{2}$.

2 Encoding and Decoding

2.1 Encoding

Encoding of a message can be done by converting parity-check matrix H into systematic form either by linear time encoding or by Gauss–Jordan elimination method. Proposed encoder design uses Gauss–Jordan elimination method. It converts parity-check matrix H into systematic form as follows:

$$H = [I|P] \tag{2}$$

where I is the identity matrix of size $(n-k) \times (n-k)$, where n is code length and k is the message length. A P is the matrix of parity bits, which is transposed and concated with the identity matrix of size $k \times k$ and form the generator matrix G which is represented by

$$G = [P^T|I] \tag{3}$$

Generator matrix G depends on the selection of parity-check matrix H. The performance of code depends on the construction of parity-check matrix H. The less number of short cycles in the parity-check matrix H gives the less possibility of error floor in BER plot. The less degree distribution will cause the degradation of a waterfall region of BER plot. As mentioned earlier, encoding of the message is done using generator matrix G.

2.2 Decoding

A min-sum message-passing iterative decoding is used to decode the message, which is given as follows. Initial Log-likelihood ratio (LLR) is calculated as given in (4).

$$\text{LLR}(y_i) \simeq \log\left(\frac{pY/X(yi|+1)}{pY/X(yi|-1)}\right) = \frac{2yi}{\sigma^2} \tag{4}$$

Variable nodes calculation is given in (5).

$$L_{ij} = \text{LLR}(y_i) + \sum_{k\epsilon C(i)/j} R_{ki} \tag{5}$$

Check nodes calculation is given in (6).

$$R_{ij} = \left(\prod_{k\epsilon v(i)/j} \text{sign}(L_{ki})\right) \min_{k\epsilon v(i)/j} |L_{ki}| \tag{6}$$

Final calculation of *Variable node* after each iteration is given in (7).

$$y_{fi} = \text{sign}\left(\text{LLR}(yi) + \sum_{k\epsilon C(i)} R_{ki}\right) \tag{7}$$

Hard decision taken at the end of given iterations or for exit condition test is given in (8).

$$\begin{aligned} c_i &= 0 \quad \text{if} \quad y_{fi} = +1 \\ &= 1 \quad \text{if} \quad y_{fi} = -1 \end{aligned} \tag{8}$$

3 Decoder Design of Code Length 1152

General diagram of LDPC decoder shown in Fig. 1. It mainly consist of *Input network*, *Output network*, *Control unit* and *Permutation network* of *Variable nodes* and *Check nodes*. It is a fully parallel design where all the *Variable nodes* and *Check nodes* are mapped. Heart of this decoder is *Control unit* which controls all the block on specific count which was synchronized with the system clock. Figure 1 shows that the encoded message in the form of 4-bit LLR's comes through AWGN channel and given to the input network then decoding process started in the form of *Variable nodes* and *Check nodes* process for specific iterations given to LDPC decoder. Decoded output comes through the *Output network*. Proposed decoder take 96 simulteneous inputs at a time inside the LDPC decoder. Within 12 clock cycles, all inputs (1152) are fed inside the *Input network* of the decoder. After decoding process is over, 96 simulteneous outputs comes outside of the LDPC decoder at a time through the *Output network*. Due to fully parallel architecture *Variable nodes* and *Check nodes* process are completed into one clock cycle in alternate manner so, for one iteration they required two clock cycles. The throughput of proposed decoder is approximately 2 Gbps due to fully parallel design but resource requirement is more. The throughput is calculated as follows:

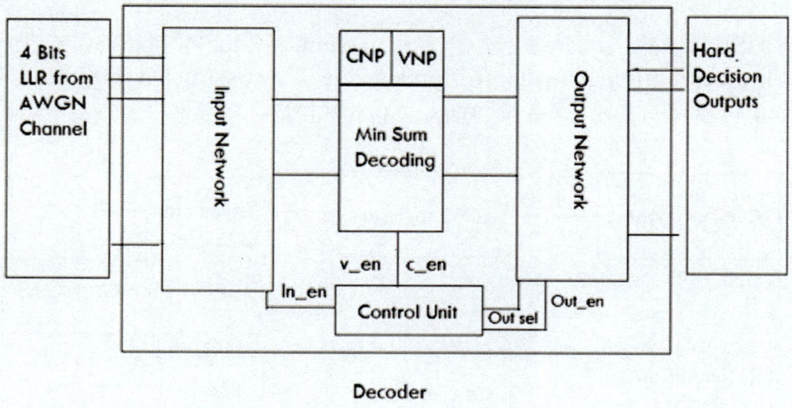

Fig. 1 General decoder architecture

$$\text{Throughput} = \frac{N \times r \times f}{itr \times \theta} \tag{9}$$

where N is code length, r is code rate, f is maximum operating frequency, itr is iteration and θ is clock cycles require to complete one iteration.

4 Image Communication Using QC-LDPC Code

The proposed work was done using MATLAB R2014a and Xilinx ISE Design Suite 14.2. The flow of Image communication using the proposed design is shown in Fig. 2. Image is read and converted into binary data which is then encoded using Gauss–Jordan elimination method on MATLAB R2014a. After encoding, image is modulated with BPSK and passed through the AWGN channel generating text files for various SNR values. Noisy data is sent to FPGA through UART and further decoded using min-sum iterative message-passing decoding algorithm on hardware. A decoded data is further sent to PC through UART. Then a decoded data is reconstructed using conversion from binary to image data on MATLAB R2014a. The proposed LDPC decoder is used to reconstruct the image at 4 dB similar to original image which can be observed from the peak signal-to-noise ratio (PSNR) value of reconstructed image. The quality of decoded image at different BER is also analyzed by computing peak signal-to-noise ratio (PSNR) with respect to original image. The PSNR is calculated as follows [5],

$$\text{PSNR (dB)} = 10 \times \log \frac{P_{\max}^2}{\text{MSE}} \tag{10}$$

$$\text{MSE} = \sum_{i=1}^{x} \sum_{j=1}^{y} \frac{(|A_{ij} - B_{ij}|)^2}{x \times y} \tag{11}$$

where MSE is mean-square error, $P_{\max}$ is maximum value of pixel in the image, A is pixel value of original image, B is pixel value of reconstructed image, x is height of image in pixels, y is width of image in pixels [5].

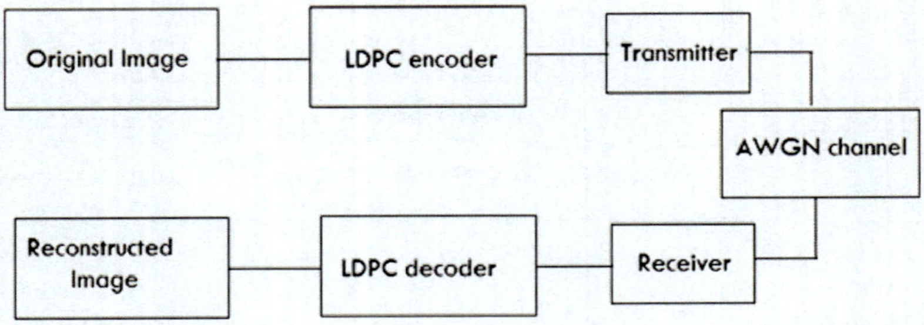

Fig. 2 Image communication

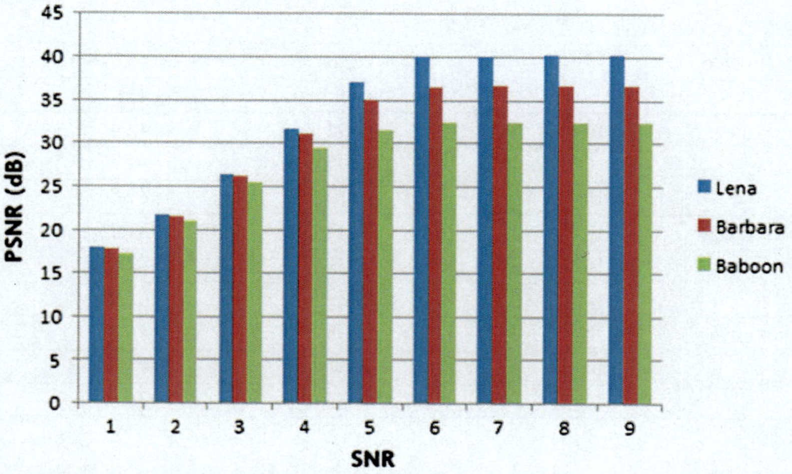

Fig. 3 PSNR versus SNR plot for different decoded image

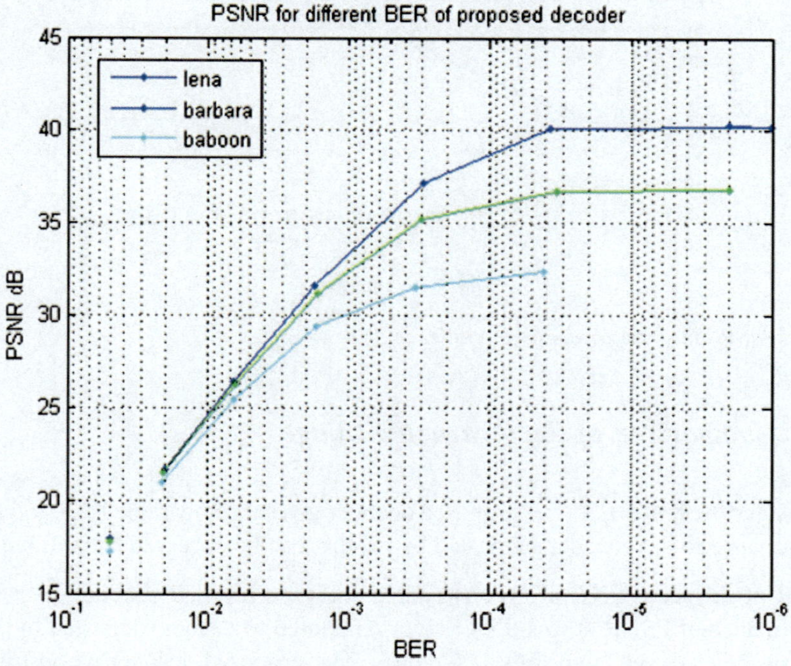

Fig. 4 PSNR versus BER plot for different Decoded image

The PSNR values for different SNR can be observed from Fig. 3. We can observe the plot of BER and PSNR from Fig. 4. We can observe that PSNR is increased as the BER is improved. We got maximum PSNR for the first image shown in Table 1 which was around 40 dB.

Table 1 Decoded images at different SNR

Reconstructed Image	BER	PSNR	
		Proposed	[5]
Lena image	10^{-4}	37.02	42.44
	10^{-5}	40.04	64.70
	10^{-6}	40.10	81.86
Barbara image	10^{-4}	31.43	41.21
	10^{-5}	32.31	52.83
	10^{-6}	32.39	62.08
Baboon image	10^{-4}	35.08	44.73
	10^{-5}	36.60	75.77
	10^{-6}	36.73	78.35

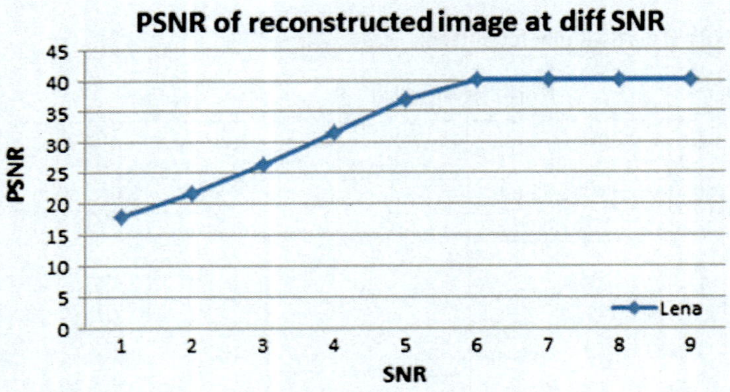

Fig. 5 PSNR of reconstructed image at different SNR

4.1 Comparison of Reconstructed Image

PSNR at different BER: The proposed decoder gives a comparable BER and PSNR performance to the decoder suggested by [5] that can be observed from Table 1.

PSNR at different SNR of reconstructed image: Proposed decoder gives better performance of PSNR at less SNR values compared to design suggested in [9, 18] that can be observed from Figs. 5, 6, and 7. The proposed decoder reconstruct the image at SNR value of 4 dB while the design suggested in [9] is able to reconstruct the image at SNR value of 12 dB while the image was reconstructed at SNR value of 4 dB but still having distortion [18]. Proposed decoder efficiently do image communication compared to design suggested in [9, 18]. The reconstructed images can be observed from the comparison shown in Table 2.

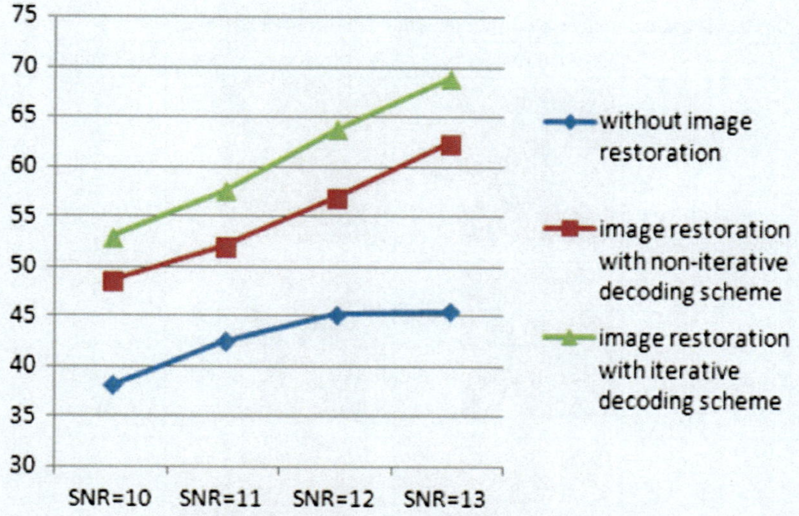

Fig. 6 PSNR of reconstructed image at different SNR [9]

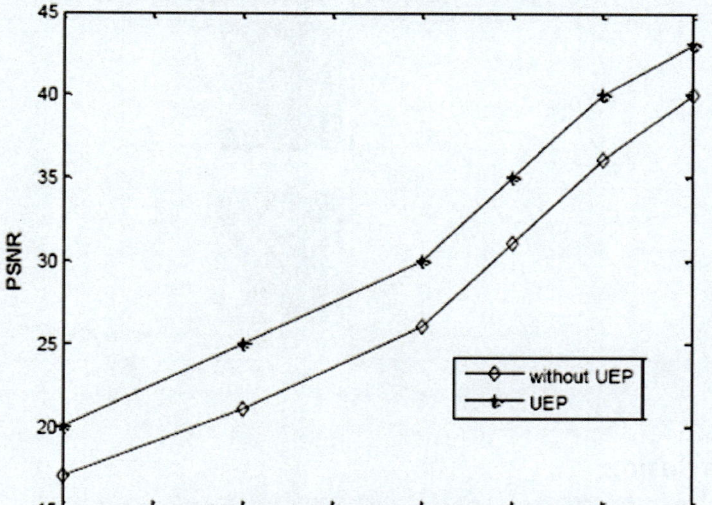

Fig. 7 PSNR of reconstructed image at different SNR [18]

Table 2 Reconstructed images by different authors at different SNR

	Original image	Reconstructed image	SNR value
Proposed			4 dB
[5]			3.5 dB
[9]			12 dB
[18]	–	 without UEP	4 dB
[18]	–	 UEP3	4 dB

5 Conclusion

Proposed decoder efficiently reconstructs the image with the good quality compared to design suggested in [9, 18]. It also gives a comparable performance of BER to the design suggested in [5]. Due to the use of min-sum decoding algorithm, the computational complexity of decoder is less, and there is not any alternate solution of a min-sum decoder which is less complex. Due to fully parallel architecture, the proposed decoder communicates an image at a high data rate.

Acknowledgments The authors would like to thank the VLSI Laboratory of Electronics engineering department, Sardar Vallabhbhai National Institute of Technology (SVNIT) surat for providing resources for this work.

References

1. Amaricai A, Boncalo O, Mot I (2015) Memory efficient FPGA implementation for flooded LDPC decoder. In: 2015 23rd telecommunications forum Telfor (TELFOR), pp 500–503
2. Anastasopoulos A (2001) A comparison between the sum-product and the min-sum iterative detection algorithms based on density evolution. In: IEEE global telecommunications conference GLOBECOM'01 (Cat. No.01CH37270), vol 2, pp 1021–1025
3. Balatsoukas-Stimming A, Dollas A (2012) FPGA-based design and implementation of a multi-GBPS LDPC decoder. In: 22nd international conference on field programmable logic and applications (FPL), pp 262–269
4. Chandrasetty V, Aziz S (2011) FPGA implementation of a LDPC decoder using a reduced complexity message passing algorithm. JNW 6:36–45
5. Chandrasetty V, Aziz S (2014) Resource efficient LDPC decoders for multimedia communication. Integr VLSI J 48:213–220
6. Chen Y, Chen X, Zhao Y, Zhou C, Wang J (2010) Design and implementation of multi-mode QC-LDPC decoder. In: 2010 IEEE 12th international conference on communication technology, pp 1145–1148
7. Fossorier MPC (2004) Quasicyclic low-density parity-check codes from circulant permutation matrices. IEEE Trans Inf Theory 50(8):1788–1793
8. Gallager R (1962) Low-density parity-check codes. IRE Trans Inform Theory 8(1):21–28
9. Jeng L-D, Chang Y-H, Lee TH (2014) Image transmission for low-density parity-check coded system under PBNJ. In: The fourth international conference on digital information processing and communications (ICDIPC2014), pp 101–106
10. Johnson SJ (2010) Introducing low-density parity-check codes
11. Kong L, Xiao Y (2008) Construction of good quasi-cyclic LDPC codes based on the row vectors of generator matrix. In: IET 2nd international conference on wireless, mobile and multimedia networks (ICWMMN 2008), pp 215–218
12. MacKay DJC, Neal RM (1997) Near Shannon limit performance of low density parity check codes. Electron Lett 33(6):457–458
13. Malema G (2007) Low-density parity-check codes: construction and implementation
14. Tanner R (1981) A recursive approach to low complexity codes. IEEE Trans Inf Theory 27(5):533–547
15. Tanner RM, Sridhara D, Sridharan A, Fuja TE, Costello DJ (2004) LDPC block and convolutional codes based on circulant matrices. IEEE Trans Inf Theory 50(12):2966–2984
16. Timakul S, Choomchuay S (2011) Construction of quasi-cyclic LDPC codes form SFT structure and cyclic shift. In: 2011 international symposium on intelligent signal processing and communications systems (ISPACS), pp 1–4
17. Wang Y, Draper S, Yedidia J (2013) Hierarchical and high-girth QC-LDPC codes. IEEE Trans Inform Theory 59:4553–4583
18. Zhang Y, Li X, Yang H (2016) Unequal error protection in image transmission based on LDPC codes. Int J Signal Process Image Process Pattern Recognit 9:1–10

Smart Soldier Health Monitoring System Incorporating Embedded Electronics

Krishna Teja, Umang Patel, Parthkumar Patel, Yash Agrawal, and Rutu Parekh

Abstract A smart soldier health monitoring system using embedded electronics and wireless communications is presented to keep real-time track of conditions of all the soldiers on the battlefield. A wireless body area network (WBAN) is created by treating each soldier as a node and each node consists of different sensors monitoring the condition of that particular soldier which could be integrated inside a uniform or a bulletproof jacket. The attributes tracked by the system are motion and physiological health of the soldier and the environment. After gaining the knowledge of these attributes, the control room can take decisions accordingly and create an effective battle strategy.

Keywords Embedded systems · Health monitoring · LoRa module · SPO2 sensor · Arduino MEGA · ECG sensor

1 Introduction

The military forces of a country play an important role in the external as well as the internal security of the country. The same is attained by activities such as patrolling, counter-insurgency operations, surgical strikes, and so on. The modus operandi of these activities is deputing a group of soldiers with a well-structured chain of command who operate as per the instructions of the designated leader. During every

K. Teja · U. Patel · P. Patel (✉) · Y. Agrawal · R. Parekh
DA-IICT, Gandhinagar, India
e-mail: parthkumarpatel101@gmail.com

K. Teja
e-mail: krishnateja87@gmail.com

U. Patel
e-mail: umangsp.1199@gmail.com

Y. Agrawal
e-mail: yash_agrawal@daiict.ac.in

R. Parekh
e-mail: rutu_parekh@daiict.ac.in

Z. Patel et al. (eds.), *Advances in VLSI and Embedded Systems*, Lecture Notes in Electrical Engineering 676, https://doi.org/10.1007/978-981-15-6229-7_18

operation, there is a controlling base station which monitors the overall activities of many such troops and keeps a track of them primarily through communication with RF devices such as walkie-talkie and communication sets. Early retrieval of information regarding the activities happening in the area of the operation plays an important role in making a decision and planning for further actions. In practice, when the soldier does not respond to the reporting medium such as walkie-talkie, communication sets, or does not turn up in providing the situational report in the stipulated time, a casualty is assumed. This procedure is timeworn and antiquated. During patrolling, the group leader generally subdivides the group into a subgroup of few soldiers, and the area to be patrolled is divided and assigned to each subgroup. A fixed time interval is decided by the group leader and the subgroups submit the situational report (called SITREP) according to the frequency of time fixed. Same is informed to the commander at the base station. In the case of action or attack during patrolling, there is no communication between the subgroups. The group leader does not know the location of the subgroup until they physically come and make the SITREP. Due to the heavy time lag in obtaining information about any kind of casualty, the group leader cannot swiftly take effective decisions such as providing backup cover, sending additional manpower to the affected area, and so on, thus, dwindling the overall fighting efficacy of the troop. The proposed model aims at the design of a prototype for a smart soldier health monitoring system. It is an incorporation of embedded systems and body area network for automatically providing information on the occurrence of any casualty in a troop. Incorporation of wearable electronics with wireless networks would rally round in improving the fighting efficiency of the troops with minimal casualty.

2 Literature Review

The functioning of a group of soldiers in the battlefield and tactical scenario can be augmented by monitoring the activities and physiological parameters. The same requires precise information about the soldier's activities, health, and position [1]. Recognition of human activities plays an essential role in such requirements and is carried out in two ways, namely through external ways and through wearable sensors, where in tactical scenario the latter is the essential requirement. A system that monitors personnel conditions includes a variety of sensors that can be put on a soldier for measuring parameters like the physiological and activity parameters. The sensors upon detection of the signals communicate with a soldier unit which can process the information. Location parameters can be obtained through a global positioning unit through which the leaders can effectively locate the soldiers and take necessary action in a tactical scenario [2]. Further, the same also helps in providing the medical help to track and treat any of the casualties in the battle. Along with that, additional sensors can provide information which may assist both the group leader as well as the medicos in understanding the area of deployment and take necessary measures before deployment to the area. The primary parameter which needs to be measured in this scenario is the motion parameter of the soldier which determines the condition of the soldier. An inertial measurement unit can be best preferred for the same. In addition to the application of accuracy and precision, this application requires high shock survivability and post-shock stability. The MEMS-based sensors can be effectively used for motion detection as the size, accuracy, and precision match the requirements for parameter measurement of a soldier. For measuring the activities of a soldier, any MEMS-based gyroscope which can measure 6–9 degrees of freedom is preferable. To cover the same, three gyroscopes and three accelerometers are essential [3].

Kozlovszky et al. in their work on IMU-based human motion tracking [4] described the usage of an inertial movement unit-based wearable rehabilitation movement tracking device which can be used for human motion tracking. A realized software, using a multisensory-based fusion solution where the data is taken from a multiple-axis accelerometer and gyroscopic sensor devices, has been implemented for detecting the motion parameters. The second parameter which is essential for the appraisal of the condition of soldiers in the battlefield or forward area is their physiological condition. The efficacy of combat medics can be greatly improved by increasing the speed and precision in which the physiological information is gathered from the soldiers who are wounded [5]. For monitoring the health of a soldier, multiple physiological parameters can aid in multiple diagnoses and help in determining the most effective forms of the treatment. The pulse rate of a soldier is calculated through measurement of heart rate and arterial oxyhemoglobin saturation (SpO2) [6]. Further, the heart rate variability plays an important role in identifying critically injured patients and is a latent predictor for mortality. The task force of the European Society of Cardiology and The North American Society of Pacing and Electrophysiology has proposed a study bringing out the various applications of heart rate variability of a human being [7]. The study brought out various measures of determining the heart rate variability, analysis, and evaluation methods. With this study, it is clear that the pulse rate measurement can be highly effective in measuring the physiological condition of the soldier. At locations where the atmospheric conditions have a deep impact on the soldiers, ECG monitoring of the soldier is preferred for detailed health monitoring.

Measurement of the atmospheric conditions in which the soldier is operating is also essential to enable the commander in control make necessary changes on the backup of troops as well as the medical help being deployed. In a field manual of US army, various weather parameters affecting the soldier in a battlefield were brought out, with the primary ones being temperature, humidity, and altitude in which the troops are operating. The measured data of the atmospheric parameters at the area of operation when transmitted to the control station gives a clear picture of the scenario and necessary corrective actions can be taken by the commander on the medical backup being deployed as well. Measurement of air quality in the area of operation would be extremely helpful for the tactical decision-makers in making effective decisions on further deploying the troops or providing additional backup. Furthermore, the data of the measured air quality would also assist in perspective health monitoring. Thus, the essential sensors and components required for making a smart soldier system have been defined. The next prerequisite for the smart soldier system is the transmission of measured values and signals as per the algorithm defined for various conditions of the soldier. The transmission module should consume less power and should be adaptable to various electronic warfare techniques, such as frequency agility and frequency hopping. Low power supply consumption is essential as the same would help in the conservation of the portable power supply, which is generally battery. Sensor connectivity plays an important role in the system. As the system is being operated in a tactical environment, the accuracy, precision, and connectivity of the sensors should be paramount, and like any sort of minute change in the activity and conditions should be detected at the immediate instinct [8]. The connectivity of sensor can either be wired or wireless. The wireless connectivity of sensors in the form of a body area network is best suited for tactical scenarios [8].

The wireless network of sensors on a human body is called as wireless body area network. A wireless body area network (WBAN) is a system that is capable of transmitting a soldier's positional information as well as physiological information to a central control unit for storage and tactical data analysis. The primary aim of WBAN is to abridge and enhance the speed, accuracy, and reliability of communication of sensors and actuators within, or on immediate proximity of a human body. Incorporating WBAN would enable continuous monitoring of physiological and environmental attributes. In the case of detection of any abnormality, the data collected by the sensors can be sent to a gateway [9]. The activities, condition, and environment affecting the soldier on the battlefield can be effectively monitored through wireless body area network. This can be carried out by integrating various sensors, GPS, and wireless networking combined with an aggregation device for communication with other soldiers and centralized monitoring [10]. WBAN, in the proposed system uses star topology in which several sensor nodes are mounted on various locations of the human body (soldier). They collect data related to various parameters and send them to the coordinator node. The coordinator node is responsible for conveying the data after aggregation to a central location for further analysis and planning [11].

The work done in this area had several constraints till now which are addressed in the presented work. One of the already published papers presented a system which used GSM module to transfer the sensor data to the control station [12]. However, the system given in the presented paper uses LoRa module which is less power-consuming than GSM, and there is no requirement of long range GSM. Furthermore, GSM signals are not available in many places, such as national borders and mountain regions. Another work used K-means clustering algorithm to determine the condition of the soldier [13]. The use of machine learning algorithm requires more computation power and more time for analyzing the data. The system presented in this paper determines the condition of the soldier based on the real-time data read from multiple sensors and makes the determining algorithm simple. One more already published paper used WiFi module to transmit the data to the base station [14]. The range of WiFi module is very less with respect to the requirement of the range for the system to work properly. Along with that, it has high power consumption than the LoRa module. Thus, the improvement in the power consumption and the range has been carried out in the presented paper. Moreover, the system also uses motion sensor which is the key point of the algorithm and along with that, various other sensors are used to get a detailed analysis of the soldier's health in real time.

3 Block Diagram and Working of Soldier Health Monitoring System

The block diagram in Fig. 1 has been conceptualized using sensors for measurement of casualty, physiological health monitoring, geographical area indication, and detection of environmental condition. The processing circuitry and transmission section form the other part of the system.

The body movement detection sensor is either an inertial measurement unit (IMU) or a gyroscopic accelerometer with suitable degrees of freedom. It would be transmitting signals as per the movement of the human body. When the soldier does not move due to a casualty or injury, this sensor would not transmit any signal to the processing circuitry. The non-receipt of the signal from the sensor can be treated as an indication of the casualty. Vibration detection sensor indicates if a soldier experiences a bullet wound. If a soldier is shot, he would experience a vibration wave above particular threshold intensity as a result of the gunshot. The same can be established on receiving a signal from the shock and vibration sensor. Information about the environment and of surrounding areas like the light intensity, temperature, humidity, and direction is provided by the respective sensors. In case the soldier intends transmitting a code during operation, the same can be done by the keypad provided. If

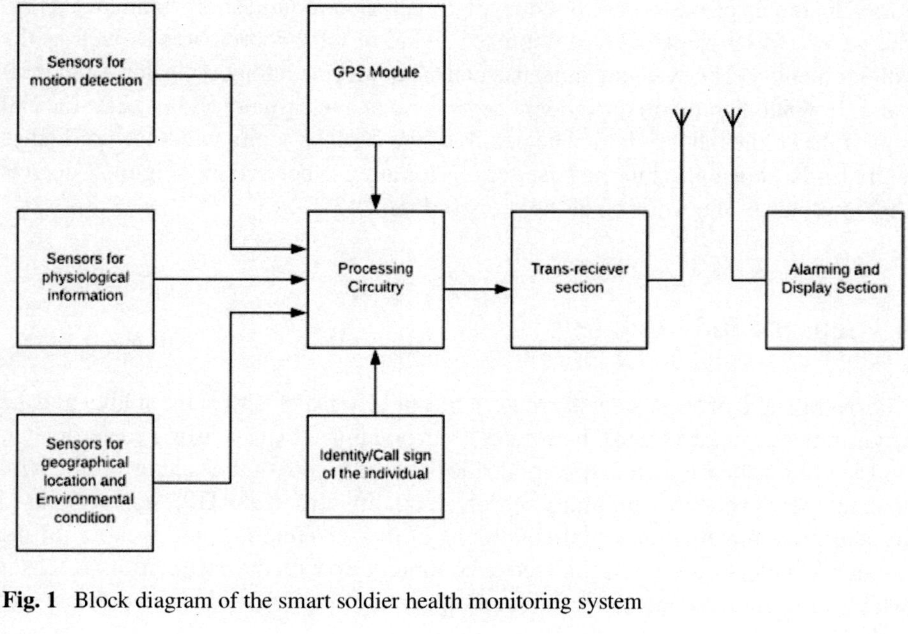

Fig. 1 Block diagram of the smart soldier health monitoring system

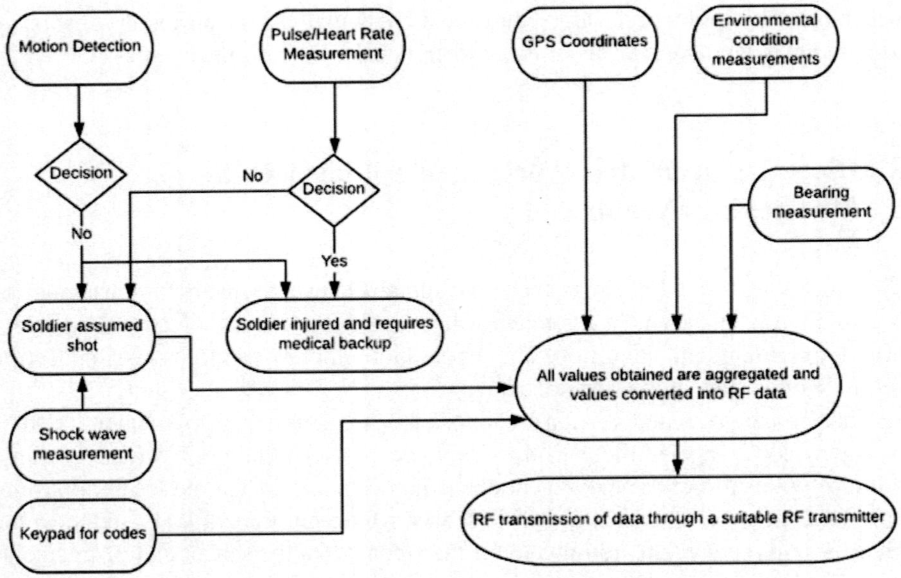

Fig. 2 Workflow diagram of the smart soldier health monitoring system

GPS coordinates and identification code of all the soldiers of a particular group are transmitted to the base station, it can be considered as a biological warfare attack. The data generated by the sensors as per the conditions are fed to the processing circuitry. The inputs from the GPS chip of the particular node transmitting would be multiplexed with the identity code of the soldier. The processed signals are converted into RF signals of the suitable frequency of IEEE communication standards. Thus, the generated RF signal is a multiplexed signal of GPS coordinates as well as the soldier identity. The system would also consist of receiver along with alarm and indicator. It would first alarm the other nodes in case of any casualty and indicate the call sign if the soldier shot or hurt. The details of the casualty along with the coordinates will also be transmitted to the base station for any further actions. Figure 2 depicts the flowchart of the working of the proposed model.

4 Sensors and Modules

The overall goal of the system is to determine the physical health of the soldier and the environmental conditions of the soldier's surroundings. ECG sensor, accelerometer, and SPO2 sensor are used to determine the physical health of the soldier. Along with that, temperature sensor, humidity sensor, air quality sensor, and GPS sensor are used to determine the environmental conditions of the surroundings and location of the soldier. Finally, LoRa module is used to communicate with the base station. A sensor which can simultaneously measure motion parameters such as body movement and vibration as well as detect high-frequency shock waves is required. In the proposed smart soldier system, due to injury such as bullet wound, this kind of sensor will be helpful. An inertial measurement unit sensor (IMU sensor) would gratify the above-mentioned requirements, as IMU sensors are a combination of accelerometer and gyroscope which indicate the position of the object in three-dimensional space as well as measure any sort of vibrations. In the real-time scenario, when a soldier is in the same position for a longer duration, it is understood that he has sustained some sort of injury and is unable to move because of it. The measurement of relative position with an IMU would indicate such a condition. Further, if a soldier is shot, the impact of a bullet would generate shockwaves throughout the body of the soldier which can be measured with a piezoelectric gyroscope of the IMU.

Three parameters, namely heartbeat, pulse rate, and stress level, are extremely essential to assess the physiological stipulation of the individual. Every parameter would have a minimum and maximum threshold value (heartbeat/pulse rate: 60–100 beats). If the heartbeat/pulse rate goes below or above the threshold level, an alarming indication would alert the group members, group leader as well as the control station. A sensor which can measure the heart rate of the soldier as well as the blood oxygen saturation can be used as the stress level measurement sensor. This sensor is called heart rate and particle sensor. In the working model of the proposed work, MAX 30105 is being used. Measurement of barometric pressure would facilitate an assessment of the atmospheric conditions in which a soldier is currently operating. Factors such as atmospheric pressure in the area of operation and altitude of the area of operation above sea level would result in better appreciation of a soldier's area of operation. The sensor being used for this purpose is BMP 280. To measure the temperature and humidity of the area of operation in which the soldier is deployed, this particular sensor can be used. The primary requirement of measuring these parameters is to appreciate the physical conditions of the area of operation where the soldiers are deployed. In the working model of the proposed system, DHT22 is used. DHT22 is a digital humidity and temperature sensor. It uses a capacitive humidity sensor and thermistor.

Arduino MEGA 2560 is used as a microcontroller in the proposed system to collect the data from the sensors, analyze the data, and display the output. In the proposed model, if a soldier gets injured upon detection by MPU 6050, the GPS indicator would be transmitting the location details. The location details would enable the group leader as well as the control station to take necessary responsive actions. The module proposed to be used is GPS module NEO 6M A6 UBLOX. To make a body area network, a wireless interconnection among the nodes is a mandate. In the real-life scenario, an earmarked frequency used by the Army can be used in making the sensor net. However, in the prototype, LoRa module is being used. The module used here is RFM95 (W) for data transmission between the hardware and computer (considered to be the controlling station). LoRa is a wireless protocol designed for long-range low-power communication. The advantage of using a LoRa module is that the frequency cannot be intercepted which makes the sensor net safer to use. Further, using LoRa technology would increase the duration of power backup. All of

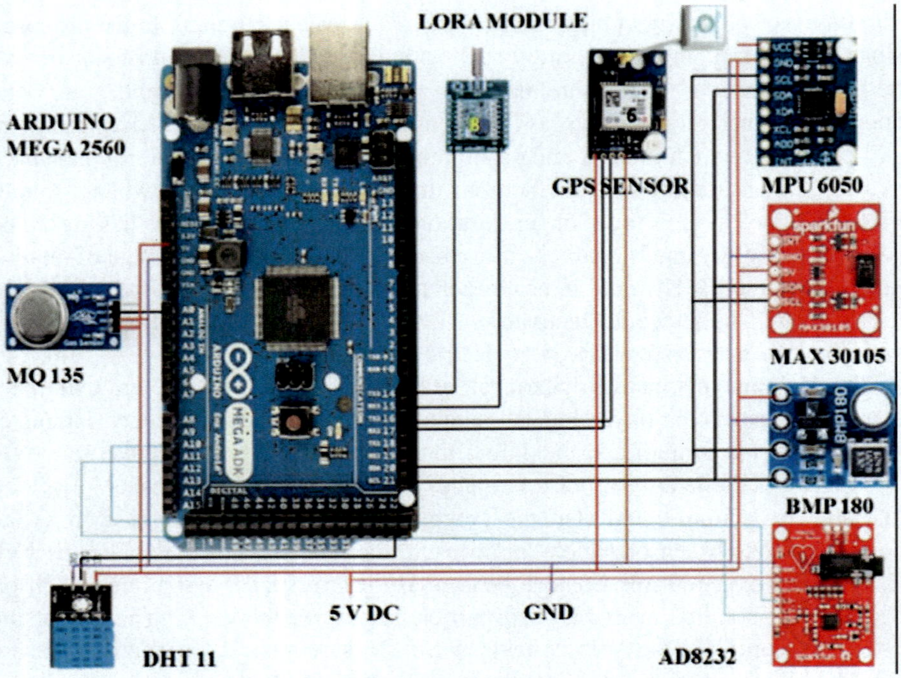

Fig. 3 Circuit diagram of the system

the sensors are selected based on the requirements of the accuracy. Figure 3 depicts the circuit diagram of the system.

5 Modeling and Results

The development, modeling, and analysis of the system are split into two categories. These two categories are namely algorithm and results. The components are first programmed as per the designed algorithm. The sensors have been individually interfaced and results have been analyzed. A graphical user interface has been created by using the concept of threading in PYTHON. A user interface has been created using QT designer editor and the real-time data of the system is displayed in the GUI module created in QT designer. The condition of the soldier along with the parameter values of the sensors is displayed in the GUI, which is assumed to be located at the central control station.

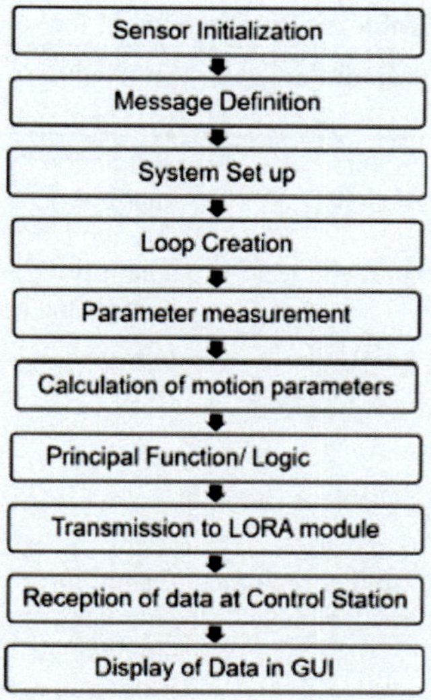

Fig. 4 Algorithm for development of the system

5.1 Algorithm

The steps involved in the development of the system are as in the algorithm shown in Fig. 4. In the first step, each of the sensors is initialized with their respective requirements. Thereafter, three messages are defined, which are: when the soldier is healthy, the soldier is wounded, and the soldier is dead. The sensors of the system are set up and a loop runs which iteratively takes the reading from the sensors. After measuring the readings from the sensors, the required calculation is done to determine the condition of the soldier using a principal logic. The data is then transmitted to LoRa module which in turn is received at the control station and the data is displayed on the GUI. In addition, Table 1 shows the range of all the parameters to determine the condition of any particular soldier.

5.2 Results

An experimentation was performed at first on a healthy human and then on a non-living thing which depicts the condition of soldier being healthy and dead. Later, to simulate the condition of a soldier being wounded, ECG and SPO2 data for a wounded person was taken and fed to the system [15]. The responses on the GUI for three cases are depicted in Fig. 5a–c from which Fig. 5a is the case when the soldier is healthy, Fig. 5b is the case when the soldier is dead, and Fig. 5c is the case when the soldier is shot or wounded. The condition of the soldier is determined on the basis of the parameters read.

Table 1 The condition of the soldier and the corresponding range of parameters

Condition	Simulation	Motion sensor (MPU 6050)	Pulse sensor (MAX 30105)	ECG sensor (AD 8232)
Soldier is healthy	Simulation by measurement through normal condition of motion and pulse rate	Acceleration X: 1–16; Y: 1–16; Z: 1–16; Pitch: 1–255; Roll: 1–255; Yaw: 1–255; Movement X: 00/01; Y: 00/01; Z: 00/01; Angle X: 0–360°; Y: 0–360°; Z: 0–360	Pulse rate between 45 and 120 (IR value from 10500 to 10600)	Heart rate between 45 and 120 resulting in the following intervals. PR interval from 0.12 to 0.20 s, QRS interval from 0.06 to 0.1 s
Soldier requires medical help	Condition is simulated by showing no motion in the body or low pulse rate through controlled respiration	Soldier assumed to be injured if any of the acceleration and movement values are read as zero	Pulse rate less than 45 (IR value less than 10500)	Heart rate less than 45; PR interval less than 0.1147; QRS interval less than 0.0585
Soldier is dead	Condition is simulated with no motion in the body and controlled respiration	Soldier is assumed to be dead if all the values are read as zero for a significant amount of time	Pulse rate is measured to be zero	Heart rate equal to 0 resulting in the no difference in time period of PR and QRST intervals

6 Conclusion

This paper presents a successfully implemented smart soldier health monitoring system which has the potential to improve the military operations substantially. The conditions and locations of all the soldiers are known to the control base station and thus swift decisions can be taken to successfully carry out the military operation. The proposed system can be embedded into soldier's uniform and if any particular group of soldiers is attacked by enemy soldiers, then immediate backup is provided to prevent high number of casualties. Hence casualties will also decrease which in turn increases the morale of the army. Moreover, the system also detects the presence of biohazards in the environment, which in turn helps the soldiers in their operations. Therefore, it can be concluded that the use of the smart soldier health monitoring system improves upon the traditional methods of executing military operations considerably.

a

ECG Pulse	Normal
Temperature:	36 *C
Humidity:	25 %
Altitude:	1.5 Mtr
Air Quality:	84 %
SPO2:	78 P/M
Air Pressure:	25.02 InHg
Gps Coordinates	
Long	69.075850
Lat	22.475690
Soldier Motion	Detected
Soldier Status:	Good

b

ECG Pulse	Not Available
Temperature:	36 *C
Humidity:	24 %
Altitude:	1.5 Mtr
Air Quality:	83 %
SPO2:	0 P/M
Air Pressure:	25.04 InHg
Gps Coordinates	
Long	69.075851
Lat	22.475696
Soldier Motion	Not Detected
Soldier Status:	soldier is dead

c

ECG Pulse	Abnormal
Temperature:	37 *C
Humidity:	24 %
Altitude:	1.5 Mtr
Air Quality:	85 %
SPO2:	39 P/M
Air Pressure:	24.98 InHg
Gps Coordinates	
Long	69.075856
Lat	22.475697
Soldier Motion	Not Detected
Soldier Status:	soldier is injured and required medical help

Fig. 5 **a** Output when soldier is healthy, **b** Output when soldier is healthy, **c** Output when soldier is wounded

References

1. Lara OD, LabradorMA (2012) A survey on human activity recognition using wearable sensors.IEEE Commun Surv Tutorials
2. Jacobson et al (2001)System for remote monitoring of personnel. United States patent number US 6198394 B
3. Habibi S, Cooper SJ, Stauffer J-M, Dutoit B (2008)Gun hard inertial measurement unit based on MEMS capacitive accelerometer and rate sensor.In: 2008 IEEE/ION position, location and navigation symposium, IEEE, pp 232–237
4. Nguyen, KD, Chen I-M, Luo Z, Yeo SH, DuhBLR (2010) A wearable sensing system for tracking and monitoring of functional arm movement.IEEE/ASME Trans Mechatron 16(2):213–220
5. Johnston W, MendelsonY (2005) Extracting heart rate variability from a wearable reflectance pulse oximeter.In: Proceedings of the IEEE 31st annual northeast bioengineering conference, 2005, IEEE, pp 157–158

6. Johnston WS, Mendelson Y (2004) Extracting breathing rate information from a wearable reflectance pulse oximeter sensor.In: The 26th annual international conference of the IEEE engineering in medicine and biology society, vol 2, IEEE, pp 5388–5391
7. Standards of Measurement, Physiological Interpretation, and Clinical Use. Task force of the european society of cardiology and the north american society of pacing and electrophysiology
8. Montgomery RR, Anderson YL (2016) Battlefield medical network: biosensors in a tactical environment. Naval Postgraduate School Monterey United States
9. Movassaghi S, Abolhasan M, Lipman J, Smith D, Jamalipour A (2014) Wireless body area networks: a survey. IEEE Commun Surv Tutorials 16(3)
10. Movassaghi S, Abolhasan M, Lipman J, Smith D, Jamalipour A (2014) Wireless body area networks: a survey. IEEE Commun Surv Tutorials 16(3):1658–1686
11. Arai H (2016) Smart suit for body area network. IEEE J Asia Pac Conf Appl Electromagnet
12. Shweta et al (2015) Soldier tracking and health monitoring systems. In: International journal of soft computing and artificial intelligence, vol 3, Issue 1
13. Aashay et al (2018) iot-based healthcare monitoring system for war soldiers using machine learning. In: International conference on robotics and smart manufacturing (RoSMa2018)
14. Pawar P, Desai A (2018) Soldier position tracking and health monitoring system: a review. In: International journal of innovative research in computer and communication engineering, vol 6, Issue 3
15. Physionet, the research resource for complex physiological signals. https://archive.physionet.org/

A 3–7 GHz CMOS Power Amplifier Design for Ultra-Wide-Band Applications

Vishakha Purnanand Bhale and Upena Dalal

Abstract In the proposed work, 3–7 GHz two-stage Ultra-Wide-Band (UWB) power amplifier has been designed and simulated using UMC 0.18 μm CMOS technology. The proposed Power Amplifier (PA) design exhibits the simplest topology to fulfill the requirement of low gain ripple, good linearity, and small group delay variations all simultaneously for Ultra-Wide-Band (UWB) application and covering only 0.75 mm^2 area of silicon. In order to obtain low gain ripple, good group delay variations and linearity at the same time, the inductive peaking and class-AB operation are employed at both the stages. The post-layout simulation results show lowest gain ripple of 19 $\pm$ 0.3 dB, good input and output matching (<-7.5 and <-6 dB), moderate group delay ($\pm$200 ps), unconditional stability, good linearity, and low power performance simultaneously for the 3–7 GHz UWB band. The design of power amplifier with these features is found to be suitable for commercial mobile or Group 1–3 of MB-OFDM UWB transmitters.

Keywords CMOS RF integrated circuits · Cascode amplifier · Feedback · Lowest gain ripple · Shunt peaking

1 Introduction

In the past few years, the UWB communication technique is considered to be a promising wideband communication technique for short-range and high-speed wireless mobile communication applications. However, it is found that the circuit level implementation of some of the key RF building blocks is very critical while showing its performance at Giga-Hertz frequency range for the application like UWB. Few important building blocks that are previously designed includes Power Amplifier (PA) [1], Low Noise Amplifier (LNA) [2–5], Voltage Controlled Oscillators (VCO)

V. P. Bhale (✉) · U. Dalal
Sardar Vallabhbhai National Institute of Technology, Ichchhanath, Surat, Gujarat, India
e-mail: v.bhale@rediffmail.com

Z. Patel et al. (eds.), *Advances in VLSI and Embedded Systems*, Lecture Notes in Electrical Engineering 676, https://doi.org/10.1007/978-981-15-6229-7_19

[6, 7], and Mixers [8] in an UWB transceiver. While designing PA on UWB transmitter side it must meet some of the stringent requirements such as achieving broadband input and output matching, achieving high power gain, reasonable efficiency, and low power consumption since PA is the power-hungry block on UWB transmitter side.

In the literature, several implementations of wideband amplifiers are already reported such as the resistive shunt feedback topology [9], the RLC matching topology [10–13], distributed amplifier [14]. Among them, the distributed amplifiers can provide good matching and linearity over a wide band of frequencies but it dissipates more power and occupies more area on chip. Whereas, RLC matching technique can provide both wideband and low power but it needs a number of reactive elements to form the wideband and thus occupies large chip area and its layout design is also complicated. In contrast, resistive shunt feedback technique can provide not only wideband matching at both input and output port but provides added advantage of providing stability in spite of occupying less area on chip [9]. The proposed PA thus uses resistive shunt feedback in its second stage.

The UWB LNAs reported previously [15, 16] proposed Current-Reused (CR) technique to enhance the gain of an amplifier at the upper end of the desired bandwidth. The same CR technique has been incorporated in the designing of UWB PA design in the present work to achieve the required gain for 3–7 GHz bandwidth. Further, the PA designs reported previously could achieve a minimum chip area of 1 mm^2 only, which needs to be further reduced to reduce the design cost of the overall transmitter design. The phase linearity is another important performance parameter that needs special attention in PA designs. These PA designs reported so far could achieve the lowest gain ripple of about ±0.6 dB [17] within the entire bandwidth. The phase ripples in the desired operating frequency band should be reduced below ±0.6 dB in designing UWB PA. Furthermore, the previous UWB PAs consumed chip area nearly 1 mm^2, which could increase the cost for transmitter design. There is a trade-off between efficiency, linearity, and low gain ripple requirement in designing UWB PA.

For the proposed PA design, this trade-off is solved with moderate efficiency, good linearity, and low gain ripple all fulfilling simultaneously without using any special techniques for gain flatness, linearity, and low gain ripple. In [15, 16], the principle of Current-Reused (CR) is used to gain flatness and matching while stagger tuning is used in [18] for the same. While for the proposed PA design, instead of using special techniques as used in [15, 16, 18]; effort has been taken in designing the optimum R, L, C components to achieve all above parameters simultaneously and it further makes proposed design more simpler than previously reported UWB PA designs [17–20]. As a result, the proposed PA design exhibit good linearity, low gain ripple of about $\pm$ 0.3 dB, moderate efficiency simultaneously.

2 Circuit Design

As a novelty, the proposed PA design exhibits the simplest design topology to fulfill the requirement of low gain ripple, good linearity, and small group delay variations all simultaneously. Murad et al. [17, 18] in his work, as depicted in Fig. 1a, used a Current Reused (CR) cascoded Common Source (CS) structure as the first stage to boost the gain since the input signal gets amplified twice with this technique for higher gain [18]. Also, inter stage inductor is used to further improve the broadband operation toward higher frequency of 7 GHz. Figure 1b shows the proposed 3–7 GHz UWB PA which used less number of lump components for simplicity and those components have been carefully designed and optimized to achieve performance improvement over [17, 18] for low gain ripple, good linearity, and efficiency simultaneously. The proposed power amplifier employs a cascode topology with inductive shunt peaking on the first stage and Common Source (CS) with resistive feedback as the second stage. The feedback resistance, R_3, here has been designed carefully to obtain resistive shunt feedback, wideband output matching, and to bias the transistor M_3. At the input stage, resistive shunt feedback R_1 is used to achieve good input impedance matching over the required bandwidth. This feedback resistance can provide excellent matching to the input by selecting small value of R_1. However, the gain of the amplifier will drop due to significant signal feedback through the path. A large value of R_1 can provide good gain but reduces the effective feedback.

Assuming a current of 10 mA to be drawn by transistor M_1, the calculated size for NMOS transistor M_1 is approximately 105 μm under saturation:

$$I_{D1} = \frac{1}{2}\mu_n C_{ox}\frac{W_1}{L}(V_{GS1} - V_T)^2 \quad (1)$$

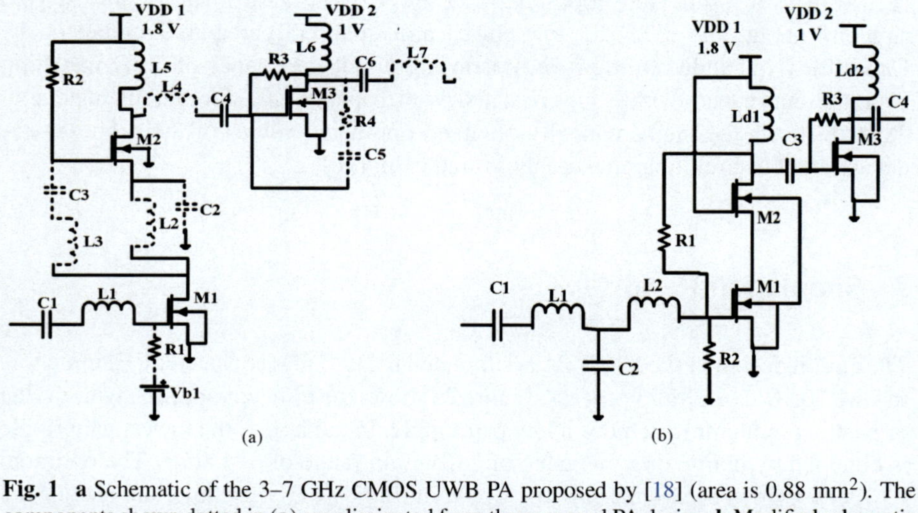

Fig. 1 **a** Schematic of the 3–7 GHz CMOS UWB PA proposed by [18] (area is 0.88 mm^2). The components shown dotted in (**a**) are eliminated from the proposed PA design. **b** Modified schematic for 3–7 GHz CMOS UWB PA proposed in the present work (area is 0.75 mm^2)

Table 1 Component values and device sizes for proposed UWB PA and PA reported in [18]

	L	C	R	M1 (W/L)	M2 (W/L)	M3 (W/L)
Proposed UWB PA	320 pH to 4.12 nH	360 fF to 2.112 pF	560 Ω to 10.4 kΩ	105 μm/0.18 μm	105 μm/0.18 μm	105 μm/0.18 μm
PA proposed in [18]	0.4 pH to 2.2 nH	0.4 pF to 1.6 pF	0.9 kΩ to 2.4 kΩ	512 μm/0.18 μm	96 μm/0.18 μm	128 μm/0.18 μm

where $\mu_n = 327.4$ cm^2/Vs, $C_{ox} = 8.42 \times 10^{-3}$ pF/μm^2, the threshold voltage, $V_t =$ 0.45 V, for a typical 0.18 μm silicon CMOS process. In general, a large transistor size M_1 is needed to provide high gain and output power of the amplifier at high frequency. The sizes for the transistors M_2 and M_3 have been designed with the same mathematical approach. Hence, in this design, optimum value of feedback resistance R_1 is chosen to meet input matching and gain flatness over the frequency band of interest. Capacitor C_1 is a DC blocking and inductors L_1, L_2, and Capacitor C_2 is a part of input T-matching. Transistors M_1, M_2, and M_3 are operated in saturation region of operation and transistor M_1 is biased into class-AB to achieve sufficient linearity and efficiency with the size of 105 μm/0.18 μm. All the shunt inductive peaking techniques with Ld_1 and Ld_2, in the proposed PA is needed to compensate for bandwidth limitation, to increase the gain flatness and small group delay variations.

As an improvement over design proposed by Murad et al. in [17, 18], more effort has been taken in well designing of RF components L_{d1}, L_{d2}, feedback resistance R_1 and R_3, and well designing of input T-matching network to prove more robustness and performance improvement compared to previously reported power amplifier design [17, 18]. As a result, the proposed PA design exhibit good linearity, lowest gain ripple of about only ±0.3 dB, low power, and small chip area all simultaneously. Capacitor C_3 is added to improve broadband output impedance of 50 Ω matching over the entire bandwidth. L_{d2} is tuned with capacitor C_4 for output matching. Table 1 shows the mathematically obtained optimized values of the proposed PA design and the circuit parameters by Murad et al. [18].

3 Simulation Results

The circuit design of the UWB PA is simulated using Cadence Specter RF simulators in UMC 0.18-μm CMOS process. Figure 2a shows the effect of optimizing the value of peaking inductor to achieve a low gain ripple. In the figure, the lowest gain ripple is obtained by optimizing the value of L_{d1} within range of 3–4.4 nH. The optimum L_{d1} is chosen to be 3.4 nH for low ripple during 3–7 GHz. While, the size of the transistor M_1 is designed for low Noise Figure (NF).

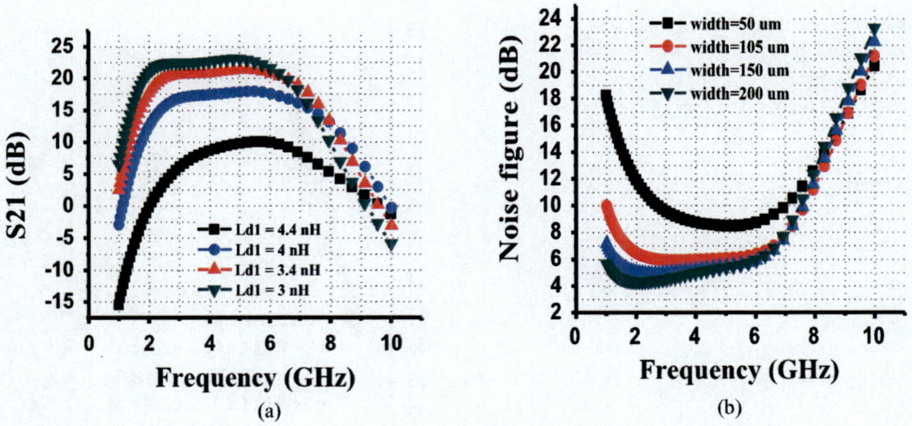

Fig. 2 **a** Simulation results for effect of L_{d1} on Gain. **b** Simulation results for effect of size of M_1 on NF

In Fig. 2b, the lowest NF is obtained by optimizing the value of width of M_1 from 50 to 200 μm. The optimum size is chosen to be 100 μm for low NF. The simulated pre- and post-layout simulation results for gain are in good agreement with each other with excellent gain flatness of 19 ± 0.3 dB as can be seen in Fig. 3. For present work, gain flatness is achieved without using the special gain flatness improvement techniques like Current Reused (CR) and stagger tuning as mentioned in [17–19].

Figure 4 shows the simulation results for the comparison of gain as obtained by the proposed PA design and previously reported PA [17] with CR technique for 3.1–4.8 GHz band. The proposed UWB PA shows very low gain ripple of ±0.3 dB for 3–7 GHz band. The simulated pre-layout and post-layout input and output return loss, S_{11} and S_{22} are plotted in Fig. 5.

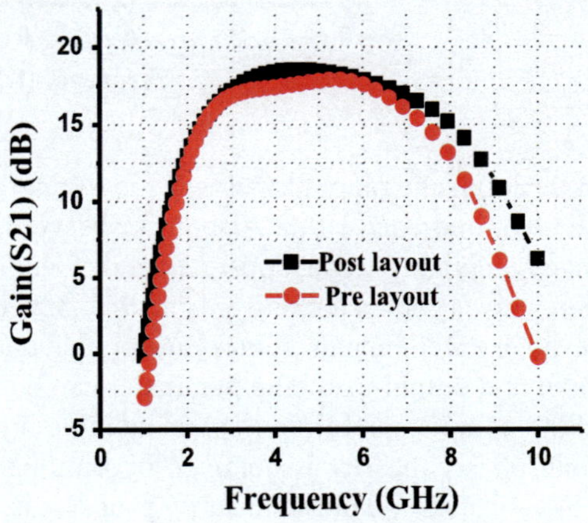

Fig. 3 Simulated gain versus frequency (pre-layout and post-layout) for proposed 3–7 GHz UWB PA

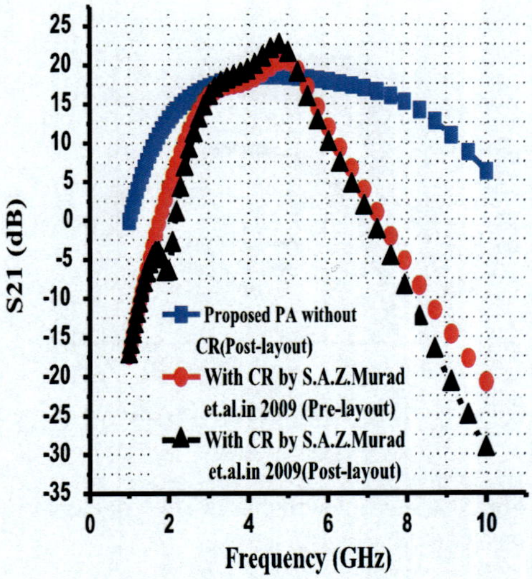

Fig. 4 Comparison of simulated gain performance for proposed 3–7 GHz PA and previously reported PA

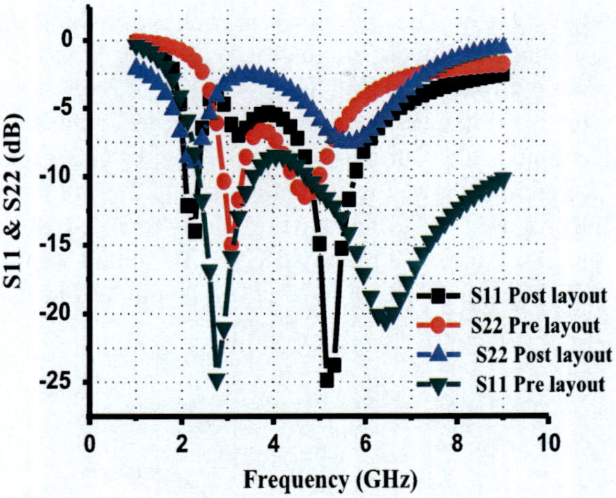

Fig. 5 Simulated input and output return loss of UWB PA (pre- and post-layout)

The return loss at the input and output is below −7.5 dB and −6 dB, respectively, is obtained within the required bandwidth. Also, it is observed from Fig. 5 that the matching performance is degrading at the upper end of desired frequency (7 GHz) so that the output matching gives good matching only up to 3–5.5 GHz. This may be due to the tradeoffs involved in excellent gain flatness and perfect matching for wideband applications. The input and output matching parameter can be further improved by using off-chip matching component during implementation, if required.

In Fig. 7, the small phase linearity is obtained by optimizing the value of L_{d1} within the range of 3–4.4 nH. Good phase linearity property is achieved, i.e., group delay variation is ±200 ps across the whole band. The operating DC supply voltage directly affects the performance of PA with respect to Power Added Efficiency (PAE) as can be seen from Fig. 6a. It is observed that with the increase in V_{DD} for 1, 1.5, and 2 V; efficiency improves for 3–7 GHz band. For UMC −0.18 μm CMOS optimum V_{DD} of 1.8 V has been chosen and accordingly PAE of 29% is obtained at −5 dBm input power from Fig. 6b. Figure 7 shows the method of optimizing the value of L_{d1} to achieve good phase linearity.

The pre- and post-layout simulation results for the Noise Figure (NF) are in good agreement with each other as can be seen in Fig. 8. A good linearity is observed from Fig. 9a, b with input and output 1-dB compression point of −3.495 dBm and 8.2 dBm, respectively, at 5 GHz. The input and output third-order intermodulation point obtained is 17.339 dBm from Fig. 9c. Stability analysis shows that $K_f > 1$ and $B_f < 1$ across the frequency band of 3–7 GHz, stating that designed PA is unconditionally stable as shown in Fig. 10. The layout of the proposed PA is a symmetric one as shown in Fig. 11. All the signal paths are designed with high-level metal 6, while metal 1 is used for the ground plane. The dc pads are used for biasing and supply voltages. RF MIM capacitors and spiral inductors are used due to its low losses. The distance between the two metal wires and in between metal 6 to component is kept as 0.8 μm.

The dimension of via's is 0.28 μm × 0. 28 μm. All pin contacts are made with metal 6. For inductor contact with other components an additional metal plate of dimension 16.1 μm × 3. 5 μm is used. Table 2 shows the simulated summary and performance comparison with other literature. As can be seen that the proposed UWB PA has obtained very low gain ripple of about ±0.3 dB, average group delay, good input and output matching, unconditional stability, and small chip area as compared to other works reported on 3–7 GHz band. However, this PA suffers from more power consumption to achieve wide bandwidth and good linearity simultaneously.

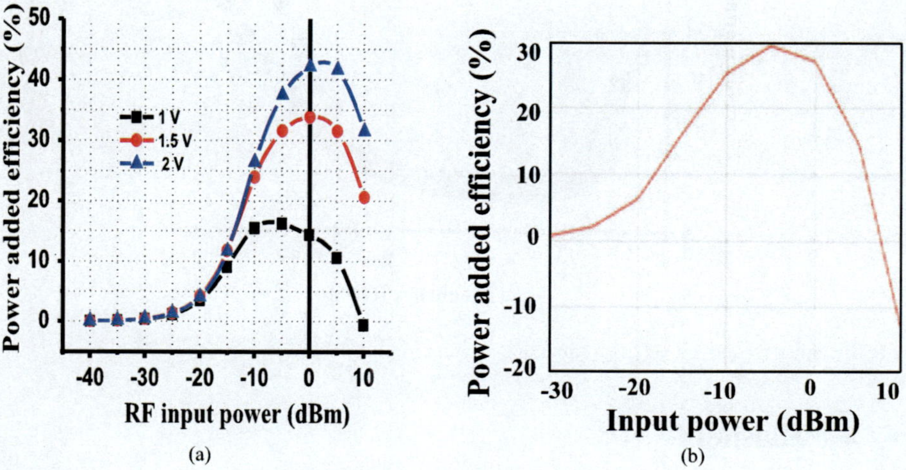

Fig. 6 **a** Variation of power added efficiency (PAE) with respect to supply voltage (V_{DD}) = 1, 1.6 and 2 V. **b** PAE for proposed PA design

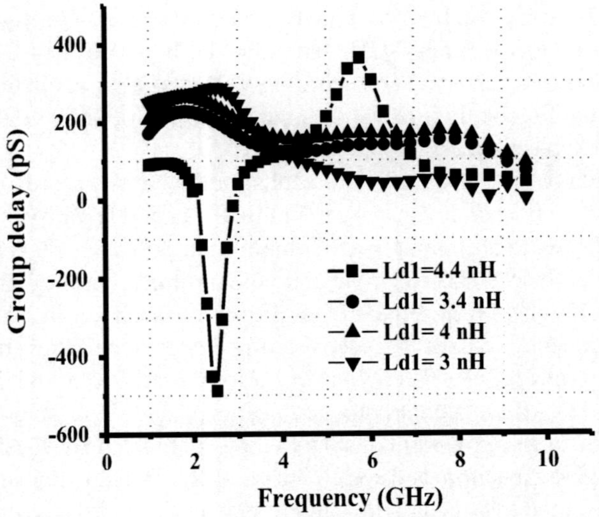

Fig. 7 Effect of variation of shunt peaking inductor on group delay

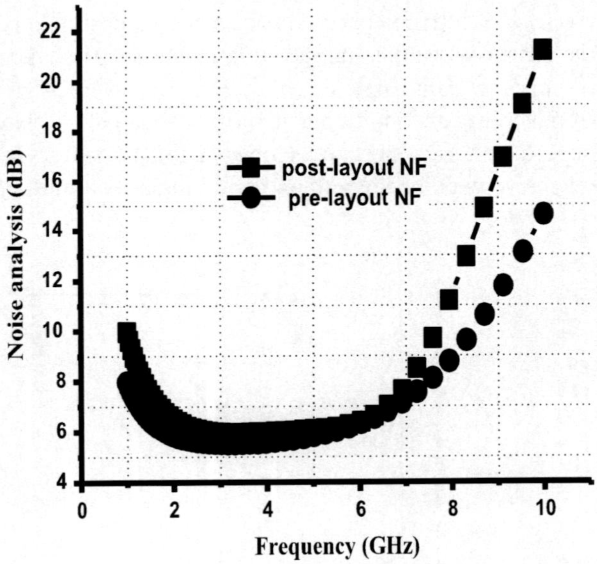

Fig. 8 Noise frequency versus frequency

4 Conclusion

For the designed 3–7 GHz UWB power amplifier the pre-layout and post-layout simulation results are found to be in great agreement with each other. By employing two stages of power amplifier, resistive shunt feedback, and inductive shunt peaking,

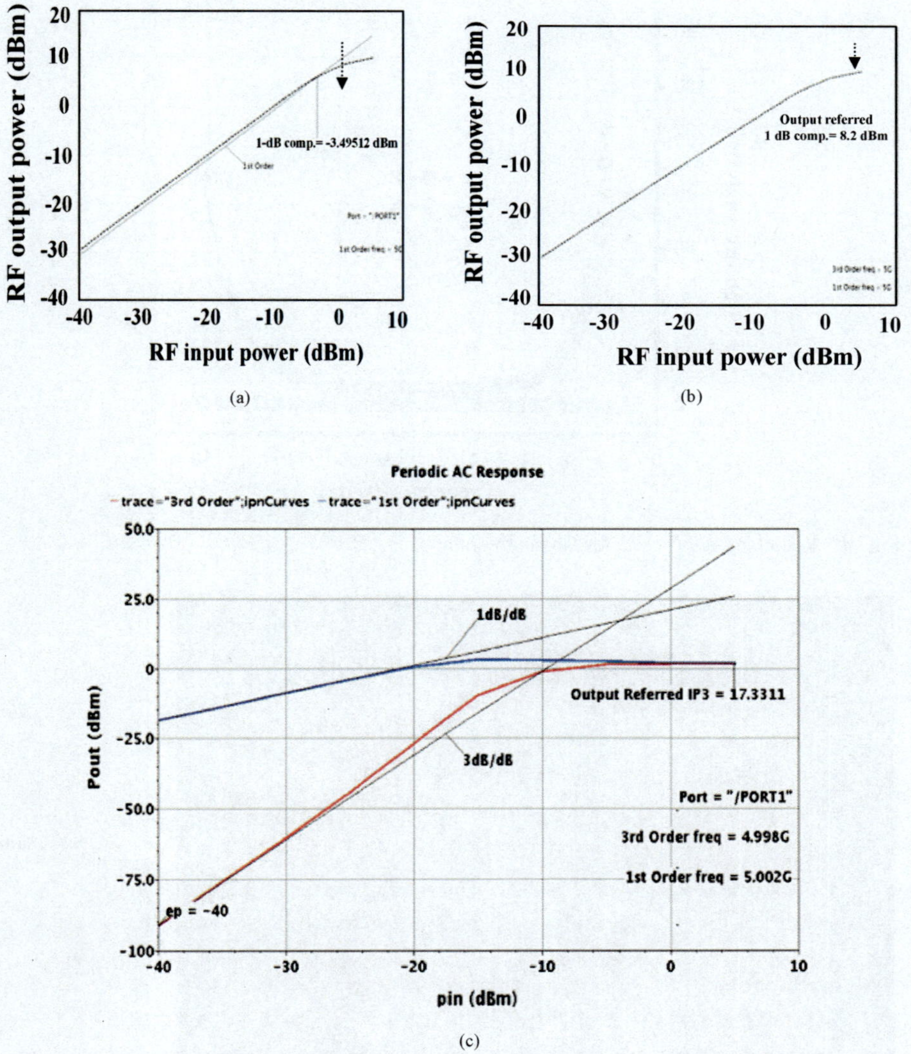

Fig. 9 Linearity analysis for proposed 3–7 GHz PA. **a** Input referred 1-dB compression point. **b** Output referred 1-dB compression point. **c** Third-order intermodulation point

the proposed design achieves 20% reduction in gain ripple and 6.37% reduction in layout area while simultaneously achieving good linearity, small group delay variation, and unconditional stability as compared to previously reported power amplifier design as shown in Tables 1 and 2 across the entire band of interest.

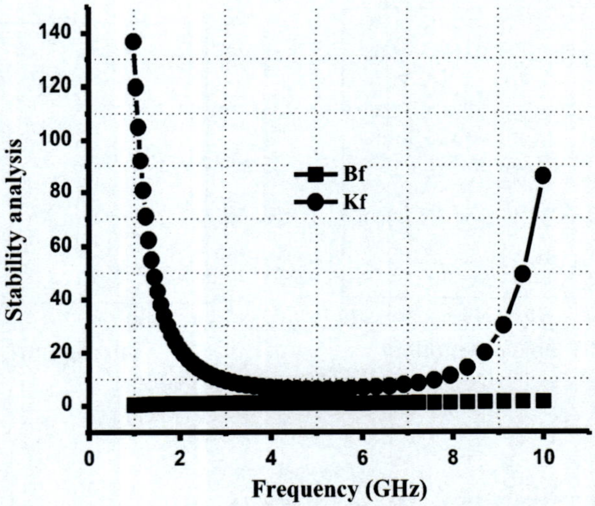

Fig. 10 Roulette's stability factor versus frequency

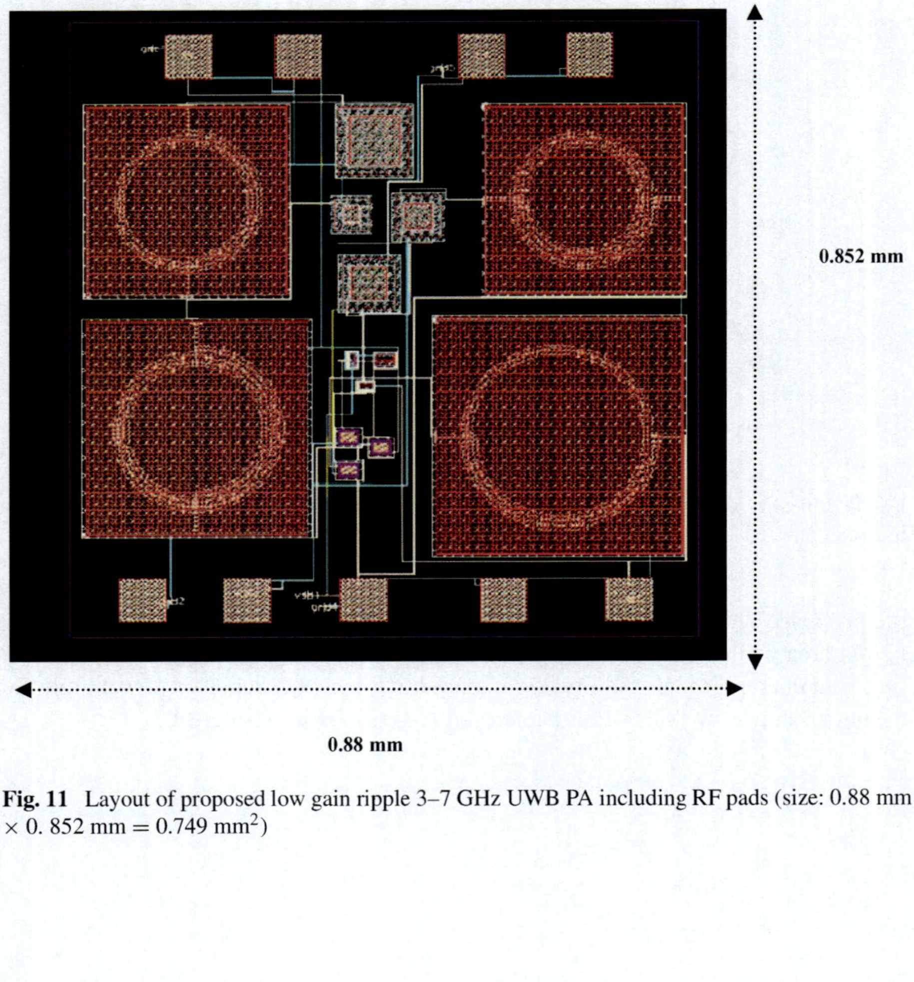

Fig. 11 Layout of proposed low gain ripple 3–7 GHz UWB PA including RF pads (size: 0.88 mm $\times$ 0. 852 mm $=$ 0.749 mm^2)

Table 2 Comparison of wideband CMOS power amplifiers performances: published and present work

Parameters	References				Proposed PA design
	[17]	[18]	[19]	[20]	
Band width (GHz)	3.1–4.8	3–7	3.1–10.6	3.1–10.6	3–7
Technology (μm)	0.18	0.18	0.18	0.18	0.18
Supply voltage (V_{DD})	1	1.8	1.8	1.8	1.8
Input matching S_{11}(dB)	<−5	<−6	<−10	<−10	<−7.5
Output matching S_{22} (dB)	<−5	<−7	<−10	<−14	<−6
Gain (dB)	18.4 ± 1	14.5 ± 0.5	10.46 ± 0.8	11.48 ± 0.6	19 ± 0.3
Input P_1dB (dBm)	−10.6 @ 4 GHz	N/A	N/A	N/A	−3.5 @ 5 GHz
Output P_1dB (dBm)	6.0 @ 4 GHz	7	5.6	5	8.2 @ 5 GHz
PAE (%)	18 @ 4 GHz	N/A	N/A	N/A	29 @ 5 GHz
Power (mW)	22	24	84	100	28
Group delay (pS)	N/A	±178.5	±	±85.8	±200
Stability	N/A				Unconditionally stable
Area (mm^2)	0.97	0.88	1.76	0.69	0.749
Remarks	Cascode with CR technique and additional CS stage to achieve high power gain	Two-stage cascode CR PA	Cascode CS	Three-stage cascode CS with stagger tuning for broadband	Two-stage cascoded PA with CS as second stage

CR—current reused, CS—common source, N/A—not mentioned

References

1. Makarov DG, Krizhanovskii VV, Shu C, Krizhanovskii VG (2008) CMOS 0.18-μm integrated power amplifier for UWB systems. In: 4th international conference on ultra wide band and ultra short impulse signals, Sevastopol, Ukraine, Sept 2008, pp 153–155
2. Blaakmeer SC, Klumbering EAM, Leenaerts DMW, Nauta B (2007) An inductor less wide-band balun-LNA in 65 nm CMOS with balanced output. In: 33rd European solid state circuit conference, Munich, 11–13 Sept 2007, pp 364–367

3. Blaakmeer SC, Klumperink EAM, Leenaerts DMW, Nauta B (2007) Wideband balun-LNA with simultaneous output balancing, noise-cancelling and distortion-canceling. In: IEEE international solid state circuit conference, pp 1341–1350
4. Tian MS, Mikklensen JH, Larsen OJT (2011) Design and implementation of a 1–5 GHz low noise amplifier in 0.18 μm CMOS. Analog Integr Circuits Signal Process 67(1):41–48
5. Duan J-H, Han X, Li S (2009) A wideband CMOS LNA for 3–5 GHz UWB systems. In: IEEE conference
6. Yoo S, Kim JJ, Choi J (2013) A 2–8 GHz wideband dually frequency-tuned ring VCO with a scalable K_{VCO}. IEEE Microwave Wirel Compon Lett 1–2
7. Kuan-Chung Lu, Wang F-K, Horng T-S (2013) Ultralow phase noise and wideband CMOS VCO using symmetrical body-bias PMOS varactors. IEEE Microwave Wirel Compon Lett 23(2):90–92
8. Yoon J, Kim H, Park C, Yang J, Song H, Lee S, Kim B (2008) A new RF CMOS Gilbert mixer with improved noise figure and linearity. IEEE Trans Microwave Theory Tech 56(3)
9. Kim CW, Kahn MS, Anh PT, Kim HT, Lee SG (2005) An ultra wideband CMOS low noise amplifier for 3–5 GHz UWB system. IEEE J Solid State Circuits 40(2):544–547
10. Jose S, Lee H-J, Ha D, Choi SS (2005) A low-power CMOS power amplifier for ultra-wideband (UWB) applications. In: Proceedings of the IEEE international symposium on circuits and systems conference, pp 5111–5114
11. Han CH, Zhi WW, Gin KM (2005) A low power CMOS full-band UWB power amplifier using wideband RLC matching method. In: Proceedings of the IEEE electron devices and solid-state circuit conference, pp 223–236
12. Lu C, Pham A-V, Shaw M (2006) A CMOS power amplifier for full-band UWB transmitters. In: Proceedings of the California, IEEE RFIC symposium, pp 397–400
13. Wang R-L, Su Y-K, Liu C-H (2006) 3–5 GHz cascoded UWB power amplifier. In: Proceedings of the IEEE Asia Pasific conference on circuits and systems, pp 367–369
14. Grewing C, Winterberg K, Waasen SV (2004) Fully integrated distributed power amplifier in CMOS technology, optimized for UWB transmitter. In: IEEE RF IC symposium, Aug 2004, pp 87–88
15. Huang ZD, Lin ZM, Lai HC (2007) A high gain low noise amplifier with current-reused technique for UWB applications. In: IEEE conference on electron devices and solid-state circuits, Tainan, 20–22 Dec 2007, pp 977–980
16. Lin Y-J, Hsu SSH, Jin J-D, Chan CY (2007) A 3.1–10.6 GHz ultra-wideband CMOS low noise amplifier with current-reused technique. IEEE Microwave Wirel Compon Lett 17(3): 232–234
17. Murad SAZ, Pokharel RK, Kanaya H, Yoshida K (2009) A 3.1–4.8 GHz CMOS power amplifier using current reused technique. In: IEEE 5th international conference on wireless communication sand, networking and mobile computing (WiCOM) 24–26 Sept 2009, Beijing, China.
18. Murad SAZ, Pokharel RK, Galal AIA, Sapawi R, Kanaya H, Yoshida K (2010) An excellent gain flatness 3.0–7.0 GHz CMOS PA for UWB applications. IEEE Microwave Wirel Compon Lett 20(9)
19. Chen C-Z, Lee J-H, Chen C-C, Lin Y-S (2007) An excellent phase-linearity 3.1-10.6 GHz CMOS UWB LNA uses standard 0.18 μm CMOS technology. IEEE Microwave Wirel Compon Lett 17(3):232–234
20. Sapawi R, Pokharel RK, Murad SAZ, Anand A, Koirala N, Yoshida K (2012) Low group delay 3.1–10.6 GHz CMOS power amplifier for UWB applications. IEEE Microwave Wirel Compon Lett 22(1)

An Approach to Detect and Prevent Distributed Denial of Service Attacks Using Blockchain Technology in Cloud Environment

Vishwani Patel, Fenil Khatiwala, and Yashi Choksi

Abstract Cloud is a novel platform providing on-demand services, web applications, network, server and storage. In spite of having problems associated with security and privacy, the cloud brings out a wide range of benefits. The distributed denial of service (DDoS) attack remains a security challenge in achieving a secure and guaranteed service and resources in cloud computing. The users have to face a problem regarding resources while having a DDoS attack. Therefore, it results in significant downtime and financial losses. Blockchain technology provides an enhanced security and privacy in many domains, like IoT, cloud, data mining and networking. It also provides a decentralized security framework, therefore, allowing sharing of attack information in a completely distributed and automated manner. Blockchain adds an additional layer of security and ensures that bad actors don't take control of systems for their own ends in cloud environment. The system proposed here offers a solution to secure data on cloud environment built on blockchain technology that detects and prevents DDoS attack.

Keywords Blockchain technology · Cloud computing · DDoS attack · Smart contract

1 Introduction

In today's technological world, cloud computing has become an essential part of human life. It is evolved from various existing technologies, like utility, computing and service orientation. Without investing a large amount of money, commercial firms can scale up using a cloud. Therefore, expenses like building large data centers, purchasing software license and infrastructure can be decreased. Distributed denial of service (DDoS) attack stands as the single largest threat to the internet. The attacker can extremely reduce the quality or completely breakdown the user's network. The main tenacity of a DDoS attack is to make the target to block the access of network

V. Patel (✉) · F. Khatiwala · Y. Choksi
Uka Tarsadia University, Bardoli, Gujarat, India
e-mail: vishwanipatel494@gmail.com

Z. Patel et al. (eds.), *Advances in VLSI and Embedded Systems*, Lecture Notes in Electrical Engineering 676, https://doi.org/10.1007/978-981-15-6229-7_20

and the resources. A DDoS attack on the cloud is executed by creating malicious traffic that exhaust resources by creating a large group of agents called botnets as discussed in [1]. In cloud environment DDoS attack can also degrade the working ability of cloud service significantly by disrupting the virtual servers.

As per the author of [2], blockchain technology is a revolutionary technology and is nowadays having significant applications in financial sectors, healthcare and government. The blockchain technology is becoming the greatest among favorable technologies of cyber security. Blockchain stores data in distributed manner that retains a list of records that goes on increasing continuously known as blocks that are secured from unauthorized alterations and reformation. These blocks contain the cryptographically hashed data as discussed in [3]. The hashed blocks are based on previous blocks so as to prevent tampering of data in blockchain. Blockchain provides protection of machines and resources, availability and reliability of networks and systems.

A smart contract is "An arithmetical function that is written in source code/platform and executed by computers that used to integrates the tamper-proof mechanism of blockchain" discussed in [4]. Smart contract defines the rules and agreements as traditional contract does. It also automatically enforces a legal bond. According to the developer's need, they are able to develop a smart contract by specifying any instructions in it. Smart contracts give authorization to trustworthy transactions and agreements to be executed among different, unidentified parties without a requirement of a central expert system, legal authority or external enforcement mechanism. As blockchain has become popular, smart contracts have received attention increasingly. Due to the promising features of blockchain, it ensures enhanced security and privacy in many domains, like IoT, cloud and data mining, as discussed in [5].

There are many techniques that can be adapted to identify and stop the DDoS attack in the cloud environment, such as classification techniques, encryption techniques, machine-learning-based techniques and entropy-based system, as discussed in [6]. Cloud computing needs more secure and efficient methods to identify and stop the cloud resources from a DDoS attack. Therefore, this research work will be helpful in the blockchain technology-based system that detects and prevents the DDoS attack in a cloud environment.

This paper is organized as follows: In Sect. 2, a detailed summary of the recent research works to stop and identify DDoS attack on the cloud is given. In Sect. 3 we show the general scenario of DDoS attack in the cloud. Section 4 explains the proposed work with blockchain technology, and we enclosed our research work with conclusion and the future directions in Sect. 5.

2 Related Works

In recent times, many researchers have been working on the identification and prevention of DDoS attack on the cloud. The history of this growing attack mechanism was detected and that the internet control message protocol (ICMP) packets make the network complex for the users and are marked as the concept of DDoS attack discussed in [7]. The consortium of network of researchers first discovered the DDoS attack in June 1998.

In early work about DDoS detection and prevention in a cloud, the research of [7], proposed fuzzy-based mechanism which is helpful for the detection of DDoS attack. The system uses three parameters, which are entropy of source IP, port address and packet arrival rate, and using these parameters, the authors analyze the incoming data packets behavior by fuzzy logic-based IF–THEN rules. It is a cost-effective, reliable and easy method for the cloud system.

As discussed in [8], the authors have proposed a system that uses practical dynamic resource allocation mechanism to face DDoS attacks. They use the intrusion prevention system (IPS) to monitor incoming packets. The proposed system will automatically and dynamically allocate extra resources from available cloud resources. IPS works simultaneously to filter out attack packets when a host cloud server is under DDoS attack. When the volume of DDoS attack packets is reduced, it releases additional resources back to the available cloud resource pool.

Machine learning techniques are being used for DDoS attack detection system discussed in [9], and are proposed to prevent attacks on the source side in the cloud. The authors analyze the statistical feature of DDoS attacks. They had developed a proof-of-concept model and tested it real cloud. They had evaluated and compared nine different types of supervised and unsupervised machine learning algorithms.

To identify and stop the DDoS TCP flood attack in public cloud using classifier, a new classifier system CS_DDoS is proposed in [10]. It provides an explanation to securing kept records. For this, it classifies the incoming packets and makes a decision on the basis of the classification results. In the detection phase, four different classifiers are used to identify and determine whether a packet is regular or irregular. In the prevention phase, it will deny access of cloud service to the malicious packets and it will add the source IP in the blacklist table.

To identify and stop/avoid the DDoS attack, Updated Snort (Usnort) model is proposed, as discussed in [11]. It was based on a Snort tool to detect DDoS implemented in a distributed virtual environment. Usnort gives an alert message about the intrusion with all relevant data associated with the attack to the network admin. Admin drops the connection with that IP for certain time. This proposed system favors to degrade the effect of DDoS by detecting the attack at very prior stages.

TCP/IP packet header features classification of connecting clients, according to their OS, as proposed in [12]. It is used to identify the original source of an incoming packet during the spoofed DDoS attack. They deployed an open source Xen Cloud Platform (XCP) test-bed. To determine the original source of attacker, they match the final TTL value of active and passive OS detection.

A method of incorporation with HTTP GET flooding among DDoS attacks and Map Reduce processing for quick attack detection is proposed in [13]. This method performs signature-based detection with low errors and in less time. It shows that the proposed pattern detection system is better compared to SNORT detection.

An entropy-based system discussed in [14], is used to provide a multilevel detection. Systems handle flow of a massive amount of file packets by multithreaded IDS approach in a cloud. It would give the observed alerts to a monitoring service. Monitoring service notifies the cloud user about their system under attack.

As discussed in [15], it shows a detailed review on DDoS attack, and classification of DDoS attack includes two types: bandwidth-based and resource-based attacks. It also shows different techniques for detection of DDoS attacks, like anomaly detection, NOMAD, packet selection and filtering method with congestion, D-WARD, MULTOPS, and misuse detection and prevention techniques like path-based circulated packet filtering, ingress and egress filtering and secure overlay services.

The research of [16], states that to identify the attacks two main research avenues need to be followed. First, the intrusion tries to harm VMs to launch a DDoS attack against a target that is external to the cloud. Second, develop defense tools for the more traditional cloud disturbance, where the goal of the attack is the cloud or a part of the cloud itself. These resources are applicable to those situations that will be constructed on a Eucalyptus cloud system.

The method proposed in [17], is used to provide security for web services against the DDoS attack. This approach protects HTTP flooding, oversized XML, coercive parsing, oversized encryption and WS, addressing spoofing vulnerabilities at the application layer. Outlier detection technique could be used for effective filtering out of malicious requests.

The Honeypot system proposed in [18], enables cloud servers to detect the unknown attacks and prevent DDoS attacks. Honeypot network blocks the entire IP address after the home IP address was detected to deal with attackers as prevention. It also shows the difference between traditional systems and the proposed system.

To avoid and identify the DDoS attack on a cloud environment, many techniques are being proposed, and most of these techniques provide less accuracy and results in slow down in the system performance.

Therefore, cloud computing is used as an efficient DDoS detection and prevention mechanism that can provide more accuracy and detect and prevent attacks accurately on a cloud environment.

3 General Scenario of DDoS Attack in Cloud

Figure 1, shows the scenario of DDoS TCP overflow (flood) attack that is performed by an attacker sending constant requests to the service provider that can affect the cloud server performance within a small time and even stop the services completely. Here, an attack is performed using three attackers and one is the genuine user. In

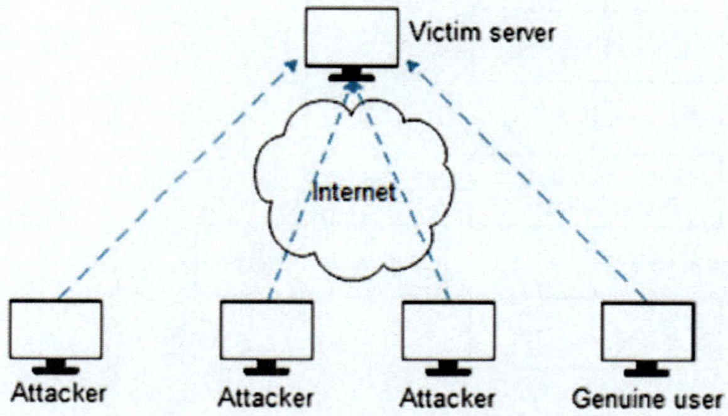

Fig. 1 DDoS attack

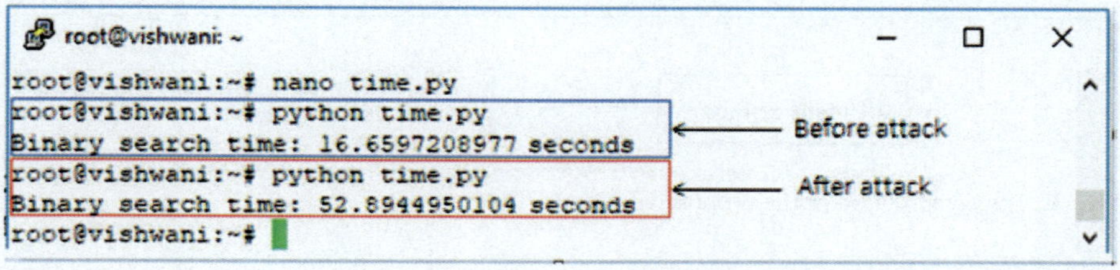

Fig. 2 Performance analysis

this scenario, the genuine user accesses cloud service to perform the execution of the python program.

A TCP flood attack was performed as a DDoS attack. This is carried out by manually sending constant requests to the victim server.

3.1 *Before Attack*

First, a genuine user is executed, and it takes 16.66 s to get the result, as shown in Fig. 2.

3.2 *During Attack*

An attack was launched by sending constant requests that perform DDoS TCP overflow (flood) attack on victim server. As shown in Fig. 2, when an attack was performed, the genuine user got to result in 52.89 s, that is, more than double times to get the result.

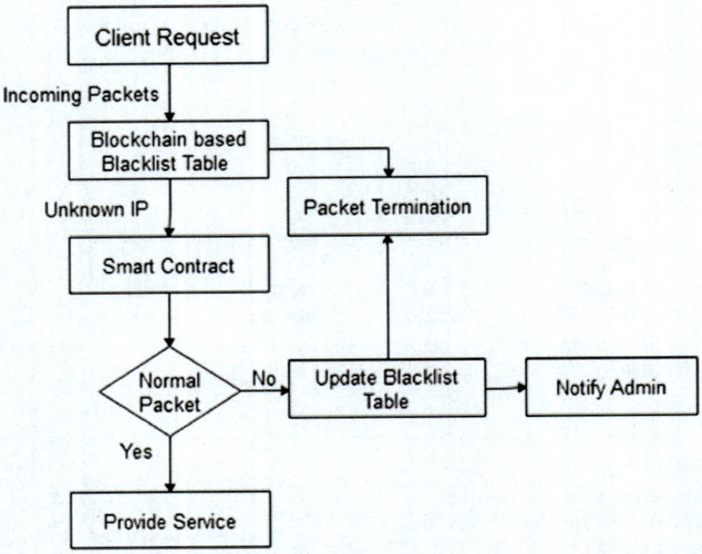

Fig. 3 The construction of the proposed system

4 Experimental Setup

For the implementation scenario, Linode service discussed in [19], is used to work in a cloud that provides platform-as-a-service, with Linux operating system, 4 GB RAM, 2 CPU cores and 80 GB storage. The practicals are being performed on Python language with the help of PuTTY emulator to connect with the cloud, while MongoDB is used for data storage in the cloud environment. Packet analyzer software Tcpdump is used to collect incoming packets in the network.

5 Proposed Work

In this section, the proposed security support to identify and stop/avoid the DDoS attack in a cloud environment is described.

There are two phases of the proposed system on which it is defined:

(i) Detection Phase
(ii) Prevention Phase.

5.1 Detection Phase

- The detection system fetches the incoming packets within a timeframe as a client request, for example, 1 min.
- The received packets are processed to find whether packet sources are blacklisted as an attacker of the system or not. For this, the system makes use of blockchain-based blacklist table, which contains blacklisted IPs.
- If source of packet is already in blacklisted IPs, the detection phase will send packets directly to the prevention phase without any processing.
- If source of packet is not blacklisted, the incoming packet will be passed through smart contract to check whether the packets are normal or abnormal.
- The smart contract contains rules that are used to classify abnormal packets. For example, predefined threshold.
- If one user makes requests more frequently than predefined threshold, it will be detected as attacker by the system. Threshold can be manually adjusted by the system admin.
- If a packet is defined as normal, the detection phase will send it to the cloud service provider. Otherwise, the detection phase will update the blacklisted table and send to the prevention phase.
- All arriving requests are stored into blocks. A new block will be stored in blockchain using the consensus algorithm known as proof of work.
- Blocks will be encrypted for making chains, and each block appended to the chain will be in a sequential manner.
- Smart contract is also executed as a part of blockchain.

5.2 Prevention Phase

- When the packet reaches to prevention phase, first it will alert the system administrator.
- Then, it will add the attacking source address to the blacklist table, if it is not already available in the blacklist table.
- Finally, the packet will be dropped.

6 Implementation Scenario in Blockchain-Based Cloud

As described earlier, the proposed system can be used to identify and stop the DDoS attack on cloud environment using blockchain technology. To perform DDoS attack on cloud environment, first blockchain-based cloud is created. Figure 4, shows three implementation scenarios in blockchain-based cloud.

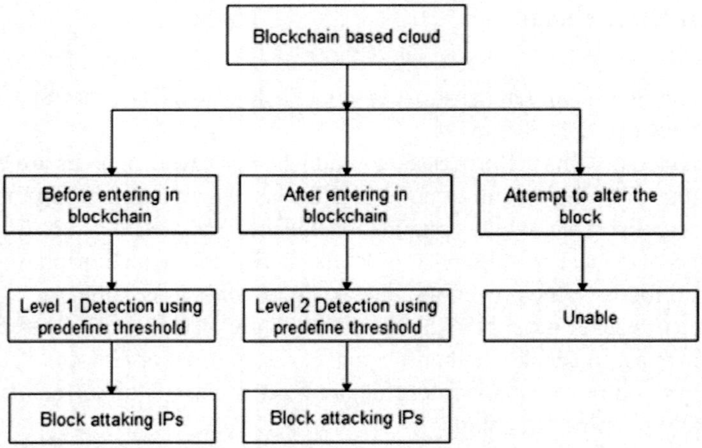

Fig. 4 Blockchain-based cloud scenarios

6.1 *Before Entering in Blockchain*

First, the detection system fetches the incoming packets within a time bound, for example 1 min. All requests are stored into IP table. If source of incoming packets is blacklisted as an attacker, then it will directly send to the prevention system. If source of packet is not blacklisted, the incoming packet will be passed through smart contract to choose whether the packets are normal or abnormal. Smart contract defines the rule like predefined threshold in it. For example, in 1 min the threshold value is defined as 10. If the incoming requests exceed from the predefined point, it is known as attacking IP.

Proof of work consensus algorithm of blockchain is used to generate block that contains data like current hash key, previous hash key and nonce, as well as the current block key. As block is generated, it is encrypted with SHA-256 algorithm and added to the blockchain.

As blockchain is created, it also identifies the attacking IPs using smart contract. Here smart contract performs Level 1 detection of DDoS attack before entering into blockchain-based cloud. If smart contract considers the incoming IP as attacking IP, then the system will send incoming packets to prevention system. If the incoming request is considered as genuine user, then portion of the blockchain and services are provided to the user. As attacking IPs are detected by smart contract, prevention phase blocks the attacking IPs.

6.2 After Entering into Blockchain

After becoming part of the blockchain if genuine user tries to perform DDoS attack, then Level 2 detection is performed using smart contract. Smart contract filters out attacker using rule of predefined threshold within a timeframe. For example, if any genuine user makes more than 30 requests in 2 min, then the user is considered as an attacker by the smart contract.

6.3 Attempt to Alter the Data

In this scenario, if any genuine user tries to alter the data of block or tries to increase the threshold value, then blockchain does not allow to alter the data because of the tamper-proof mechanism. If any user tries to alter the block, then when counting the hash key different key is generated from the already stored in the database. Therefore, it gives match false and system founds that blockchain is altered and it allows adding an updated block in blockchain because of its tamper-proof mechanism.

7 Performance Analysis

In this section, the performance of the proposed system is evaluated. The accuracy represents the rate of correctly identified attackers and genuine users in network. The accuracy of the proposed system is measured by the following equation.

$$\text{Accuracy} = \frac{TP + TN}{TP + TN + FP + FN}$$

where

True positive (TP) is correctly identified abnormal packets.
True negative (TN) is correctly identified normal packets.
False positive (FP) is incorrectly identified abnormal packets.
False negative (FN) is incorrectly identified normal packets.

Table 1, shows that proposed system is more efficient, but it requires more execution time to identify the DDoS attack because it generates blockchain that requires more time to be generated.

Table 2, shows the evaluation of the CS_DDoS system with the proposed method with respect to a number of genuine users and the attacks that are performed on the cloud environment. CS_DDoS is known as a classifier system that is used in cloud environment to detect and prevent the DDoS attack. Results show that when the number of attackers is increased with respect to genuine users, both systems will provide 100% accuracy. As soon as the number of attackers is increased (e.g. 6, 8),

Table 1 Detection accuracy based on the number of attackers

Attackers	Genuine user	TP	TN	FP	FN	Execution time (s)	Accuracy (%)
1	1	1	1	0	0	64	100
2	1	2	1	0	0	64	100
4	2	4	2	0	0	64	100
6	3	6	3	0	0	64	100
8	4	8	4	0	0	64	100
10	4	10	3	0	1	65	92
12	4	12	3	0	1	65	93.75
14	6	14	5	0	1	66	95
18	6	18	5	0	1	66	96

Table 2 Detection accuracy based on the number of attackers

Attackers	Genuine users	Accuracy of CS_DDoS system (%)	Accuracy of proposed system (%)
1	1	100	100
2	1	100	100
4	2	100	100
6	3	88	100
8	4	83	100
10	4	85	92
12	4	81	93.75

the change in the accuracy of the CS_DDoS system get changed while the proposed system shows the highest accuracy. When the number of attackers is further increased (e.g. 10, 12), the proposed system attains high accuracy as compared to the CS_DDoS system as 92 and 93.75%, respectively.

8 Conclusion and Future Work

DDoS attacks are quickly becoming the most prevalent type of cyber threat, growing rapidly in the past years, both in numbers and volumes. Many detection techniques are used against DDoS attack on a cloud, but these techniques including CS_DDoS have disadvantages like sometimes attacks are not properly detected, it slows down the performance of the system or stops the services completely. The proposed system shows that DDoS attack is properly identified by the detection phase and blocked by the prevention phase. It also shows that blockchain provides tamper-proof mechanism.

The system can be tested and analyzed with real server and clients on a large network. The system can be checked with different threshold values for performance analysis. The proposed system can be optimized in terms of execution time.

References

1. Singh N, Hans A, Kumar K, Birdi MPS (2015) Comprehensive study of various techniques for detecting DDoS attacks in cloud environment. Int J Grid Distrib Comput 8(3):119–126
2. Tama BA, Kweka BJ, Park Y, Rhee K-H (2017) A critical review of blockchain and its current applications. In: 2017 international conference on electrical engineering and computer science (ICECOS). IEEE, pp 109–113
3. Zheng Z, Xie S, Dai H, Chen X, Wang H (2017) An overview of blockchain technology: architecture, consensus, and future trends. In: 2017 IEEE international congress on big data (BigData congress). IEEE, pp 557–564
4. Cheng J, Lee N, Chi C, Chen Y (2018) Blockchain and smart contract for digital certificate. In: 2018 IEEE international conference on applied system invention (ICASI), Chiba, pp 1046–1051
5. Miraz, MH, Ali M (2018) Applications of blockchain technology beyond cryptocurrency
6. Devi BSK, Subbulakshmi T (2017) DDoS attack detection and mitigation techniques in cloud computing environment. In: 2017 international conference on intelligent sustainable systems (ICISS), Palladam, pp 512–517
7. Mondal HS, Hasan MT, Hossain MB, Rahaman ME, Hasan R (2017) Enhancing secure cloud computing environment by detecting DDoS attack using fuzzy logic. In: 2017 3rd international conference on electrical information and communication technology (EICT), Khulna, pp 1–4
8. Yu S, Tian Y, Guo S, Wu DO (2014) Can we beat DDoS attacks in clouds? IEEE Trans Parallel Distrib Syst 25(9):2245–2254
9. He Z, Zhang T, Lee RB (2017) Machine learning based DDoS attack detection from source side in cloud. In: 2017 IEEE 4th international conference on cyber security and cloud computing (CSCloud), New York, NY, pp 114–120
10. Sahi A, Lai D, Li Y, Diykh M (2017) An efficient DDoS TCP flood attack detection and prevention system in a cloud environment. IEEE Access 5:6036–6048
11. Khadka B, Withana C, Alsadoon A, Elchouemi A (2015) Distributed denial of service attack on cloud: detection and prevention. In: 2015 international conference and workshop on computing and communication (IEMCON). Vancouver, BC, pp 1–6
12. Osanaiye OA, Dlodlo M (2015) TCP/IP header classification for detecting spoofed DDoS attack in Cloud environment. In: IEEE EUROCON 2015—International conference on computer as a tool (EUROCON), Salamanca, pp 1–6
13. Choi J, Choi C, Ko B, Kim P (2014) A method of DDoS attack detection using HTTP packet pattern and rule engine in cloud computing environment. Soft Comput 18(9):1697–1703
14. Navaz AS, Sangeetha V, Prabhadevi C (2014) Entropy based anomoly detection system to prevent DDoS attacks in cloud. Soft Comput 18(9):1697–1703
15. Deshmukh RV, Devadkar K (2015) Understanding DDoS attack and Its effect in cloud environment. In: 2015 international conference on advances in computing, communication and control, pp 203–210
16. Carlin A, Hammoudeh M, Aldabbas O (2015) Defence for distributed denial of service attacks in cloud computing. Procedia Comput Sci 73:490–497

17. Vissers T, Somasundaram TS, Pieters L, Govindarajan K, Hellinckx P (2014) DDoS defense system for web services in a cloud environment. Future Generat Comput Syst 37(2014):37–45
18. Manoja I, Sk NS, Rani DR (2017) Prevention of DDoS attacks in cloud environment. In: 2017 international conference on big data analytics and computational intelligence (ICBDAC), Chirala, pp 235–239
19. Linode Service. https://www.linode.com/

Variability Analysis of On-Chip Interconnect System Using Prospective Neural Network

Ajita Misra, Yash Agrawal, Vinay Palaparthy, and Rutu Parekh

Abstract Interconnects are densely laid as multiple layers in ICs and play a dominant role in characterizing the system performance. At nanodimensions, variability and reliability in interconnects become a dominant concerning issue. Variability analysis can be accomplished using several statistical mathematical techniques, such as Monte-Carlo, parametric, process corner, ANOVA and rank table analyses. However, these conventional techniques are becoming obsolete and deficient in determining the severe variability effects of the sophisticated, densely packed and enormously large on-chip interconnects. Moreover, traditional techniques are tremendously computational-expensive. Henceforth, in the present paper, prospective neural network-based back-propagation technique is incorporated to capture the effect of highly variable parameters that are existing in on-chip interconnects. Back-propagation neural network (BPNN) is used along with Levenberg–Marquardt (LM) algorithm to adjust weights of hidden layer in order to minimize error. The proposed model is efficient and versatile such that it predicts the reliability of the circuit performance based on input parameters, such as interconnect resistance, inductance and capacitance. The performance parameter considered for the developed model is mean square error. The obtained results exhibit high accuracy and adaptability. The accuracy of the proposed model is assessed through regression, error and histogram plots.

A. Misra (✉) · Y. Agrawal · V. Palaparthy · R. Parekh
VLSI and Embedded Systems Research Group, Dhirubhai Ambani Institute of Information and Communication Technology, Gandhinagar, Gujarat, India
e-mail: ajita24101996@gmail.com

Y. Agrawal
e-mail: mr.yashagrawal@gmail.com

V. Palaparthy
e-mail: vinay_shrinivas@daiict.ac.in

R. Parekh
e-mail: rutu_parekh@daiict.ac.in

Z. Patel et al. (eds.), *Advances in VLSI and Embedded Systems*, Lecture Notes in Electrical Engineering 676, https://doi.org/10.1007/978-981-15-6229-7_21

Keywords Back-propagation neural network (BPNN) · Integrated circuits (ICs) · Levenberg–Marquardt (LM) · On-chip interconnects · Very large-scale integration (VLSI)

1 Introduction

With the magnificent advancements in VLSI technology, the demand for high-end electronic appliances has increased manifolds. These technology developments have been achievable due to continuous scaling of technology. Stepping with the miniaturization of technology, the device performance increases. However, at the same time, the on-chip interconnects become more dense and huge at lower technology nodes and consequently proven to be a major bottleneck for IC designs [1]. At scaled technology nodes, interconnects are the major governing parameter in determining the efficiency of electronic devices.

At lower technology nodes, the non-ideal effect such as variability also aggravates substantially. The variability refers to the change in output system responses due to deviation in certain input parameters. Variability can be induced during any of the manufacturing steps such as fabrication and packaging. As technology advances toward sub-micron regime, interconnects dominate the model behavior and become more prone to process variations in such scaled down environment. Such variations can result in output that may be quite contrary to the expected results. Therefore, circuit performance is hampered by variability effects and it becomes a necessity to consider them in order to construct a robust model.

Interconnect variability can occur in any of its physical parameters such as spacing between interconnects (s), thickness of wire (t), interlayer dielectric thickness (h) and width of the wire (w). These defects arise due to imperfections in fabrication process, thereby significantly affecting the parasitic of interconnect, namely resistance, inductance and capacitance [2]. These in turn affects the circuit performance parameters as delay and power dissipation. Till date, majority of analysis of such variability have been performed using traditional techniques, such as process corner, parametric sensitivity, Monte-Carlo method, Taguchi, ANOVA and rank table [3, 4]. In process corner analysis, typical region (formed by corners) is defined within which it is expected that the system can withstand noise and works well. The corners are basically formed based on the combination of fast and slow transistors, namely NMOS and PMOS. These corners enclose an imaginary area, and it is expected that the circuit performance is executed correctly within this area. Parametric analysis studies the effect of variation in each process parameter individually over circuit performance, whereas Monte-Carlo method investigates the cumulative effect of all parameters simultaneously over circuit performance. Taguchi and ANOVA methods indicate the influence of each factor on the system in terms of percentage, whereas rank table determines the impact of parameters by assigning ranks to each factor. The higher rank or percentage implies stronger parameter influence on the system performance. Monte-Carlo has been a widely used variability technique. Henceforth, in the ongoing paper, Monte-Carlo analysis has been incorporated for creating initial dataset for on-chip neural network-based model.

In order to design a system, determination of desired output performance parameters values is needed. This task cannot be solely dependent on the traditional methods as the designer might have to iterate multiple numbers of times to re-optimize the circuit in accordance to a new performance requirement. In order to increase performance efficiency of a system, prediction of output parameters is essentially needed. This can be attained by developing initial training dataset. The development of these datasets, and subsequently its system prediction is based on regression models that can be developed using prospective machine learning and neural network (NN) techniques [5–7]. Neural networks have been employed as regression and classification algorithm in various fields, such as speech processing, image processing, control system and more, as an optimization technique [5]. They mimic the biological functioning of neurons. The algorithm has two main aspects: training and testing. Training is used so as to make the system learn, while testing is used to check the accuracy of the designed structure. In order to minimize the error, back-propagation techniques are used as training algorithm [5–7]. Various training algorithms are used in neural network such as gradient descent, conjugate gradient, quasi-Newton, Levenberg–Marquardt (LM) and so on. These algorithms are classified on the basis of speed and memory usage. This paper incorporates LM algorithm while generating weights for NN model [8]. Back-propagation neural network (BPNN) is a powerful predictive and optimization techniques which can be prospectively used in performance analysis of interconnect. BPNN replaces the manual time taking process that is used in traditional variability analysis method and speeds up the process of optimal designing of the system.

The present paper prominently attempts to bridge the gap between VLSI design and neural network modeling approach. In the paper, a robust approach to integrate faster optimization technique for prediction of performance with variable parasitics is presented.

2 Model Development Using Neural Network Approach

Variability is one of the major contributors for anomaly in circuit performance. These variations could be result of varying processes used in manufacturing such as etching, lithography, doping, polishing and proximity. The variations in several manufacturing process lead to process-induced variations in parameters such as oxide thickness, effective gate length and threshold voltage that further deviates electrical parasitic of interconnect, namely resistance, inductance and capacitance. In order to predict system response and analyze the effect of variability, a model is developed using NN approach in the present paper. The proposed model is developed using three-step processes as depicted in Fig. 1.

The detailed process steps for the development of the proposed system are depicted in Fig. 2.

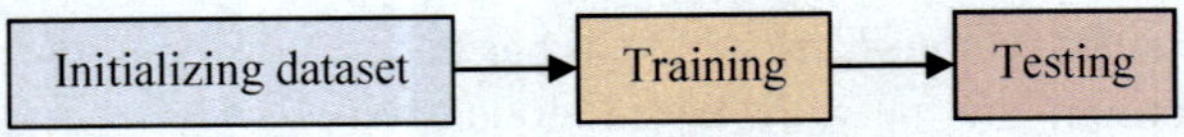

Fig. 1 Block diagram representation of sequential steps for NN-based model development

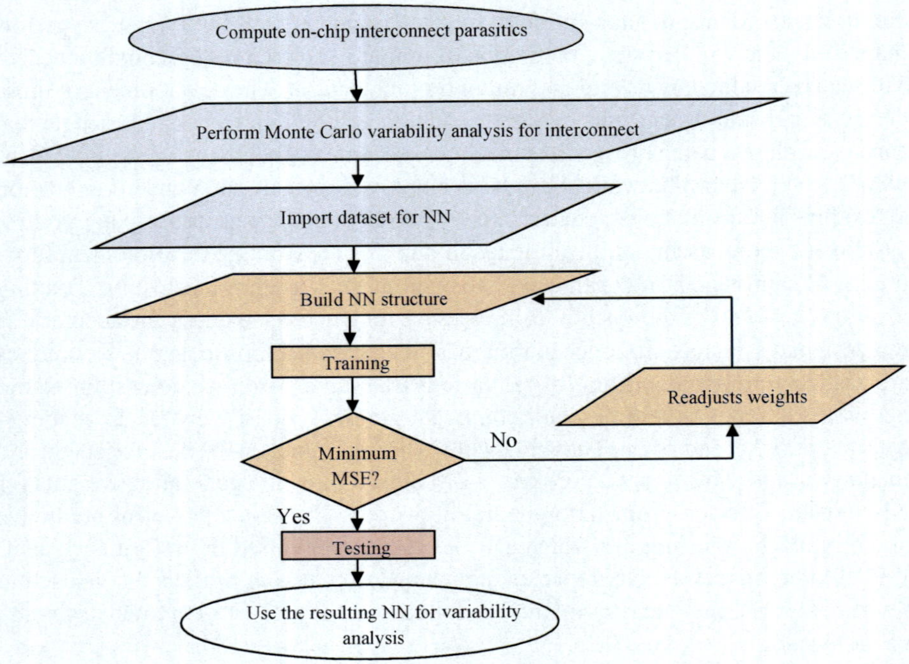

Fig. 2 Process steps for creation of NN-based on-chip interconnect system

The first step involves creation of initial dataset. This is performed using Monte-Carlo analysis. Monte-Carlo analysis is a statistical-based technique used for assessment of variation impact of multiple parameters on the circuit performance. According to this method, multiple iterations are performed for response evaluation based on random set of varying inputs specified within a certain range. The system chosen in the current paper is a driver-interconnect-load (DIL) where a CMOS inverter is considered as driver and a capacitor as load [1]. This is shown in Fig. 3.

Fig. 3 Driver-interconnect-load (DIL) system

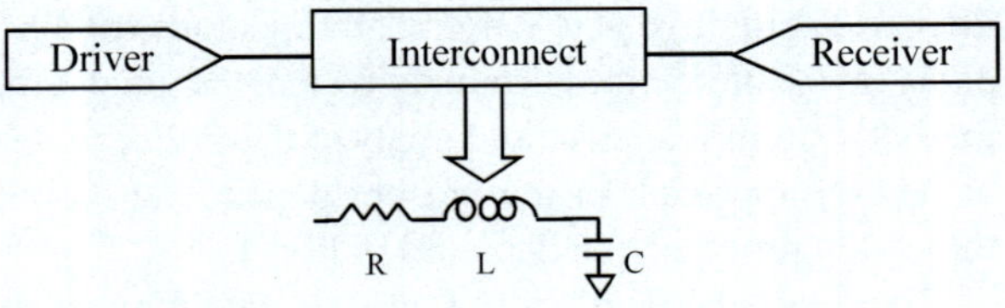

The second step involves building the NN-based model for training. To implement this, it is necessary to be acquainted with the NN approach. Algorithms based on neural network (NN) are inspired by biological neural network. These are basically mathematical interpretation of human neuron model that attempts to emulate the associative and parallel pattern of the human brain. Neural network-based algorithms are being widely researched presently and employed for various regression and classification problems. The structure of any neural network consists of input, hidden and output layers. Each layer consists of a set of neurons that are defined by user. Multilayer perceptron (MLP) is the most frequent representation of NN network, as depicted in Fig. 4a.

The middle layer, which is the hidden layer, transforms the obtained weighted linear function to a nonlinear function by using activation function [9]. Based on the weights or strength of each connection of the nodes, the data is passed on to the

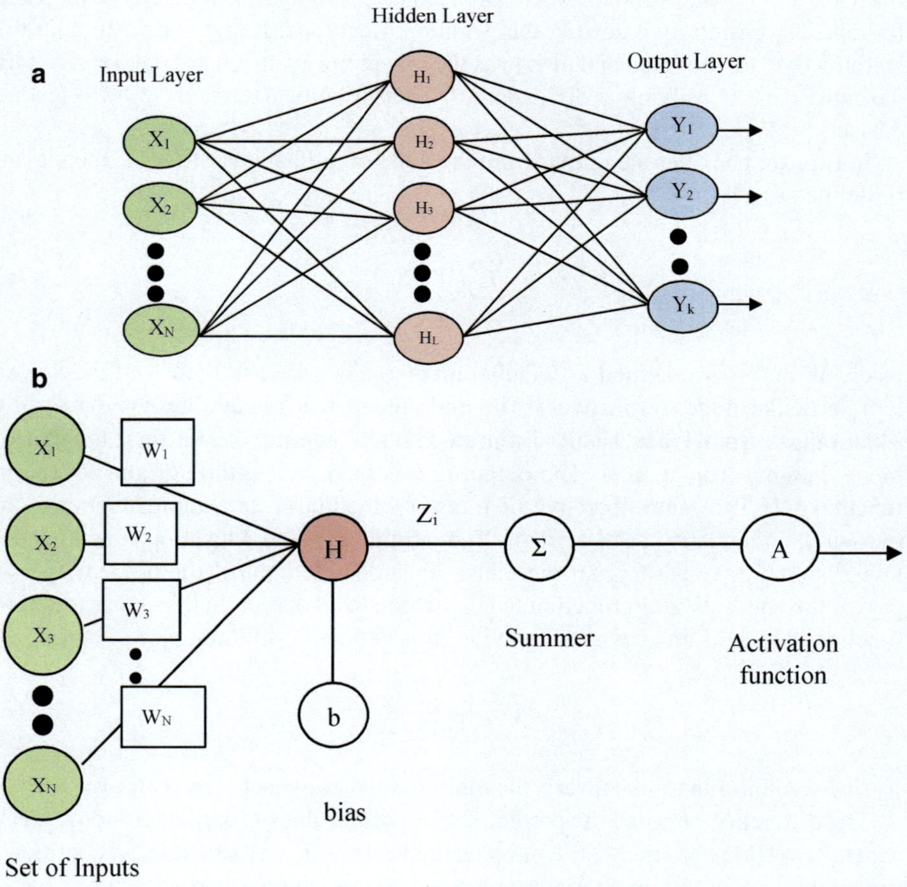

Fig. 4 **a** Neural network architecture. **b** Detailed architecture corresponding to a node of the hidden layer

other hidden layer or the output layer. The input layer consists of neurons which are equal to the number of input variables. The input nodes accept the input independent dataset. The number of hidden layers can be more than one depending upon the complexity of the problem. A detailed structure corresponding to single neuron of the hidden layer (of Fig. 4a) is shown in Fig. 4b. In the figure all inputs are connected to each node of the hidden layer with some strength known as weight. Also, a bias has been added in order to incorporate shifting.

Consider a function such that

$$Y = f(X) \tag{1}$$

where X is the independent variable given to the input layer and Y is the dependent variable which is obtained through output layer. X and Y are variables that contain set of values arranged in the form of array. Even if the exact function which computes Y from X is unknown, neural network approach has the magnificent ability to map and form a relationship by adjusting the weights of neurons. It first learns through the learning data set provided and then tests the system by using the test dataset. Usually the entire dataset available is divided into 70% as training dataset and 30% as testing dataset.

In Fig. 4b, consider the output from the node of hidden layer to be Z, then it can be defined as

$$Zi = \sum_{j=1}^{N} (W_j.X_j) + b \tag{2}$$

where W and b are defined as weight corresponding to each input and bias given to a particular node, respectively. The node number of hidden layer is defined by i that ranges from 1 to L, while j represents the count number of total number of inputs ranging from 1 to N. The output (Z_i) is then passed through the activation function 'A'. The main objective of transfer function or activation function is to introduce nonlinearity in the circuit. The commonly used activation functions are linear, sigmoid, Gaussian and piecewise linear. The output of the hidden layer is passed through activation function to get a normalized output. In this paper, sigmoid function is used as an activation function. It is defined as follows:

$$\sigma(Z) = \frac{1}{1 + e^{-z}} \tag{3}$$

The sigmoid function converts the output in the range between 0 and 1.

The difference between the predicted and actual values is used to compute mean square error (MSE). The MSE is propagated backwards so that each layer readjusts its weights so that minimum error is obtained. This method is hence known as back-propagation algorithm [7]. There are many NN back-propagation algorithms that have been proposed by different researchers [10]. Levenberg–Marquardt (LM) algorithm is one of the best NNBP algorithms as it considerably lowers computational time [8]. LM algorithm is stable and faster than other algorithms when it comes to weights updation. Weight updation using LM algorithm is defined as:

$$W_{k+1} = W_k - (J_k^T J_k + \mu I)^{-1} J_k e_k \tag{4}$$

where W is the weight of the NN network, J is the Jacobian matrix, I is the identity matrix and e is the error matrix. The computational complexity of LM algorithm depends on the Jacobian matrix. Thus, the complexity of the proposed system depends on the number of input, hidden, output layers and on the number of neurons in the hidden layer. This complexity is defined only for training phase as in testing phase we use the NN network which has been created using LM algorithm.

The LM algorithm has been used in the current paper. The training process is executed until minimum MSE is achieved. Once the desired MSE is obtained, the training process is completed and last step for model formation is executed. The final step comprises testing process. Here, the reserved testing dataset is provided as input and the output is analyzed using developed NN-based model.

3 Results and Discussion

This section presents the results that are obtained using the developed model that is based on NN approach. The precision of the developed model is also compared with the traditional analytical model. The NN-based model is developed for 32 nm technology. The on-chip interconnect dimensions are defined in Table 1 [11, 12].

The neural network is trained using dataset that is generated from Monte-Carlo simulations. These have been performed using HSPICE. Interconnect parasitics, namely resistance, inductance and capacitance, are varied. The output performance parameters are delay and power dissipated. The interconnect length considered is 0.5 mm. The Monte-Carlo simulations have been performed for 1000 iterations. The process parameters are varied with deviation of $\pm 10\%$ simultaneously. Further training, testing and creation of neural network structure have been implemented

Table 1 Interconnect dimensions for 32 nm technology

Interconnect dimension	Value
Width (w)	48 nm
Space (s)	48 nm
Thickness (t)	144 nm
Height (h)	110.4 nm
Aspect ratio	3
Vdd	0.9 V
Length	0.5 mm
Dielectric constant (ε)	2.25

in MATLAB with the help of nntraintool. Training dataset consists of 70% of the total dataset. Further, the remaining dataset is divided into validation set and testing dataset, each with 15% of the total dataset. Input layer is connected to hidden layer. Number of hidden layers and neurons in it depend on the complexity of the problem. Since feed forward neural network does not have any prior knowledge regarding the weights or hidden layers, henceforth back-propagation algorithm is used to compute weights. These weights are iteratively updated in order to fit with respect to the performance parameter. For this problem, two hidden layers have been employed with count of neuron number as 10 and 8, respectively. The LM training algorithm is used with 1000 epochs (iterations) and mean square error as a performance parameter. The activation function used is sigmoid in order to introduce nonlinearity to the hidden layer.

3.1 Validation and Accuracy Assessment of the Developed NN-Based Model

Regression curves are plotted and mean square error has been computed as depicted in Fig. 5. Regression is a statistical modeling technique used for setting up the relationship between input and output variables. It also defines the level of closeness of the developed model with initial dataset. Figure 5a–c indicates regression plots for output data with the target values. The correlation factor 'R' indicates the linearity between the actual obtained and expected output values. The closer the value of R to 1, the greater is the linear relationship between targets and output. The fitting of the dataset and predicted output is plotted using regression plots. From Fig. 5, it can be seen that the best fit line for the clustered dataset lies on the target line. This shows that the developed NN-based model works very optimally.

To compute the accuracy and confirming the efficiency of the NN-based developed model, mean square error is computed. Figure 6 indicates the mean square error, which in this case is low, thereby indicating the network is working very closely to the analytical model. It can be observed that best performance or least error is obtained after 32 epochs. Hence, a system stabilizes and adjusts its parameters for best performance in 32 iterations.

In order to analyze each sample's accuracy in the total data set, error histogram is plotted as shown in Fig. 7. The error histogram curve indicates that around 700 training samples, 150 validation samples and 150 testing samples have error near to 0. Therefore, these results confirm that NN works closely when compared to SPICE simulation, so NN can be employed as an alternative predication method.

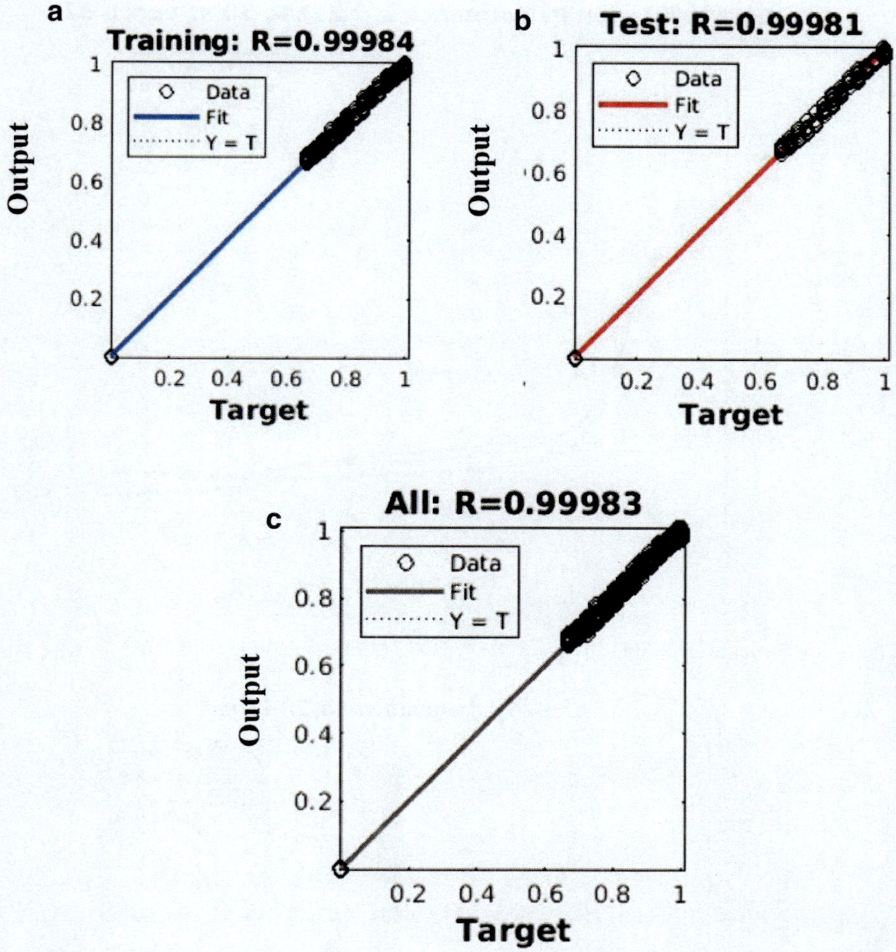

Fig. 5 **a** Training regression plot. **b** Testing regression plot. **c** Regression plot for entire data

3.2 Variability Analysis Using Developed Neural Network Model

In this section, variability analysis is performed using NN approach. This is compared with the Monte-Carlo simulations. The trained neural network is utilized to compute the average power dissipation and delay for different interconnect lengths. Therefore, a new dataset is created for 32 nm technology and variation of $\pm 10\%$ is considered [11, 12]. The training dataset consists of 750 samples for each interconnect length. The number of samples taken in Monte-Carlo and for testing dataset in NN is 125 for each interconnect length. An analysis methodology using proposed NN is depicted

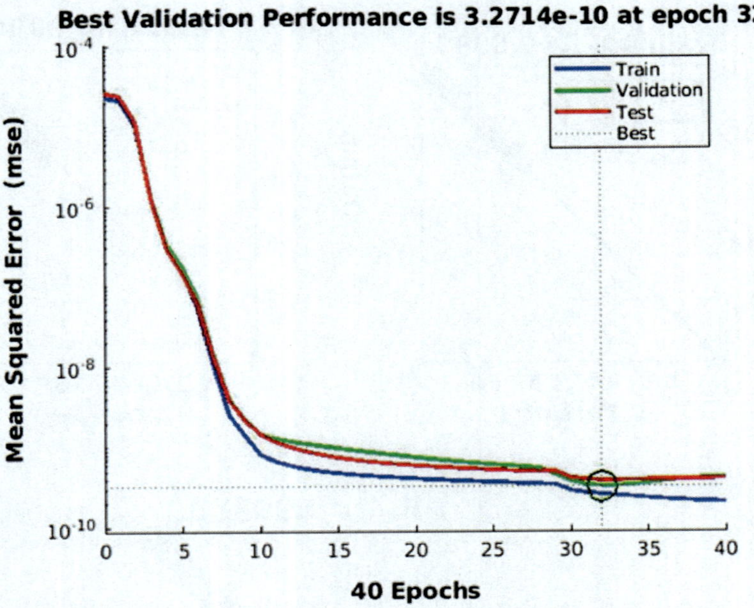

Fig. 6 Mean square error plot

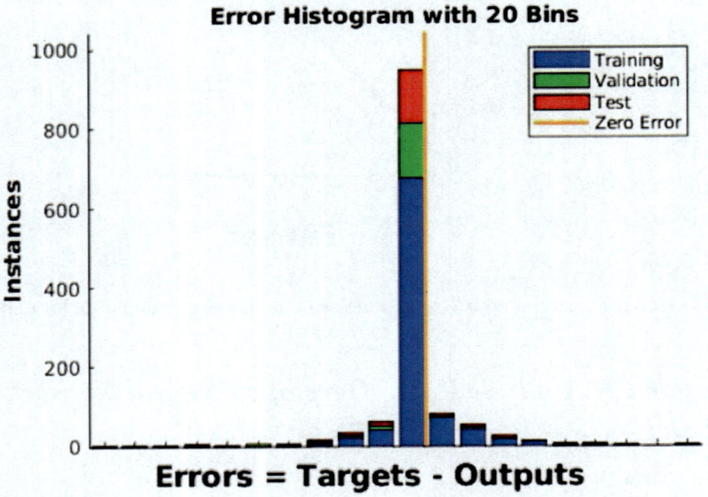

Fig. 7 Error histogram plot

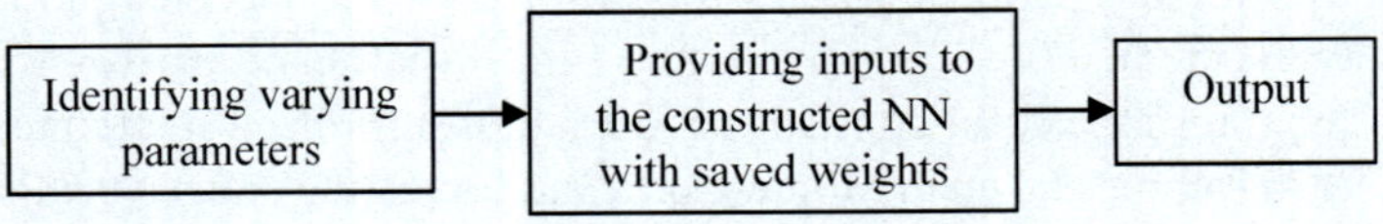

Fig. 8 Variability analysis methodology using NN

in Fig. 8. Comparison of output performance parameter, viz., power dissipation and delay between the proposed NN and existing analytical method is shown in Fig. 9.

From Fig. 9, it is observed that delay and power dissipation increases with interconnect length. This is due to increase in interconnect parasitics with length of the

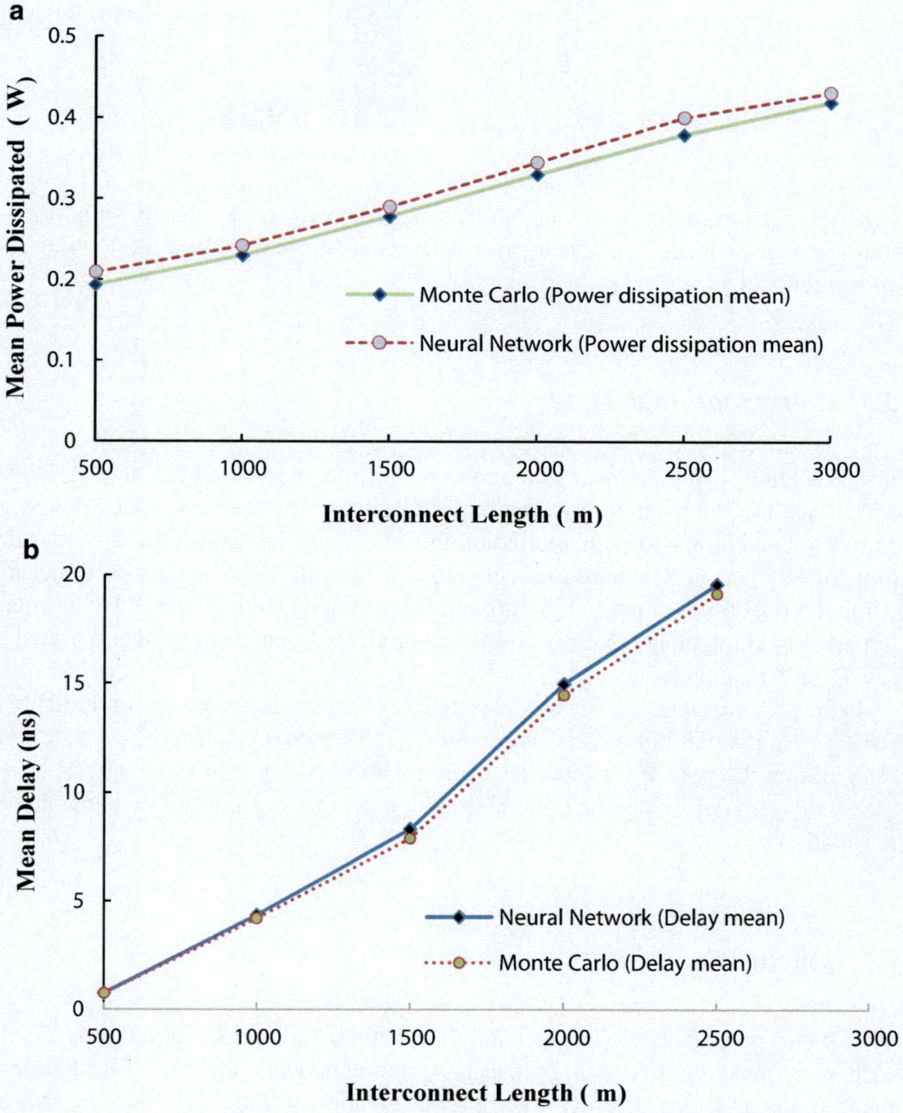

Fig. 9 **a** Monte-Carlo and NN comparison plot based on power dissipation. **b** Monte-Carlo and NN comparison plot based on delay

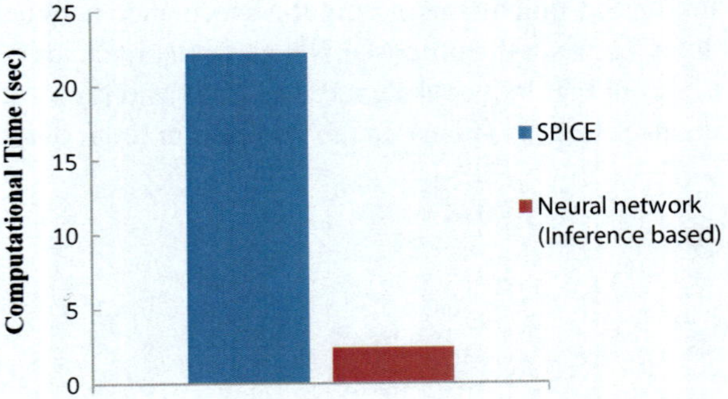

Fig. 10 Monte-Carlo and NN total time elapsed plot

wire. It is seen from the figure that NN-based model works very closely with Monte-Carlo analysis. Hence, there is a good agreement of the proposed model with the existing model.

3.3 *Computational Time*

In this section, computational time required for inference-based NN model, which considers the time taken to predict the performance after the network is trained, is analyzed and compared with simulation model. Figure 10 depicts the time elapsed plot for NN and SPICE-based model. From the figure it can be observed that the computational time for prediction through NN is much less as compared to its counterpart simulation model. Hence, NN-based model is fast and can be incorporated in several other analyses.

In the present paper, the NN model is used to compute delay and power dissipation for assessing variability effects. Subsequently, this model can also be used to optimize/design the system for low-power dissipation system. Hence, using NN-based model, system design can be done optimally such that it produces very low power dissipation.

4 Conclusion

This paper presents a novel and effective method for variability analysis of on-chip interconnects system using prospective neural network approach. The approach incorporates a systematic process comprising creation of initial dataset, then training and building the NN-based model. This is followed by testing of the model. The developed neural network-based model is computationally efficient. The validity and accuracy of developed model is ensured using regression plots and analyzing mean square error. Further, variability analysis has been performed using NN-based model. This has also been validated using conventional Monte-Carlo method. It has been analyzed that NN-based model is 9.2 times faster as compared to conventional Monte-Carlo method. Since neural network does not have a prior knowledge, it has the capacity to grasp from abstract dataset, and thereby it is not limited to a particular interconnect dimension. Also the proposed model can handle high dimension datasets which have unknown non-linear relationship with each other. The neural network-based model is versatile, dynamic and can be effectively incorporated to implement, analyze and design efficient circuits and systems.

References

1. Lin Q, Wu HF, Jia GQ (2018) Review of the global trend of interconnect reliability for integrated circuit. Circuits Syst 9:9-21Y
2. Scheffer L (2006) An overview of on-chip interconnect variation. In: ACM international workshop on system-level interconnect prediction, pp 27–28
3. Agrawal R, Chandel, Dhiman R (2016) Variability analysis of stochastic parameters on the electrical performance of on-chip current-mode interconnect system. IETE J Res (Taylor and Francis) 63(2):268–280
4. Agrawal Y, Parekh R, Chandel R (2018) Performance analysis of current-mode interconnect system in presence of process, voltage, and temperature variations. In: Garg A, Bhoi A, Sanjeevikumar P, Kamani K (eds) Advances in power systems and energy management. Lecture notes in electrical engineering, vol 436. Springer, Singapore
5. Tu TY, Chao PC-P (2018) Continuous blood pressure measurement based on a neural network scheme applied with a cuffless sensor. Springer, Berlin Heidelberg
6. Trinchero R, Manfredi P, Stievano IS, Canavero FG (2018) Machine learning for the performance assessment of high-speed links. IEEE Trans Electromagn Compat
7. Zhang L, Wang F, Sun T et al (2018) A constrained optimization method based on BP neural network. Neural Comput Appl 29
8. Lourakis MIA (2005): A brief description of the Levenberg–Marquardt algorithm implemented by Levmar. Proc Found Res Technol 1–6
9. Sibi P, Bordes A, Jones SA, Siddarth P (2013): Analysis of different activation functions using back propagation neural network. J Theor Appl Inf Technol 47(3)
10. Saduf MAW (2013) Comparative study of back propagation learning algorithms for neural networks. Int J Adv Res Comput Sci Softw Eng 3(12)
11. Predictive Technology Models. https://ptm.asu.edu (2016)
12. International Technology Roadmap for Semiconductors. https://public.itrs.net (2013)

Prospective Incorporation of Booster in Carbon Interconnects for High-Speed Integrated Circuits

Takshashila Pathade, Yash Agrawal, Rutu Parekh, and Mekala Girish Kumar

Abstract VLSI technology has eulogistically grown over the years. This has been made feasible due to continuous scaling of technology. Miniaturization of technology and conscientious demand of high-speed applications have resulted in dense and compact packaging of on-chip interconnects and devices. Copper has been widely used as an on-chip interconnect material in VLSI chips. However, copper is constrained by the grain and surface boundary scattering effects at scaled technology nodes that limit its utility and further incorporation in futuristic integrated circuit (IC) designs. Subsequently, carbon-based on-chip interconnects have been investigated and proven to be one of the effective alternatives to conventional copper interconnects. The present work vividly explores and investigates carbon nano-materials that can be used for high-speed VLSI interconnects. The other important entanglement with on-chip interconnect is that its parasitic increases and henceforth output performance degrades as the length of wiring over IC increases. This can be eminently alleviated using prospective booster insertion in between on-chip interconnects. The elegant booster insertion technique for prosperous carbon interconnects has been marginally explored till date and henceforth taken up in this work. This paper meticulously investigates various system characteristics such as delay, power dissipation, crosstalk, signal integrity, eye diagram for different interconnect materials such as copper, carbon-based CNTs and GNRs. It is observed that carbon interconnects impressively outperform in terms of delay, power, crosstalk and high-speed operation than conventional copper interconnects.

T. Pathade (✉) · Y. Agrawal · R. Parekh
Dhirubhai Ambani Institute of Information and Communication Technology, Gandhinagar 382007, Gujarat, India
e-mail: takshupathade@gmail.com

Y. Agrawal
e-mail: mr.yashagrawal@gmail.com

R. Parekh
e-mail: rutuparek@gmail.com

M. Girish Kumar
Vidya Jyoti Institute of Technology, Hyderabad 500075, Telangana, India
e-mail: girishkumarece@vjit.ac.in

Z. Patel et al. (eds.), *Advances in VLSI and Embedded Systems*, Lecture Notes in Electrical Engineering 676, https://doi.org/10.1007/978-981-15-6229-7_22

Keywords Booster insertion · Carbon nanotube (CNT) · Eye diagram · Graphene nanoribbon (GNR) · Nanomaterials · Very large-scale integration (VLSI)

1 Introduction

Progressive advancements in analog, digital and mixed signal systems and their realization in integrated circuits (ICs) have created need for high-speed requirements, and necessity to attain more functionality over same silicon chip area. This requires scaling of technology that leads to compact integration of millions of transistors and interconnections. In the early era of VLSI development, when the technology dimensions were large, the performance of IC was dominantly determined by transistors. However, as the technology miniaturized, on-chip interconnects have become more vital factor and dictate in characterizing system performance of nano-chips [1–10]. Hence, performance analysis of on-chip interconnects to enhance efficiency of ICs has become very essential. Incorporation of advanced integration techniques, like 3D and multicore ICs, to attain high speed and enhanced performance leads to reliability concerns like thermal reliability, crosstalk, electromigration and scattering effects [2, 3, 11]. As the devices are miniaturized below submicron technologies, new nanomaterials are explored and introduced to meet high-performance requirements. International Technology Roadmap for Semiconductors (ITRS) published in its reports of year 1994 about the necessity to incorporate new on-chip conducting materials for attaining projected overall technology requirement [1]. And later from year 2001 onwards, ITRS emphasized on prospective new materials and highlighted the problems that are existing in conventional copper on-chip interconnects. It is reported that copper resistivity increases at scaled technology nodes due to downscaling of physical dimensions of interconnects. Nanomaterials that can be aptly used in high-speed on-chip interconnects are carbon nanomaterials that are briefly discussed in subsequent sections.

The vital system performance is majorly evaluated and analyzed using propagation delay and power dissipation. Both propagation delay and power dissipation are function of on-chip interconnect length. The increase in the length of interconnect has deteriorating effects on these output performance parameters. The global on-chip interconnects refer to top layer of on-chip interconnect system and are of longest length in IC designs. These global interconnect provides power, ground and control signals to entire chip modules. The length of global interconnect increases in proportion with the die size and is prime evildoer for propagation delay and signal degradation in miniaturized ICs [2, 3, 10, 12]. Booster insertion is one of the efficient techniques, to mitigate the signal integrity issues in long interconnects [10, 12, 13]. In this technique, boosters (buffers) are placed in between long wires by segmenting it into smaller sections. Since propagation delay increases with the length of interconnect, dividing the line into smaller sections is an efficient scheme to limit system latency due to interconnects.

2 Carbon Nanomaterials

Carbon is a very versatile element, distinguished by various forms of allotropes and structures. Their extraordinary physical and electromechanical properties are because of occurrences of sp, sp^2 or sp^3 hybridization in different allotropic forms [6–8]. Among the variety of carbon nanomaterials, graphene nanoribbons and carbon nanotubes are good electrical conductors, hence can be used as promising alternative to copper as on-chip interconnects.

Carbon nanomaterials have all preferable properties such as very long mean free paths up to a few micrometers, less electro-migration effect, high thermal conductivity, high current carrying capability and ability to work effectively at high frequencies, carrier mobility and so on. The various important properties of carbon nanomaterials are summarized in Table 1.

The values of the CNT parameters as described in Table 1 are function of CNT diameter, tube density, width of graphene sheet and also fabrication process. It can be observed from the table that properties of carbon nanomaterials are significantly superior than copper. These makes carbon nanomaterials as one of the best suited materials for on-chip interconnects [2–9, 11, 17–20].

Graphene nanoribbons (GNRs) and carbon nanotubes (CNTs) are the most widely researched and investigated carbon nanomaterials for VLSI applications. Both the forms are derivatives of graphene that comprises single atom thick layer of carbon atoms arranged in honeycomb lattice structure. CNT is a sheet of graphene rolled up as hollow cylinder. However, GNR is obtained by patterning graphene as a planar structure [12, 13, 16].

From physical structure point of view, GNRs can be categorized as armchair (*ac*) or zigzag (*zz*), as depicted in Fig. 1. Depending on the edge pattern of GNR, it can act as either semiconducting or metallic. The armchair edge-patterned GNRs can be used as either semiconductor or metal. This depends on the number of carbon atoms residing across the GNR width. While GNR with zigzag edge always attributes metallic properties [4–7], GNR with one layer is termed as single-layer

Table 1 Electrical properties of carbon nanomaterials against copper interconnects [8–16]

Parameter	Copper	SLGNR/MLGNR	SWCNT	MWCNT	CNT Bundle
Mean free path at room temperature (nm)	40	1000	>1000	$\approx 25 \times 10^3$	$>25 \times 10^3$
Thermal conductivity ($\times 10^3$ W/m K)	≈0.39	≈3–5	≈1.8–5.8	≈3	>3
Current density (A/cm^2)	$\approx 10^7$	$>10^8$	$>10^9$	$>10^9$	$>10^9$
Carrier mobility (cm^2/V s)	5–9	$\approx 10^5$	$>10^5$	$>10^5$	$>10^5$

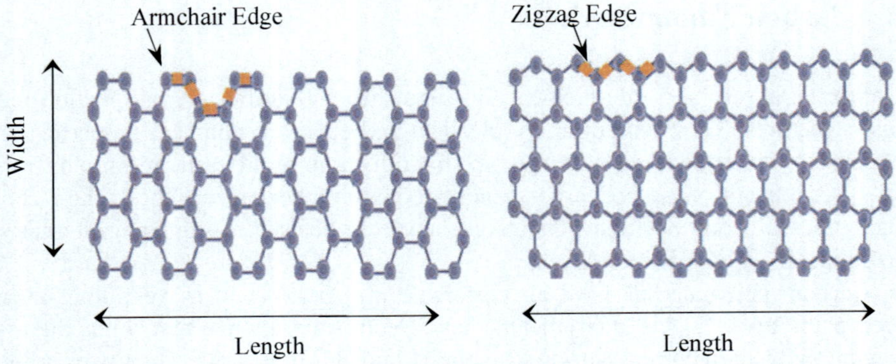

Fig. 1 Representation of armchair (left) and zigzag (right) structures of GNR

GNR (SLGNR). Likewise, when multiple layers of GNR are stacked one above the other maintaining minimum distance of van der Waal's gap (≈0.34 nm) between the intermediate layers, it is called as multilayer GNR (MLGNR). MLGNRs are most preferred for on-interconnect applications. This is because MLGNR possesses better current conduction and lesser impedance than SLGNR structure [5–7]. The geometric representation of MLGNR interconnect is demonstrated in Fig. 2, where N represents the number of graphene layers. These are separated by van der Waal's gap (δ). Width (w) and thickness (t) of GNR correspond to dimension of interconnects [4–8].

The total number of layers (N_{layer}) in MLGNR is computed as [4–7].

$$N_{\text{layer}} = 1 + \text{Integer}\left(\frac{t}{\delta}\right) \tag{1}$$

Alike GNR structures, CNTs are also categorized into single-wall carbonnanotube (SWCNT), which is a single tubular structure, and multiwall carbon nanotube (MWCNT) which is a CNT with multiple shells concentric in it. In addition to these, potential research has been performed considering CNTs in bundle form. Depending on the type of CNTs packed in bundle structure, they are correspondingly defined and nomenclatured. Bundle consisting of only SWCNT is called SWCNT bundle while bundle of MWCNT is termed as MWCNT bundle. Later researchers have investigated

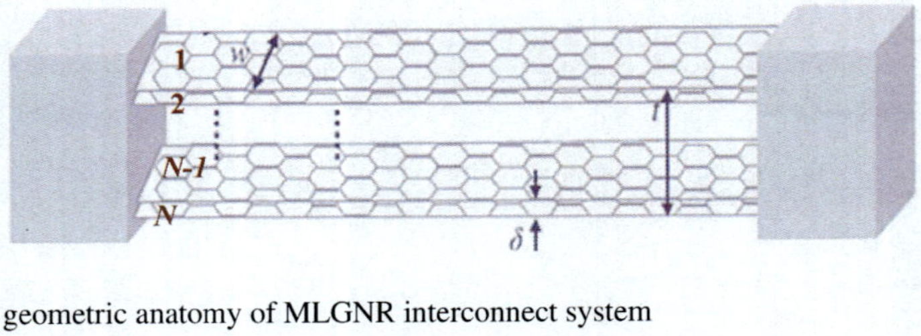

Fig. 2 The geometric anatomy of MLGNR interconnect system

that during development of CNTs using chemical vapor deposition process (CVD), carbon nanotubes are formed in a bundle structure comprising both SWCNTs and MWCNTs. This structure is referred to as mixed CNT bundle (MCB). The structures of SWCNT, MWCNT and mixed CNT bundle are shown in Fig. 3.

Figure 3 also depicts the spatial arrangement of SWCNTs, MWCNTs and mixed CNTs that are packed in a bundle structure. It is seen that the centers of SWCNTs, MWCNTs and mixed wall CNTs in a bundle are arranged in triangular placement form. It has been analyzed that by placing centers of CNTs in the shown triangular formation gives the high package density and incorporation of maximum CNTs in the rectangular area. In any CNT bundle with given cross-sectional area, total number of CNTs is a critical factor to compute equivalent circuit parasitical parameters. Total number of CNTs (SWCNTs/MWCNTs) in a bundle are determined as [10, 11, 14, 16, 21, 22].

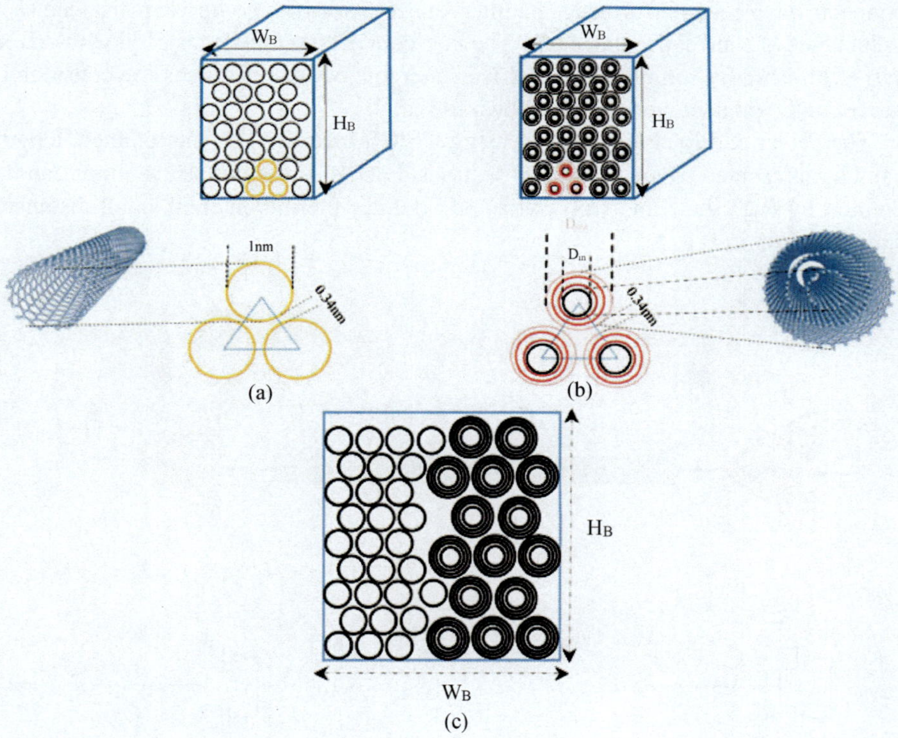

Fig. 3 Packing arrangement of **a** SWCNT, **b** MWCNT and **c** Mixed CNT bundles

$$N_{CNT} = \begin{cases} N_{W_B} N_{H_B} - \frac{N_{H_B}}{2}, \text{ if } N_{H_B} \text{ is even} \\ N_W N_{H_B} - \frac{N_{H_B}-1}{2}, \text{ if } N_{H_B} \text{ is odd} \end{cases} \quad (2)$$

$$N_{W_B} = \left\lfloor \frac{W_B - D}{D + \delta} \right\rfloor + 1 \text{ and } N_{H_B} = \left\lfloor \frac{H_B - D}{\frac{\sqrt{3}}{2}(D + \delta)} \right\rfloor + 1 \quad (3)$$

where D is diameter of SWCNT (D_{OUT} in case of MWCNT), W_{B} and H_{B} are width and height CNT bundle, respectively, N_{WB} and N_{HB} are number of CNTs present along the width and height, respectively. $\lfloor x \rfloor$ is used to denote a largest integer number, that is, less than or equal to x.

3 Electrical Modeling of Copper and Carbon Interconnects

In this section, electrical modeling, its model description and parasitic extraction of conventional copper and advanced carbon (MLGNR and CNT bundles) based interconnects have been performed. The interconnect parasitics comprise resistance (R), inductance (L) and capacitance (C). The interconnect has to be energized with driver and terminated by some load. The driver-interconnect-load (DIL) is the considered model in the present work and is shown in Fig. 4.

The driver and load are realized using CMOS inverter. The interconnect length can be represented by small lumped sections. Each of these lumped sections can be defined by $R\Delta z$, $L\Delta z$ and $C\Delta z$, where Δz defines the incremental small distance. This is shown in Fig. 4.

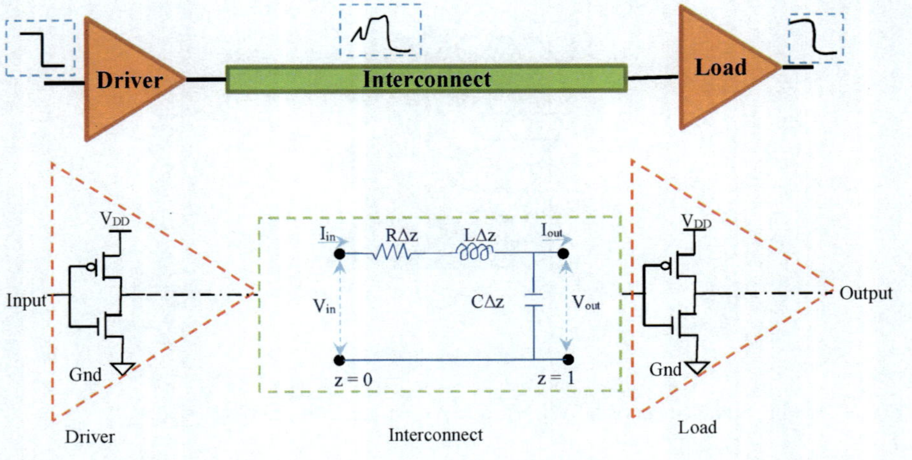

Fig. 4 On-chip interconnect system represented in terms of driver-interconnect-load (DIL) model

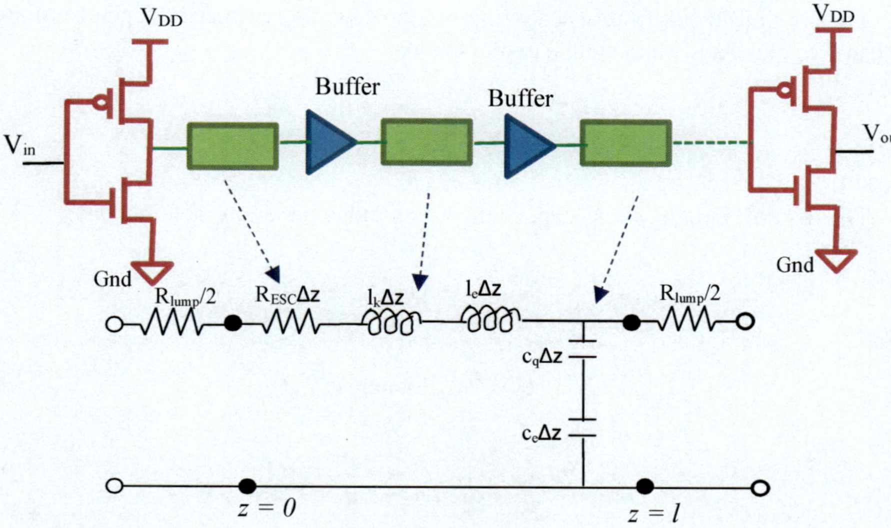

Fig. 5 Representation of buffer insertion technique in DIL model

As discussed in Sect. 1, long on-chip interconnects deteriorate the system performance. This can be effectively mitigated using booster (buffer) insertion technique. The DIL system along with boosters inserted in between interconnects is shown in Fig. 5.

3.1 Modeling of MLGNR Interconnect

MLGNR interconnect is electrically represented as equivalent single conductor (ESC) model. This comprises lumped resistance (R_{lump}), scattering resistance (r_{ECS}), kinetic inductance (l_{k}), magnetic inductance (l_{e}), quantum (c_{q}) and electrostatic (c_{e}) capacitances. These parameters are extracted using (5)–(9) [4–8].

$$R_{\text{lump}} = R_q + R_{\text{mc}} \tag{4}$$

where

$$R_q = \frac{h}{2 \cdot e^2 N_{\text{ch}}} \tag{5}$$

In Eq. (5), N_{ch} defines the number of conduction channels and it is evaluated on the basis of spin and sub-lattice degeneracy of carbon atoms that is proportional to the width, Fermi energy (E_{f}) and temperature (T) [4–7].

The per unit length (*p.u.l.*) scattering resistance (r_{ECS}) primarily depends on the mean free path (λ_{mfp}) and can be expressed as

$$r_{ESC} = \frac{R_q}{N_{layer} \cdot \lambda_{mfp}} \tag{6}$$

The per unit length (*p.u.l.*) capacitances and inductances are obtained as

$$l_k = \frac{l_{k0}}{2N_{ch} \cdot N_{layer}}; \text{ where } l_{k0} = \frac{h}{2e^2 v_F} \tag{7}$$

$$c_{q,n} = 2c_{q0} N_{ch} \cdot N_{layer}; \text{ where } c_{q0} = \frac{2e^2}{h v_F} \tag{8}$$

$$l_{e,n} = \frac{\mu_0 d}{w}, c_e = \frac{\varepsilon_0 w}{d} \tag{9}$$

3.2 Modeling of MWCNT and Mixed CNT Bundle Interconnect

Among various carbon-based allotropes, this work is motivated to use carbon nanotube bundle and its various structures as on-chip interconnects. This is due to higher current density, longer mean free path and higher current mobility of CNT bundle structures over GNRs. Indeed, making isolated SWCNT/MWCNT from a bundled CNT is a cumbersome task, since it is found that CNTs are developed by CVD mechanism in the form of bundle [11]. It is also inferred from the literature that mixed CNT bundles offer superior performance and reliability. Hence, the current work focuses on demonstrating and comparing performance of all the forms of bundled CNTs. MWCNTs behave metallic in nature; SWCNTs may be metallic or semiconducting depending on their chirality [11, 21]. Electrical parameters of SWCNT bundles are determined using (10)–(15).

$$R_{lump} = \frac{\frac{h}{4 \cdot e^2}}{N_{total}} + R_{mc} \text{ and } R_{ESC} = \frac{R_q}{\lambda_{mfp} \cdot N_{total}} \tag{10}$$

where

$$\lambda_{mfp} \approx 10^3 D_i \text{ and } N_{total} = \sum_{i=1}^{N_{CNT}} N_{CH} \tag{11}$$

Quantum capacitance, electrostatic capacitance and effective equivalent single-conductor capacitance are estimated as in (12)–(14).

$$C_{\text{ESC}}^{'} = \frac{2\pi\varepsilon}{\cosh^{-1}\left(\frac{h_t+D}{D}\right)} \approx 50\,\text{aF} \tag{12}$$

$$C_{q_o^{'}} = \frac{4.e^2}{h \cdot v_F} \approx 193.7\,\text{pF and } C_q = C_{q_o^{'}} \cdot N_{\text{total}} \tag{13}$$

$$C_{\text{ESC}} = C_q || C_{\text{ESC}}^{'} = \frac{C_q \cdot C_{\text{ESC}}^{'}}{C_q + C_{\text{ESC}}^{'}} \tag{14}$$

Kinetic and mutual inductances are in series, thus equivalent single-conductor inductance is evaluated using (15)

$$L_{\text{ESC}} = \frac{\frac{h}{2e^2 v_F}}{2 \cdot N_{\text{total}}} + \frac{1}{N_{\text{total}}}\left[\frac{\mu_o}{2\pi}\ln\left(\frac{h_t}{D}\right)\right] \tag{15}$$

MWCNTs and mixed CNT bundles are of greater importance due to their metallic property and reliability; their modeling differs in terms of number of shells each CNT have. Number of conducting channels is a critical parameter that determines the performance of interconnect and is estimated on the basis of how many shells each CNT possesses and it can be given as

$$N_{\text{ch}}(D_i) = \begin{cases} K_1 T D_i + K_2, \text{ when } D_i > d_{t/T} \\ \frac{2}{3}, \text{ when } D_i < d_{t/T} \end{cases} \tag{16}$$

where $T = 300$ K, $K_1 = 3.87 \times 10^{-4}\ \text{nm}^{-1}\ \text{K}^{-1}$ and $K_2 = 0.2$.

Lumped and distributed RLC parameters of MWCNT bundle and mixed CNT bundle can be derived using (17)–(22) as

$$R_{\text{lump}} = \left[\sum_{j=1}^{N_{\text{CNT}}}\left(\sum_{i=1}^{N_{\text{shells}}}\left(\frac{R_q}{N_i(D_i)} + R_{\text{mc}}\right)\right)^{-1}\right]^{-1} \tag{17}$$

$$R_{\text{ESC}} = \left[\sum_{j=1}^{N_{\text{CNT}}}\left(\sum_{i=1}^{N_{\text{shells}}}\left(\frac{N_i(D_i) \cdot \lambda_{\text{mfp},i}}{R_q}\right)\right)\right]^{-1} \tag{18}$$

$$C_{e,\text{ESC}} = \frac{2\pi\varepsilon_0\varepsilon_r}{\cosh^{-1}\left(\frac{D_{\text{shell}}+2\cdot h_t}{D_{\text{shell}}}\right)} \times N_x \tag{19}$$

$$C_{q,\text{ESC}} = 2 \cdot C_{q0} \cdot N_{\text{total}} \tag{20}$$

$$L_{k,\text{ESC}} = \frac{L_{k_o^{'}}}{2 \cdot N_{\text{total}}} \tag{21}$$

$$L_{e,\mathrm{ESC}} = \frac{1}{N_x}\left[\frac{\mu_o}{2\pi}\ln\left(\frac{D_{\mathrm{shell}} + 2\cdot h_t}{D_{\mathrm{shell}}}\right)\right] \tag{22}$$

4 Simulation and Analysis

In the present work, effectiveness of prospective booster insertion for futuristic carbon interconnects is investigated. This has been compared with conventional buffer inserted copper interconnects. Various signal integrity and reliability analyses have been performed. In the context of analyzing high-speed operation of carbon and copper interconnects, jitter and signal-to-noise ratios (SNR) are estimated and analyzed. For all the performance analyses, technology node considered is 22 nm. Interconnect materials considered for analysis are copper, MLGNR, SWCNT bundle, MWCNT bundle and mixed CNT bundle. Figure 6 depicts the effect of increasing length of interconnect on the system latency. This analysis is performed for interconnect lengths of 0.5, 1, 1.5 and 2 mm. It is observed that delay across output of DIL model goes on increasing if length of interconnect increases.

Figures 6 and 7 represent the effect of global interconnects on system performance. The efficiency of buffer insertion technique is visualized in Fig. 7. CMOS buffers are inserted in regular intervals of long interconnect. It is observed from Fig. 7 that delays can be reduced using booster insertion technique in case of all

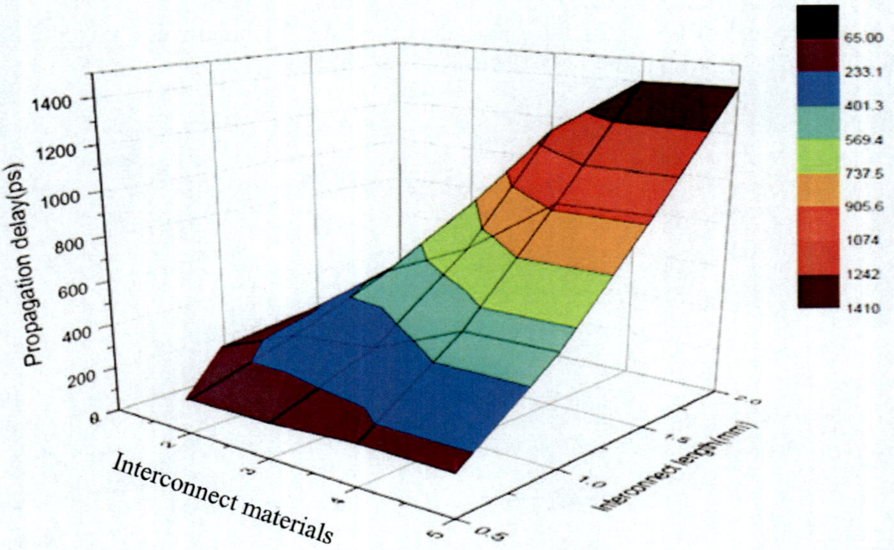

Fig. 6 Variation in propagation delay with varying interconnect length for different interconnect materials. (Notations 1–5 on x-axis of the figure represent as follows: 1. MWCNTB, 2. SWCNTB, 3. MCB, 4. MLGNR, 5. Cu)

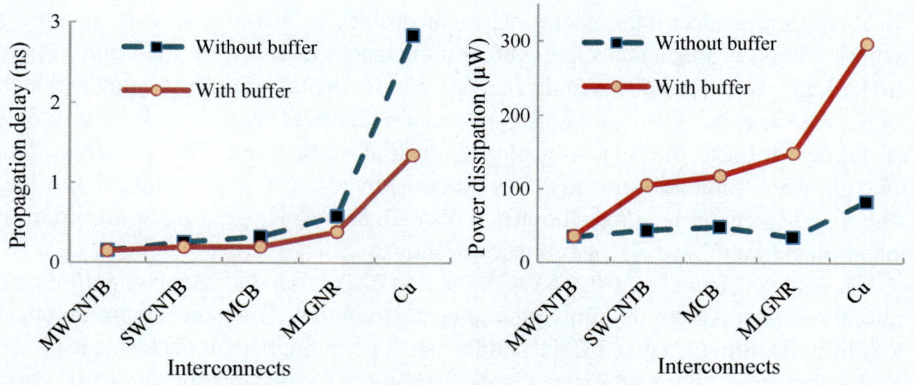

Fig. 7 Delay (left) and power dissipation (right) in varying interconnect materials for with buffer and without buffer insertion cases

the interconnect materials. But since CMOS buffers consume significant amount of energy from the power supply, these buffers are at the same time power-hungry also. So this work also gives direction to future researchers in this area for using efficient booster that consumes lesser power. It is investigated from the figures that MWCNTB and SWCNTB interconnects outperform among all the interconnect materials. But due to their development infeasibility, MCB interconnect is considered as suitably optimal performing material for interconnects since MCB possesses more number of conduction channels than SWCNT and MWCNT bundles. Hence, MCB is proposed as a prospective interconnect material in the current work.

Another critical issue in interconnect performance is coupling phenomenon, and it is modeled using the following crosstalk model. Figure 8 shows crosstalk model

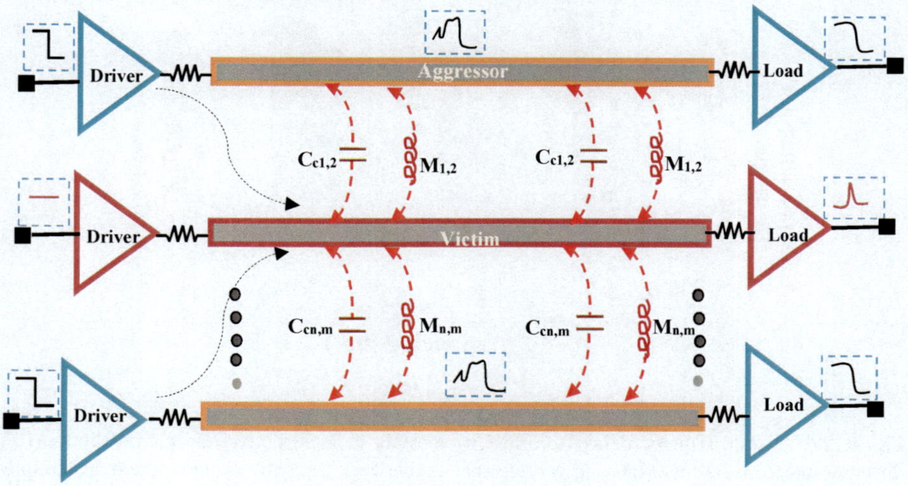

Fig. 8 Multiple line crosstalk model for interconnects

for three interconnect lines. Since interconnect lines are parallel to each other, they exhibit coupled capacitances and mutual inductances as shown in the figure. There are basically two types of crosstalk analysis. First is the functional crosstalk in which victim line is at quiescent and other lines are switching. Due to switching activities of aggressor lines, there is insertion of noise at victim line. This is depicted in the following simulation results. Due to this forced switching action, victim line signal gets undershooted/overshooted. Crosstalk analysis is performed for 5 mm of interconnect length for all interconnect structures.

It is analyzed from Fig. 9 that SWCNT and MWCNT bundles are good performing candidates over other on-chip interconnect structures. But since their practical feasibility is difficult, mixed CNT bundle can cope in the race of performance.

Another emerging critical issues in the high-speed nano-interconnects are its noise immunity and clock jitter problems. Hence, the present work also briefly illustrates

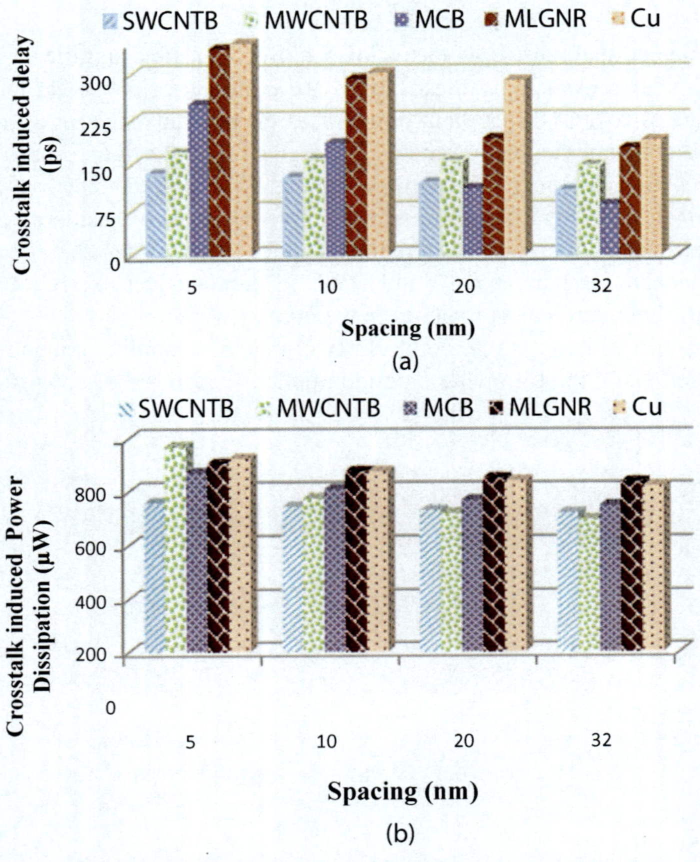

Fig. 9 **a** Crosstalk-induced delay (top) and **b** crosstalk-induced power dissipation (bottom) for different interconnect materials and varying spacing between capacitively and inductively coupled interconnect lines

signal integrity analysis in high-speed interconnects. The clock jitter and signal-to-noise ratio are the measured entities and are computed using eye diagram, as shown in Figs. 10 and 11.

An eye diagram is a powerful tool for understanding signal impairments in the high-speed digital channel system, verifying transmitter output compliance, and revealing the amplitude and time distortion elements that degrade the BER for diagnostic purposes [23]. By taking bandwidth samples of a high-speed digital signal, an eye diagram represents the sum of samples by superimposing high and low levels of the signal, and corresponding transition measurements. The eye diagram analysis is performed for different interconnect materials, viz., copper, MLGNR, SWCNTB,

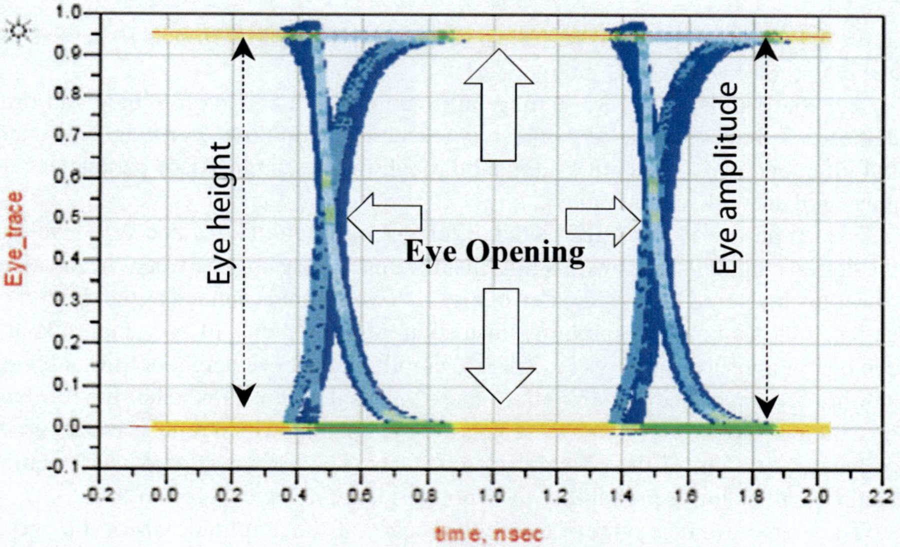

Fig. 10 Eye diagram representations and measurement entities for 1 Gbps data rate operation of MWCNTB interconnect

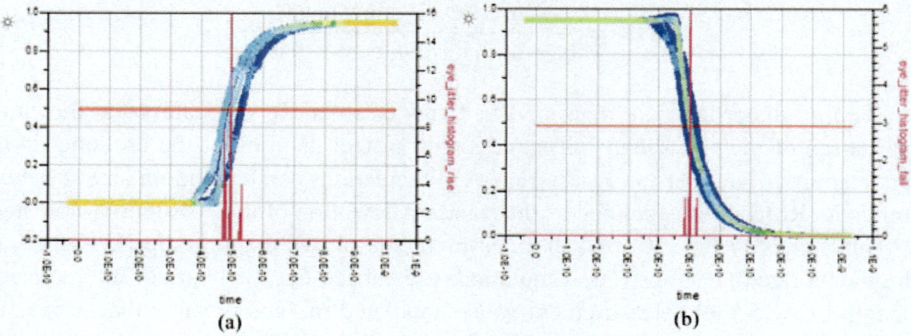

Fig. 11 **a** Rise time jitter, **b** fall time jitter representations for 1 Gbps data rate operation of MWCNTB interconnect

Table 2 Eye diagram measurement of parameters for different interconnect topologies at data rate of 1 Gbps

Parameter	Copper	MLGNR	SWCNTB	MWCNTB	MCB
Eye amplitude (V)	0.899	0.901	0.951	0.947	0.829
Eye height (db)	−0.462	−0.476	−0.219	−0.237	−1.439
Eye width (ns)	9.875E−10	9.942E−10	9.831E−10	9.46E−10	9.81E−10
Eye SNR	7.77E−12	9.012E19	9.509E19	9.47E19	13.159
Eye jitter (rms)	3.999E−11	2.238E−12	4.263E−12	1.445E−11	4.375E−12

MWCNTB and MCB. Their performance is shown in Table 2. Figure 10 depicts the eye diagram of MWCNTB interconnect.

Eye diagram is an aid for getting information about signal amplitude and time distortions, and some useful parameters relevant to amplitude, eye height, rise and fall jitter, and eye masks are computed through this research. These parameters are measured and enlisted in Table 2.

Eye amplitude is measured as the difference between the one and zero levels of the digital signal. In other words, it is the difference measured between mean values of yellow lines (red lines) indicated in Fig. 10. Eye height is defined as the difference between the variance of amplitude distortion. Measurements of eye amplitude and eye height are shown in Fig. 10. The eye amplitude and eye height entities are vital amplitude terms since the strength of these values determines whether the received data bit is a "0" or "1". Figure 10 depicts the eye opening which reflects the accuracy of receiver system of the on-chip interconnects. The higher value of eye opening factor signifies lower probability to 'wrongly identify' logical '1' and '0'.

The signal-to-noise ratio of eye is the metric of the amplitude distortion and is computed as a ratio of the desired signal level to the level of background noise, together with other existing distortions. Higher SNR values are desirable for good signal transmission. SNR is defined as,

$$\mathrm{SNR} = \frac{\text{high level} - \text{low level}}{\sigma \times \text{high level} + \sigma \times \text{low level}} \tag{23}$$

Timing distortions are measured in terms of jitter. To estimate jitter, the time variances of the rising and falling edges of an eye diagram at the crossing point are captured. These are shown in Fig. 11. Fluctuations can be random and/or deterministic. RMS jitter is defined as the standard deviation of the histogram (indicated by pink lines in Fig. 10). All the measurements of eye diagram are performed at high data rate of 1 Gbps. The simulations are carried out for 1 mm of interconnect length in ADS tool. Measurement values tabulated in Table 2 show that copper is failing to suffice with the high-speed operation while CNT bundles are frontrunners in high-speed operational metrics.

5 Conclusion

The present work investigates prospective booster technique for different interconnect materials. The various structures of on-chip interconnect considered are copper and carbon-based SWCNT bundle, MWCNT bundle, mixed CNT bundle and MLGNR. To speculate the performance of different interconnect materials and then switching to futuristic interconnects, several signal integrity and reliability analyses have been performed. It has been analyzed that with increase in length of interconnects in DIL system, the performance of interconnect system degrades. To mitigate the signal degradation issues in long interconnects, repeater insertion technique has been proposed and implemented for all interconnect structures. It is inspected that up to 20% delay can be reduced by incorporating carbon interconnects using CMOS buffer insertion. Similarly, nearly 30% reduction in delay is possible for copper interconnect. Crosstalk is a function of pitch (spacing) between interconnect lines and is one of the main parameters for inducing delay and power dissipation in the system. Hence, crosstalk analysis is performed for various values of spacing between interconnects. It is observed that by using carbon nano-interconnects, crosstalk-induced delay and power consumption reduction can be attained up to 50 and 25%, respectively. It is inferred from various analyses that bundled CNTs are leading in high performance. But due to feasibility issue of developing SWCNTB and MWCNTB, MCB is proposed as potential on-chip interconnect structure for high performance. Analysis shows that approximately 20% speed improvement and approximately 60% lesser power dissipation improvement can be achieved if carbon interconnects are used in place of copper. Furthermore, signal integrity of on-chip interconnect is assessed using eye diagram and jitter analyses. From eye diagram analysis, signal-to-noise ratio (SNR) is computed. It is investigated that graphene interconnects possess higher SNR than its counterpart copper interconnects. From different analyses performed in this work, it is envisaged that booster insertion technique along with prospective grapheme-based mixed CNT bundle interconnects is prominent solution to attain high performance for on-chip interconnects in high-speed IC designs.

References

1. International Technology Roadmap for Semiconductors (ITRS) (2007). https://www.itrs.net/.
2. Kurokawa A, Sato T, Kanamoto, Hashimoto M (2009) Interconnect modeling- A physical design perspective. IEEE Trans Electron Devices 56(9):1840–1851
3. Steinhögl W, Schindler G, Steinlesberger G, Traving M, Engelhardt M (2005) Comprehensive study of the resistivity of copper wires with lateral dimensions of 100 nm and smaller. J Appl Phys 97:023706
4. Jiang PC, Zhao WS, Yin WY (2012) Signal transmission analysis of multilayer graphene nano-ribbon (MLGNR) interconnects. IEEE Trans Electromag Compat 54(1):126–132
5. Rai MK, Chatterjee AK, Sarkar, Kaushik BK (2016) Performance analysis of multilayer graphene nanoribbon (MLGNR) interconnects. J Comput Electron 15(2):358–366.
6. Naeemi A, Meindl JD (2009) Compact physics-based circuit models for graphene nanoribbon interconnects. IEEE Trans Electron Devices 56(9):1822–1833

7. Politou M, Wu X, Asselberghs I, Contino A, Soree B, Radu I, Huyghebaert C, Tokei Z, Gendt SD, Heyns M (2017) Evaluation of multilayer graphene for advanced interconnects. Microelectron Eng 167:1–5
8. Li H, Xu C, Banerjee K (2010) Carbon nanomaterials: the ideal interconnect technology for next-generation ICs. IEEE Des Test Comput 27(4):20–31
9. Agrawal Y, Girish M, Chandel R (2018) An efficient and novel FDTD method based performance investigation in high-speed current-mode signaling SWCNT bundle interconnect. Sådhanå 43(11)
10. Lu Q, Zhu Z, Yang Y, Ding R (2016) Analysis of propagation delay and repeater insertion in single-walled carbon nanotube bundle interconnects. Microelectron J 54:85–92
11. Sathyakam U, Mallick PS (2012) Towards realisation of mixed carbon nanotube bundles as VLSI interconnects: a review. Nano Commun Netw 3:175–182
12. Alpert C, Devgan A (1997) Wire segmenting for improved buffer insertion. In: IEEE/ACM proceedings on design automation, pp 588–593
13. Tang M, Mao J (2003) Wire sizing optimization for buffered global interconnects. Int Proc Microwave Milimeter Wave Technol 2:479–482
14. Patel D, Kim B (2010) Carbon nanotube bundle interconnect—performance evaluation optimum repeater size and insertion for global wire. In: 53rd IEEE international midwest symposium on circuits and systems (MWSCAS).
15. Agrawal Y, Mekala G, Chandel R (2016) Comprehensive model for high-speed current-mode signaling in next generation MWCNT bundle interconnect using FDTD technique. IEEE Trans Nanotechnol 15(4):590–598
16. Pathade T, Shah U, Agrawal Y, Parekh R (2018) Preeminent buffer insertion technique for long advanced on-chip graphene interconnects. In: IEEE proceedings on electrical design of advanced packaging and system
17. Pu SN, Yin WY, Mao J, Liu QH (2009) Crosstalk prediction of single- and double-walled carbon-nanotube (SWCNT/DWCNT) bundle interconnects. IEEE Trans Electron Devices 56(4):560–568
18. Nieuwoudt A, Massoud Y (2006) Evaluating the impact of resistance in carbon nanotube bundles for VLSI interconnect using diameter-dependent modeling techniques. IEEE Trans Electron Devices 53(10)
19. Mekala G, Agrawal Y, Chandel R (2017) Modelling and performance analysis of dielectric inserted side contact multilayer graphene nanoribbon interconnects. IET Circuits Devices Syst 11(3):232–240
20. Majumder MK, Kaushik BK, Manhas SK (2014) Analysis of delay and dynamic crosstalk in bundled carbon nanotube interconnects. IEEE Trans Electromag Compat 56(6)
21. Singh KS, Thakur A (2019) Comparative analysis of mixed CNTs and MWCNTs as VLSI interconnects for deep sub-micron technology nodes. J Electron Mater 48(4):2543–2554
22. Karthikeyan A, Mallick PS (2017) Optimization techniques for CNT based VLSI interconnects—a review. J Circuits Syst Comput 26(3):1730002
23. Anritsu's application notes on understanding eye pattern measurements. https://rintintin.colorado.edu/~gifford/5830-AWL/.

Island Engineering of Single-Electron Transistor for Room Temperature Operation

Raj Shah, Rutu Parekh, and Rasika Dhavse

Abstract Single-electron transistors (SETs) have shown their competence to overpower MOSFETs as well as FinFETs in low power regime. However, research to exhibit room temperature operation and its demonstration for application in logic and memory circuits is still in infancy. A SET contains two ultra-thin tunnel barriers and a conductive island which work on the principle of Coulomb blockade (CB) and tunneling. Its dimensions must be planned properly to exhibit room temperature operation with CMOS compatibility. In this work, the island engineering technique is proposed for devising SET so as to observe both Coulomb blockade and quantum mechanical tunneling at room temperature. The island engineering is carried out for aluminum and copper island. The SET has been simulated in TCAD using the designed dimensions and its output as well as transfer characteristics have been plotted at room temperature. The analyses illustrate that out of all the SET devices, aluminum island SET with 4 nm tunnel barrier with 17 nm island length offers the desired CB and tunneling current in tens of nA range. To validate the results and ascertain the concept of CMOS compatible design, SET transfer and output characteristics have been simulated using Cadence Virtuoso using MIB model. The proposed technique of SET is also compared with the other SET optimization techniques.

Keywords Coulomb blockade · Island engineering · MIB model · Nano-electronics · Quantum mechanical tunneling · Single-electron transistor · TCAD

R. Shah (✉) · R. Dhavse
Sardar Vallabhbhai National Institute of Technology, Surat, Gujarat 395007, India
e-mail: shahraj91@gmail.com

R. Dhavse
e-mail: rsk@eced.svnit.ac.in

R. Parekh
Dhirubhai Ambani Institute of Information and Communication Technology, Gandhinagar, Gujarat 382421, India
e-mail: rutu_parekh@daiict.ac.in

Z. Patel et al. (eds.), *Advances in VLSI and Embedded Systems*, Lecture Notes in Electrical Engineering 676, https://doi.org/10.1007/978-981-15-6229-7_23

1 Introduction

Owing to small size and low power consumption, single-electron transistors (SETs) are becoming an attractive alternative in the post-CMOS era of semiconductor technology [1–4]. They operate on the principles of Coulomb blockade and quantum mechanical tunneling (QMT), depicting the on and the off states of the device, respectively [5–7]. SETs have proved their potential to be a promising candidate for non-volatile memory as well as logic applications [2, 3, 8]. SET-MOS hybridization is also one huge field of interest for logic development. It is envisaged that SETs can outperform CMOS in nanometer regime with its virtues like ultra-low power consumption and scalability [2, 3, 9]. Since last two decades, many researchers have worked to overcome two fundamental issues of SETs, viz., (a) mass production and (b) room temperature operation. In recent years, with advent in time and technology, many CMOS-compatible processes are demonstrated for fabrication of SETs [2, 4, 10–12]. They include nanodamascene processes [4], patterning with reactive ion etching and electromigration, chemical mechanical polishing-based fabrication [10], self-assembly and self-alignment of quantum dots [11], even optical lithography of silicon nanowire [12]. Thus, prospects of integrating SETs in a CMOS process are bright. However, room temperature operation is still a matter of experimentation.

Coulomb blockade is inherently a low energy phenomenon [6, 7]. Hence, it cannot be detected at ambient temperatures easily. So, to observe the SET operation in true sense, the device needs to be engineered carefully. The material and dimensions of island play crucial role in tunneling of electron between source and drain. Hence, the effect of island on SET drain current along with CB and QMT must be observed. The structure of island should also be compatible with the CMOS fabrication processes. In this paper, we will present the rudiments of SET design followed by a detailed analytical approach. We will then simulate the device demonstrating thorough island engineering. Results show occurrence of both, the Coulomb blockade and the quantum mechanical tunneling in the optimized device. Finally, we will present validation of our device with MIB model [3].

2 SET Design Considerations

As shown in Fig. 1, a SET consists of two tunnel junctions on source and drain sides sandwiching a quantum dot. Every junction offers a barrier—popularly known as Coulomb blockade—to flow of electrons. These junctions or barriers ensure weak coupling of source and drain with the quantum island. An electron from source, after overcoming Coulomb blockade energy level, enters the island by the process of tunneling. Simultaneously, an electron leaves the island and tunnels to the drain, causing current to flow. Thus, to ensure the charge transfer, one needs to supply Coulomb energy to the island. This energy can be supplied by an external voltage source. It is given by [5–7],

Fig. 1 Structural schematic of single-electron transistor

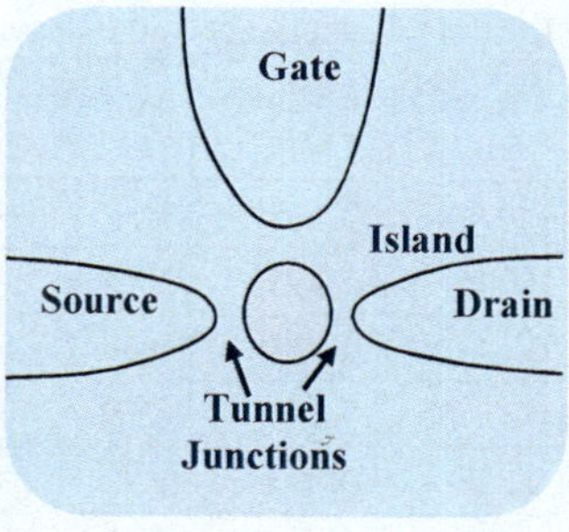

$$E_C = \frac{e^2}{2C_\Sigma} \quad (1)$$

where C_Σ is the capacitance of the island given by $(2C_J + C_G)$, C_J being the tunneling junction capacitance. Correspondingly, Coulomb voltage is given by

$$V_{CB} = \frac{e}{2C_\Sigma} \quad (2)$$

This capacitance is primarily governed by the device geometry and material parameters. With reduction in this capacitance, E_C increases, making the junction more opaque.

In addition to this capacitive coupling, the barrier offers tunnel resistance, which must be larger than the fundamental resistance $R_T \gg \frac{\hbar}{q^2}$, that is, approximately 25.8 KΩ [5–7]. It ensures localization of electron states to the island during blockade. R_T also controls the driving capability of the device. The device maximum operating frequency is $1/(R_T C)$ [9]. Conflict of speed and power can be addressed by fine tuning these two electrical parameters. This demands appropriate materials choice and fabrication of capacitors of the order of attofarads, which is possible in today's era.

For designing a rugged SET, Coulomb energy of the SET should be sufficiently higher than the ambient thermal energy. Table 1 illustrates capacitance and bias requirements with respect to thermal robustness. General trend dictates acceptable range as 20 $K_B T$.

It is evident from Table 1 that for room temperature operation of SETs, appropriate tunnel junction engineering and optimization is crucial so as to obtain: (a) very low total capacitance (aF), (b) high driving current (nA) and (c) very low thermionic current (during opaque state of the junction). A low-k dielectric like SiO_2 can offer a barrier less than 0.4 eV [13] aiding high level of transparency. But for junction opacity, high-k dielectric like TiO_2 which offers barrier greater than 1 eV is required. To resolve this conflict, Hajjam et al. proposed multi-dielectric stack to control the off current [13]. However, it involved realizing ultra-thin layers of various dielectric increasing design and process complexity. We propose that instead of tunnel

Table 1 Capacitance and bias requirements with respect to thermal robustness

C_G (aF)	C_Σ(aF)	E_C (eV)	$T = E_C/nK_B$ ($n = 1$) (K)	$T = E_C/nK_B$ ($n = 20$) (K)	$V_{CB} = e/2C_\Sigma$ (V)	$V_{BIAS} = e/2C_G$ (V)
0.023	0.115	1.39	16,000	800	0.695	3.47
0.046	0.138	1.15	13,000	650	0.575	1.73
0.14	0.232	0.68	7880	394	0.34	0.57
0.18	0.272	0.59	6840	342	0.295	0.44
0.023	0.093	1.72	19,900	9950	0.86	3.47
0.046	0.116	1.37	15,900	7950	0.685	1.73
0.14	0.21	0.76	8810	440	0.38	0.57
0.18	0.25	0.64	7420	371	0.32	0.44

engineering, one can opt for island engineering to achieve acceptable I_{ON} and I_{OFF} values. It is comparatively easy, simple to comprehend and process-friendly.

Figure 2a, b shows structure and MIB model of a single-gate SET [3]. Source, drain and gate are made up of n-polysilicon (phosphorus, 1e19 cm^{-3}), whereas the substrate is p-silicon (boron, 1e15 cm^{-3}). Tunneling junction capacitances on source as well as drain side are $C_{TD} = C_{TS} = C_J$, control gate capacitance is C_G, tunnel junction resistances are $R_{TD} = R_{TS} = R_T = 1\ M\Omega$. The total device capacitance is defined as $C_\Sigma = 2C_J + C_G$. SET design parameters compatible with BSIM predictive technology model (PTM) [14] and 22 nm node CMOS model operation at $T = 300$ K are chosen. Consequently, supply voltage is chosen to be 0.8 V. For a robust reliable design of SET logic, which can operate with the least possible error at room temperature, the charging energy E_C must be as large as possible compared with the

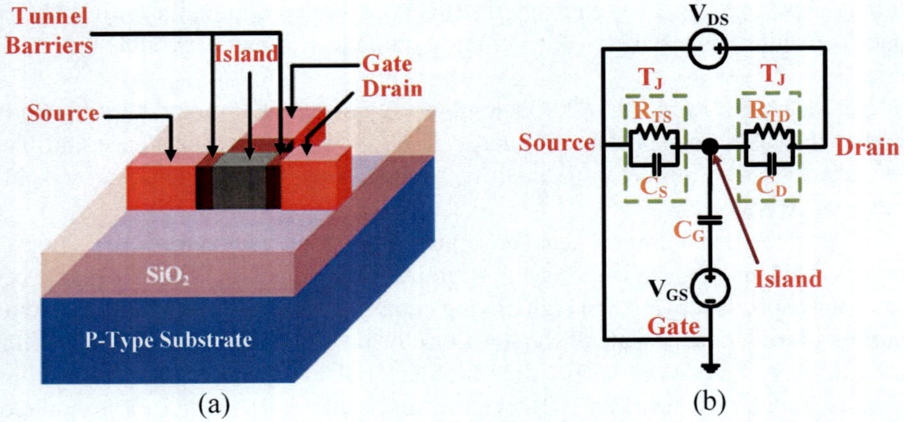

Fig. 2 **a** Single-gate SET structure, **b** equivalent MIB model where tunnel barriers are represented by parallel RC combinations ($R_{TS} \parallel C_{TS}$ and $R_{TD} \parallel C_{TD}$), gate dielectric as gate capacitance C_G, conductive island as black dot and two bias supplies, namely, V_{DS} and V_{GS} between drain—source and gatesource, respectively [3]

thermal energy. As such, Coulomb/charging energy $= e^2/C_\Sigma = 40\ \mathrm{K_B T} = 1.036$ eV. This leads to a value of 0.155 aF for C_Σ. So, Coulomb blockade voltage $= e/2C_\Sigma = 0.51$ V. Control gate capacitance $= e/2V_{DD} = 0.045$ aF. Hence, junction/tunnel capacitance $= (C_\Sigma - C_G)/2 = 0.055$ aF. Background charge is neglected.

3 Island Engineering

Figure 3 shows top view of single-gate SET under consideration.

Table 2 illustrates SET electrical parameters and their possible mapping with its physical parameters. Our strategy is to play with island and junction dimensions while maintaining overall device dimension constant. The tunnel barrier engineering plays a vital role for this work to choose the tunnel barrier thickness [15]. We have experimented with aluminum and copper as island material, due to general good conductivity. In TCAD simulations, for global device, we used Philips Unified Mobility Model because it incorporates doping and temperature dependencies. For modeling tunneling near insulator for rapidly varying field, we used band-to-band and Schenk recombination models [16]. At material interfaces, we also used direct tunneling model as the barrier is trapezoidal [15, 16]. Junction width less than 3 nm did not result in satisfactory Coulomb blockade and greater than 4 nm did not show tunneling in the neighborhood of bias voltage. Under no influence of gate, devices

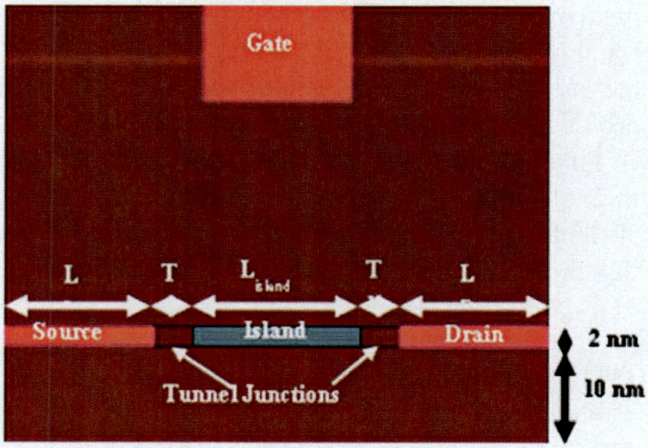

Fig. 3 Top view of single-gate SET

Table 2 Physical and electrical parameters of SET

Junction width	Physical parameters	Electrical parameters	Device name
$T_J = 4$ nm	$L_S = L_D = 15$ nm, $L_{\text{island}} = 17$ nm	$C_\Sigma = 0.11$ aF, $n = 27$	A
	$L_S = L_D = 30$ nm, $L_{\text{island}} = 32$ nm	$C_\Sigma = 0.18$ aF, $n = 17$	B
$T_J = 3$ nm	$L_S = L_D = 15$ nm, $L_{\text{island}} = 19$ nm	$C_\Sigma = 0.14$ aF, $n = 22$	C
	$L_S = L_D = 30$ nm, $L_{\text{island}} = 34$ nm	$C_\Sigma = 0.20$ aF, $n = 15$	D

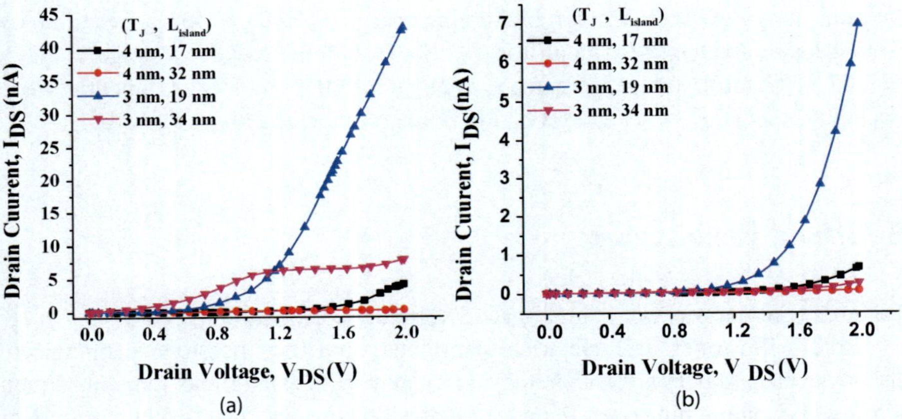

Fig. 4 V_D–I_D characteristics of single-electron devices for **a** aluminum island and **b** copper island for different island dimensions

exhibit blockade and tunneling as shown in Fig. 4a, b, for aluminum and copper island, respectively.

The devices clearly show the mechanisms of Coulomb blockade and quantum mechanical tunneling. In the limiting case of small structures, the structure resistivity becomes proportional to ρ0 * λ for any given fixed dimension [17], where ρ0 and λ are the bulk resistivity and mean free path for electron phonon scattering, respectively. Thus, the metal with the lowest product ρ0 * λ is expected to exhibit the highest conductivity. This product is 6.70e–16Ω m^2 for copper and 5.01e–16Ω m^2 for aluminum [17]. Hence, the current capacity of aluminum devices is higher than the copper devices. Coulomb voltage and SET capacitances extracted from above profiles for all the devices are listed in Table 3.

First observation here is that copper island devices offer better confinement than their aluminum counterparts. Small islands have higher Coulomb voltage. This is attributed to increased confinement and hence higher blockade. It is also observed that the extracted values have higher degree of matching as the device capacitance

Table 3 Derived and extracted parameters of SET

Device name	Analytical capacitances (aF)			Analytical V_{CB} (V)	Extracted V_{CB} (V)		Extracted total capacitance $C_{\sum}$(aF)	
	C_G	C_J	$C_{\sum} = 2C_J + C_G$	$e/C_{\sum}$	Aluminum Island	Copper Island	Aluminum Island	Copper Island
A	0.045	0.035	0.11	1.38	1.2	1.2	0.134	0.134
B	0.11	0.035	0.18	0.88	0.6	1.1	0.26	0.144
C	0.045	0.0468	0.14	1.14	0.7	1.25	0.22	0.13
D	0.10	0.0468	0.20	0.8	0.2	1.05	0.8	0.15

becomes smaller and smaller. Effect of island dimension and material can be seen in Fig. 5.

So, it is found that bigger island leads to higher Coulomb blockade and lower drain current.

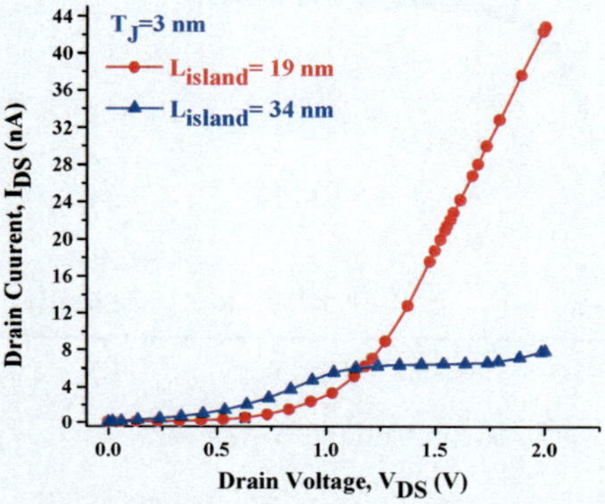

(a) Aluminum island variation for T_J= 3 nm

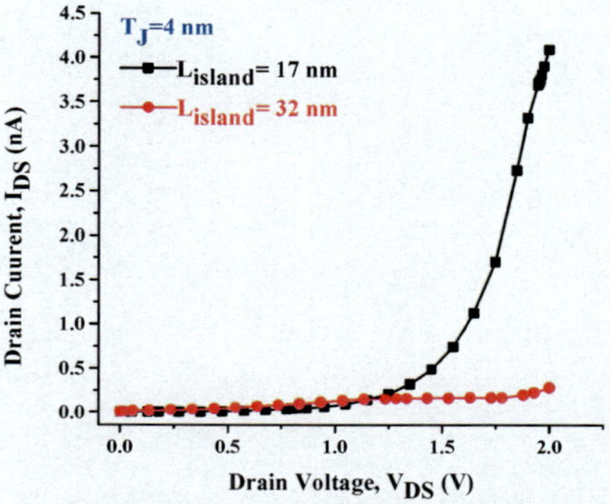

(b) Aluminum island variation for T_J= 4 nm

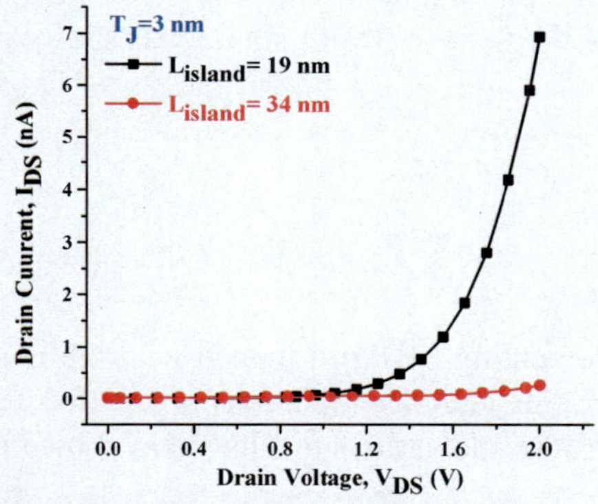

(c) Copper island variation for T_J= 3 nm

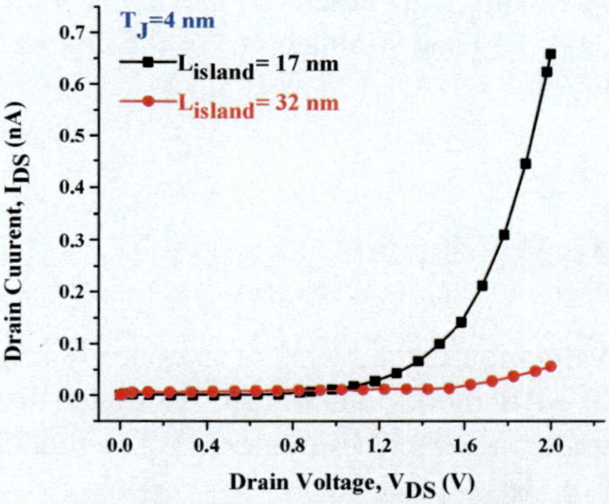

(d) Copper island variation for TJ= 4 nm

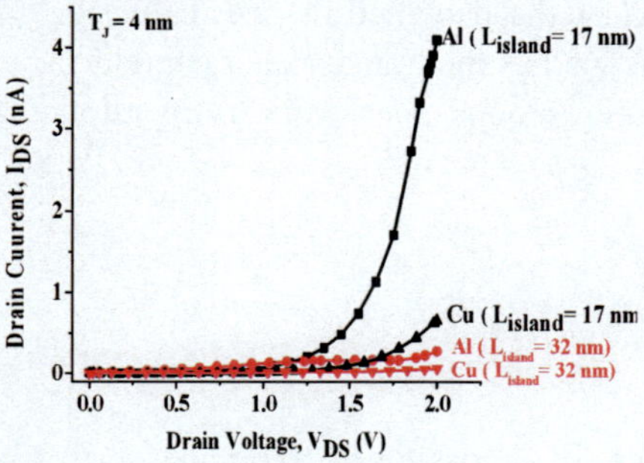

(e) Profile T_J= 4 nm

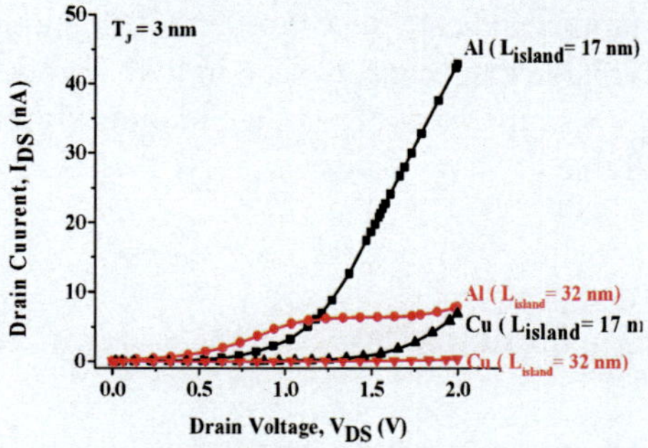

(f) Profile for T_J= 3 nm

Fig. 5 I–V profiles for variation in materials and dimensions of island and junction

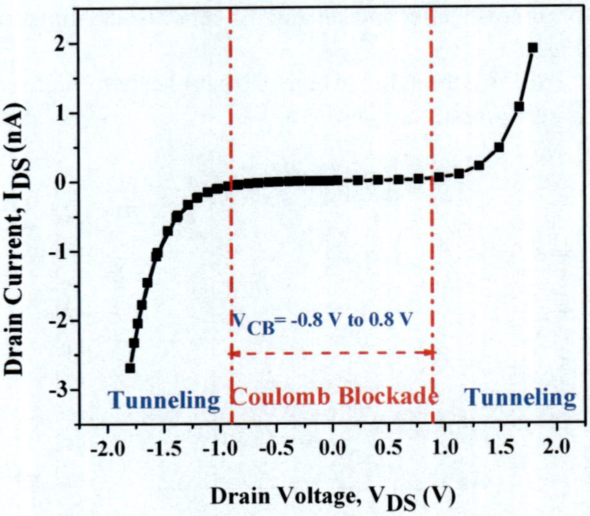

Fig. 6 Output characteristics of SET using TCAD

Finally, we settled for Device A with aluminum island ($C_G = 0.045$ aF, $C_J = 0.035$ aF) and simulated it for the bipolar I–V behavior. Result obtained is shown in Fig. 6.

4 Validation

Validation of our work was done by implementing SET in cadence with the help of MIB model. Parameters extracted from the TCAD simulations were used for implementing MIB model [3]. The model setup and various results are as shown in Fig. 7a–d.

Clear-cut manifestation of Coulomb blockade, tunneling and Coulomb oscillations indicate successful implementation of the designed device. Effect of gate voltage can be easily seen in both output as well as the transfer characteristics.

Comparison of our technique with other groups peer work is mentioned in Table 4.

5 Conclusion

Owing to modern technological inventions, it is possible to fabricate very small capacitive structures. Hence, SET has a potential to operate at room temperature in ultra-low power applications. Island engineering can be sought as simplest method for optimizing SET. The aluminum island device simulated here gives a stable room temperature operation with a current of the order of 10 nA and is compatible with

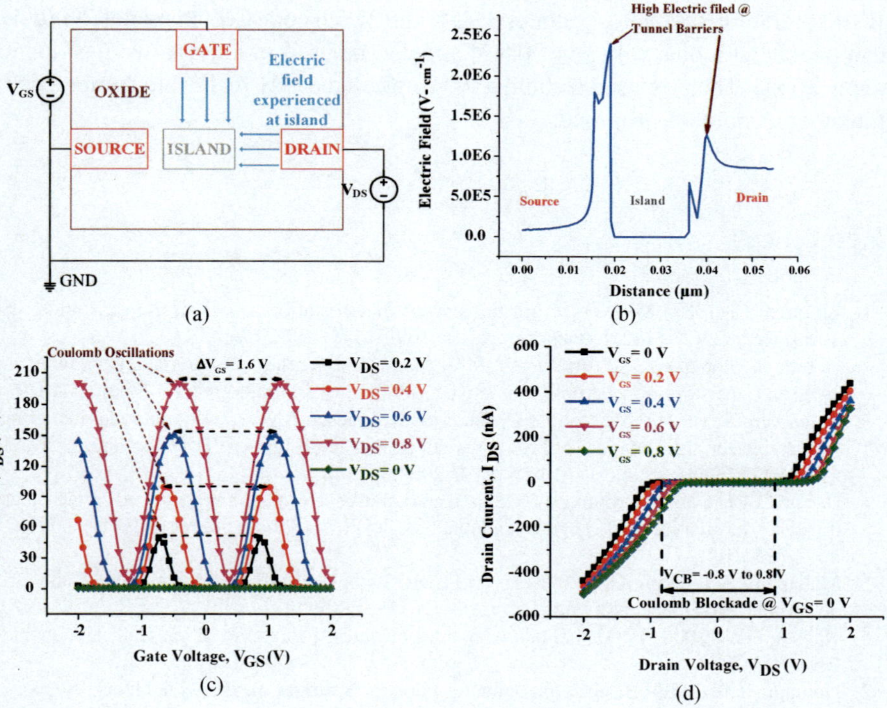

Fig. 7 **a** Electric field approaching at island, **b** simulated electric field of SET, **c** and **d** transfer and output characteristics of SET using MIB

Table 4 Comparison of proposed SET with other peer work with respect to various electrical and physical parameters

Reference	Technique	Pros	Cons
Dubuc [4, 18]	BEOL processing	CMOS BEOL process compatibility, high current drive	Required 3D integration, more back ground charge due to metallic structure
Joshi [10, 19]	Island material variation	Simple, CMOS fabrication processing	No room temperature operation
Hajjam [13]	Tunnel barrier optimization	Accurate	Time consuming, no CMOS technology compatibility
Proposed work	Island engineering (choosing proper island material and dimension)	Simple, accurate	Depends on resolution of nanoscale fabrication processes

BSIM predictive technology model (PTM) and 22 nm node CMOS model. The device can operate at a bias voltage of 0.8 V and can be used in logic as well as memory applications. The proposed technique is simple and easily realizable compared with the other techniques proposed.

References

1. Likharev K (1987) Single-electron transistors: electrostatic analogs of the dc squids. IEEE Trans Magn 23(2):1142. https://doi.org/10.1109/TMAG.1987.1065001
2. Uchida K, Matsuzawa K, Toriumi A (1999) A new design scheme for logic circuits with single electron transistors. Jpn J Appl Phys 38:4027–4032. https://doi.org/10.1143/jjap.38.4027
3. Mahapatra S, Vaish V, Wasshuber C, Banerjee K, Ionescu AM (2004) Analytical modeling of single electron transistor for hybrid cmos-set analog ic design. IEEE Trans Electron Devices 51(11):1772. https://doi.org/10.1109/TED.2004.837369
4. Dubuc C, Beauvais J, Drouin D (2008) A nanodamascene process for advanced single-electron transistor fabrication. IEEE Trans Nanotechnol 7(1):68–73. https://doi.org/10.1109/TNANO.2007.913430
5. Mahapatra S, Ionescu AM (2006) Hybrid CMOS single-electron-transistor device and circuit design. Artech House, Norwood, MA
6. Hanson GW (2008) Fundamentals of nanoelectronics. Pearson/Prentice Hall, Upper Saddle River, NJ
7. Hamaguchi C (2009) Basic semiconductor physics. Springer, Berlin, Heidelberg, Neckar
8. Durrani ZAK, Irvine AC, Ahmed H, Nakazato K (1999) A memory cell with single-electron and metal-oxide-semiconductor transistor integration. Appl Phys Lett 74:1293–1295. https://doi.org/10.1063/1.123528
9. Parekh R, Beaumont A, Beauvais J, Drouin D (2012) Simulation and design methodology for hybrid SET-CMOS integrated logic at 22-nm room-temperature operation. IEEE Trans Electron Devices 59(4):918–923. https://doi.org/10.1109/TED.2012.2183374
10. Joshi V, Orlov AO, Snider GL (2008) Silicon single-electron transistor with oxide tunnel barriers fabricated using chemical mechanical polishing. J Vac Sci Technol B Microelectron Nanometer Struct 26(6):2587–2591. https://doi.org/10.1116/1.297887
11. Ray V, Subramanian R, Bhadrachalam P, Ma L-C, Kim C-U, Koh SJ (2008) CMOS-compatible fabrication of room-temperature single-electron devices. Nat Nanotechnol 3(10):603–608. https://doi.org/10.1038/nnano.2008.267
12. Sun Y, Rusli, Singh N (2011) Room-temperature operation of silicon single-electron transistor fabricated using optical lithography. IEEE Trans Nanotechnol 10(1):96–98. https://doi.org/10.1109/TNANO.2010.2086475
13. Hajjam KGE, Bounouar MA, Baboux NA, Ecoffey S, Guilmain M, Puyoo E, Francis L, Souifi A, Drouin D, Calmon F (2015) Tunnel junction engineering for optimized metallic single-electron transistor. IEEE Trans Electron Devices 62(9):2998–3003. https://doi.org/10.1109/TED.2015.2452575
14. Predictive Technology Model (PTM). In: Predictive Technology Model (PTM). https://ptm.asu.edu/.
15. Shah R, Dhavse R (2018) Tunnel barrier optimization for room temperature operation of single electron transistors. In: Proceedings of the nanotech France 2018 international conference (nanotech France 2018), Sector, Paris, pp 37–40
16. Sentaurus Device User Guide.
17. Gall D (2016) Electron mean free path in elemental metals. J Appl Phys 119(8):085101. https://doi.org/10.1063/1.4942216

18. Beaumont A, Dubuc C, Beauvais J, Drouin D (2009) Room temperature single-electron transistor featuring gate-enhanced on-state current. IEEE Electron Device Lett 30:766–768. https://doi.org/10.1109/led.2009.2021493
19. Lee Y-C, Joshi V, Orlov AO, Snider GL (2010) Si single electron transistor fabricated by chemical mechanical polishing. J Vac Sci Technol B Nanotechnol Microelectron Mater Process Meas Phenom 28(6). https://doi.org/10.1116/1.3498748